# Cystic Fibrosis and DNA Tests: Implications of Carrier Screening
# OTA Project Staff

Roger C. Herdman, *Assistant Director, OTA Health and Life Sciences Division*

Michael Gough, *Biological Applications Program Manager*

**Robyn Y. Nishimi,** *Project Director*

Kathi E. Hanna, *Senior Analyst*

Margaret A. Anderson, *Analyst*[1] and *Contractor*

Sheryl M. Winston, *Research Analyst*

Ellen L. Goode, *Research Assistant*

Claire L. Pouncey, *Research Analyst*[2]

Alyson Giardini, *Intern*[3]

## *Administrative Staff*

Cécile Parker, *Office Administrator*

Linda Rayford-Journiette, *Administrative Secretary*

Jene Lewis, *Secretary*

## Contractors

Gary B. Ellis, *Editor*, Silver Spring, MD

Adrienne Asch, New York, NY
John M. Boyle, Schulman, Ronca & Bucuvalas, Inc., Silver Spring, MD
Ferdinand J. Chabot, Newton, IA
R. Alta Charo, University of Wisconsin, Madison, WI
Jeffrey L. Fox, Washington, DC
Bob Jones, IV, Arlington, VA
F. John Meaney, Scottsdale, AZ
Thomas H. Murray, Case Western Reserve University, Cleveland, OH
Mark V. Pauly, Leonard Davis Institute of Health Economics, Philadelphia, PA
Susan Sanford, MedSciArtCo, Washington, DC
Andrew M. Shorr, Charlottesville, VA
Ann C.M. Smith, Reston, VA
Bonnie Steinbock, University of Albany, Albany, NY
Janet Weiner, American College of Physicians, Philadelphia, PA

[1] Through December 1991

[2] Through May 1991

[3] September - December 1991

# Contents

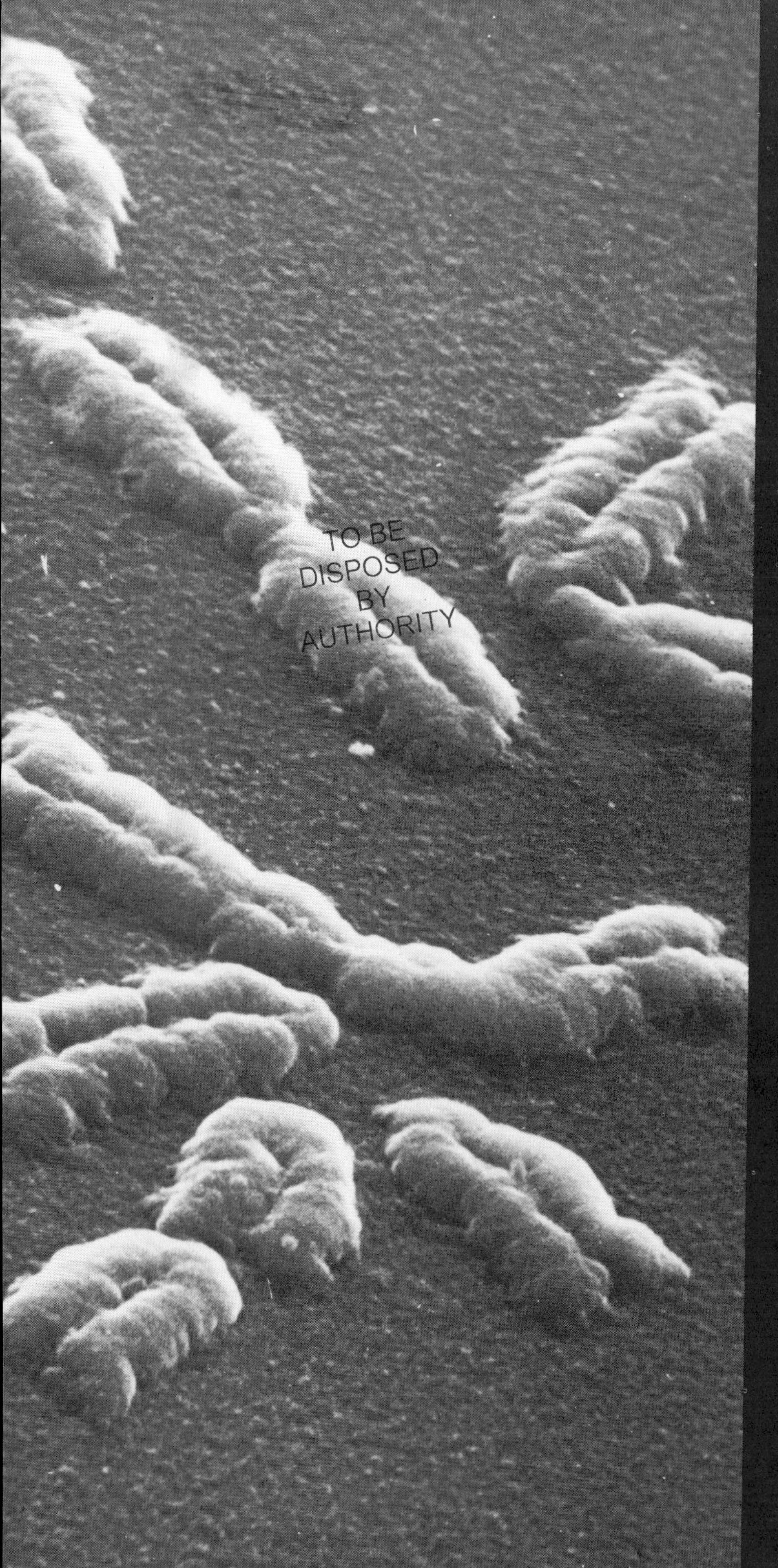



# Cystic Fibrosis and DNA Tests: Implications of Carrier Screening

Congress of the United States
Office of Technology Assessment

## Recommended Citation:

U.S. Congress, Office of Technology Assessment, *Cystic Fibrosis and DNA Tests: Implications of Carrier Screening*, OTA-BA-532 (Washington, DC: U.S. Government Printing Office, August 1992).

For sale by the U.S. Government Printing Office
Superintendent of Documents, Mail Stop: SSOP, Washington, DC 20402-9328
ISBN 0-16-037986-5

# Foreword



Nearly 10 years ago, the President's Commission for the Study of Ethical Problems in Medicine and Biomedical and Behavioral Research speculated about the potential ethical, legal, and social consequences that might occur if a test were available to identify carriers for cystic fibrosis (CF)[1], the most common, life-shortening, recessive genetic disease in American Caucasians. Time and technology have moved forward. The mysteries of biological inheritance—first explored by Austrian monk Gregor Mendel over a century ago—are yielding to modern science. A CF carrier test is no longer a prospect; it is now reality. The test's existence raises broad societal questions about the use of genetic information. And beyond CF tests, expectations of scores of additional genetic tests loom on the horizon as scientists in the United States and abroad pursue an ambitious mission to map and sequence the entire human genetic blueprint, or genome.

Ongoing interest in the Human Genome Project, as well as concern about the potential magnitude and effects of routine CF carrier screening, led the House Committee on Science, Space, and Technology and the House Committee on Energy and Commerce to request an evaluation of the scientific, clinical, legal, economic, and social considerations of widespread carrier screening for CF. The study was also endorsed by Representative David R. Obey. *Cystic Fibrosis and DNA Tests: Implications of Carrier Screening* presents a range of options for action by the U.S. Congress in six broad policy areas:

- genetics education and the public,
- genetics training and education of health care professionals,
- discrimination,
- clinical laboratory and medical device regulation,
- instrumentation to automate DNA diagnostics, and
- integration of DNA assays into routine clinical practice.

OTA prepared this report with the assistance of a panel of advisors and reviewers selected for their expertise and diverse points of view. Additionally, hundreds of individuals cooperated with OTA staff through interviews or by providing written material. These authorities were drawn from academia, industry, and professional societies, as well as Federal and State agencies. OTA gratefully acknowledges the contribution of each of these individuals. As with all OTA reports, however, responsibility for the content is OTA's alone.

In publishing this report, OTA concludes that the value of the CF carrier test is the information it provides. No one can estimate in common terms what it means to an individual to possess information about his or her genetic status, especially when the value concerns reproductive decisionmaking. As our knowledge of the human genome increases, what we do with information such as CF carrier status will depend on the perceptions and beliefs of all Americans. We believe that public understanding of this new knowledge and its implications is necessary for its wise and thoughtful application.

JOHN H. GIBBONS
*Director*

---

[1]President's Commission for the Study of Ethical Problems in Medicine and Biomedical and Behavioral Research, *Screening and Counseling for Genetic Conditions: The Ethical, Social, and Legal Implications of Genetic Screening, Counseling, and Education Programs* (Washington, DC: U.S. Government Printing Office, 1983).

## Cystic Fibrosis and DNA Tests: Implications of Carrier Screening Advisory Panel



NOTE: OTA is grateful for the valuable assistance and thoughtful critiques provided by the advisory panel members. The panel does not, however, necessarily approve, disapprove, or endorse this report. OTA assumes full responsibility for the report and the accuracy of its contents.

Chapter 1

# Summary, Policy Issues, and Congressional Options

# Contents



## *Boxes*



## *Figures*



## *Tables*

# Chapter 1
# Summary, Policy Issues, and Congressional Options

Seeking to learn what the future holds is an enduring human quality. What will happen? When will it happen? How will it happen? People have always pondered such questions about their health and that of their families. Folk ways once enjoyed wide favor in medicine, but over the years technology has increasingly eclipsed such methods of divination. Today, medical technology includes genetic tools that can deliver predictive information with ever-increasing accuracy. This report is about one of those tools: a test that can tell people about their potential to pass to their offspring a genetic condition called cystic fibrosis (CF). Some people want and seek this information; others do not.

CF is the most common, life-shortening, recessive disorder affecting Caucasians of European descent. Between 1,700 and 2,000 babies with CF are born annually in the United States. As in many genetic conditions, the diagnosis of an infant with CF often reveals the first clue that the genetic trait exists in the family. In fact, four of five individuals with CF are born to families with no previous history of the illness. In such cases, the parents—as well as their siblings, parents, and other relatives—do not have CF. These individuals, referred to as CF carriers, have no symptoms of CF and might not even have heard of the condition.

In 1989, scientists identified the most common change, or mutation, in the genetic material, deoxyribonucleic acid (DNA), that causes CF. Hard on the heels of this discovery, scientists developed tests to detect mutations in the area of DNA—the CF gene—that is responsible for the disease. This report focuses on using these DNA tests to screen and identify CF carriers before they have a child with CF (box 1-A). Beyond the approximately 30,000 Americans who have CF, as many as 8 million individuals could be CF carriers. The report concentrates on these millions of CF carriers, who are, today, largely unidentified.

Concern about the scientific, legal, economic, ethical, and social implications of the prospect that large numbers of people might be screened for their CF carrier status led the House Committee on Science, Space, and Technology and the House Committee on Energy and Commerce to request, and Representative David R. Obey to endorse, this Office of Technology Assessment (OTA) report.[1] CF carrier screening also commands the attention of Congress because of Congress' interest in the Human Genome Project (box 1-B).

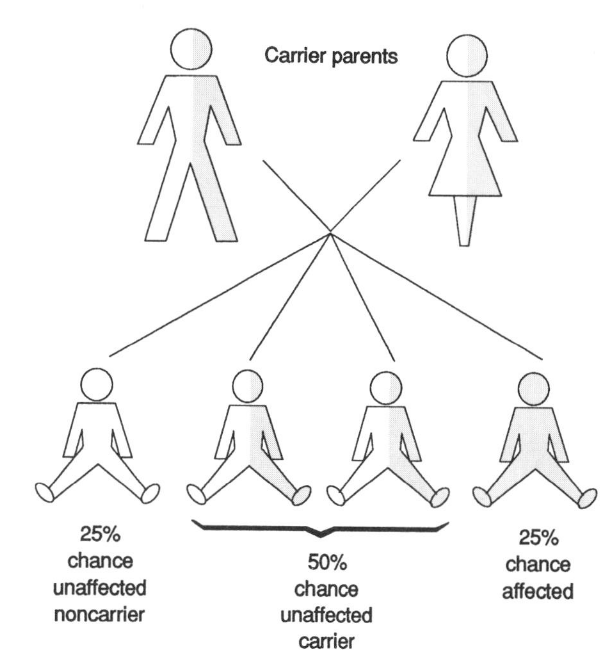


*Photo credit: Office of Technology Assessment, 1992*

Inheritance of cystic fibrosis.

## WHAT IS CYSTIC FIBROSIS?

CF is not a new disease. First described in 17th century folklore, medical literature has long documented that CF compromises many functions throughout the body—chiefly the sweat glands and the respiratory, gastrointestinal, and reproductive systems. It occurs in all racial and ethnic groups, although more frequently in some than in others (table 1-1). In fiscal year 1991, public and private

[1] Specific analysis of several topics related to CF carrier screening have been assessed in previous OTA reports, including: newborn screening for CF; genetic monitoring and screening in the workplace; the Human Genome Project; the commercial development of tests for human genetic disorders; safety and efficacy of amniocentesis, prenatal care, and pregnancy management; and reproductive technologies and assisted conception.

### *Box 1-A—Terminology*

OTA defines *genetic testing* as the use of specific assays to determine the genetic status of individuals already suspected to be at high risk for a particular inherited condition. While any individual can be considered "at high risk" for a particular unknown trait, and hence be "tested," "at high risk" in this report denotes the presence of a family history or clinical symptoms. The terms *genetic test, genetic assay*, and *genetic analysis* are used interchangeably to mean the actual laboratory examination of samples.

*Genetic screening* usually uses the same assays employed for genetic testing, but it is distinguished from genetic testing by its target population. OTA uses the term "screening" selectively. In this report, it refers to analyzing samples from individuals without a family history of the disorder, groups of these individuals, or populations. *Carrier screening for CF* (or *CF carrier screening*), then, involves performing tests on persons for whom no family history of the disorder exists to determine whether they have one normal and one aberrant copy of the CF gene, but not the disorder (which results from having two aberrant CF genes).*

Many individuals are CF carriers but do not have a positive family history. In fact, 80 percent of babies born with CF each year are cases where there was no known family history for CF. Thus, a person contemplating procreation could inquire about the availability of an assay to determine the probability that he or she could have a child affected with CF. If there are no relatives with the disorder, the individual could be informed that a test would provide information about his or her genetic status for CF. The person could then elect to be *screened* to determine whether he or she is a carrier for CF. If, however, there is a family history of the disease, a practitioner would ideally inform the individual and his or her partner about CF carrier assays and they might choose to be *tested* to determine if they are both carriers.

*Genetic counseling* is a clinical service that includes providing an individual (and sometimes his or her family) with information about heritable conditions and their risks. When centered around genetic testing or screening, it involves both education and psychological counseling to convey information about the ramifications of possible test outcomes, prepare the client for possible positive or negative analyses, and discuss the implications of the actual test results. Many types of health professionals perform genetic counseling. OTA reserves the term *genetic counselor* specifically for master's-level individuals to clarify the legal distinctions in licensing and third-party reimbursement among the different types of practitioners. But, OTA uses the term *genetic counseling* generically to refer to the educational and informational process performed by genetic specialists, including physicians, Ph.D. clinical geneticists, genetic counselors, nurses, and social workers.

OTA avoids using the term "program" in discussing CF carrier screening in the United States. For some, the term connotes a formal public health effort led or sanctioned by Federal, State, or local governments. In analyzing CF carrier screening, OTA's premise is only that large numbers of Americans could—or will—be screened for their CF carrier status. OTA remains neutral on whether the assays will be a component of a fixed, regulated scheme or another facet of general medical practice.

*In contrast, OTA uses the term *CF screening* (or *screening for CF*), to mean screening individuals to diagnose the presence or absence of the actual disorder, in the absence of medical indications of the disease or a family history of CF. This type of diagnostic screening usually involves newborns, but is rarely done for CF except in Colorado and Wisconsin. CF testing of newborns is common if a family history of the condition exists.

SOURCE: Office of Technology Assessment, 1992.

institutions spent more than $55 million studying medical and genetic aspects of CF. This section provides a brief overview of what this—and past—research has revealed, providing context for the policy aspects of CF carrier screening that follow.

### *Pathology, Diagnosis, and Prognosis*

Many affected babies are not immediately diagnosed as having CF. Although the disease is always present at birth in affected individuals, the onset of recognizable clinical symptoms varies widely; about 10 percent of cases show symptoms at birth. Other childhood ailments often share symptoms with CF, which contributes to diagnostic difficulties. In general, most diagnoses occur by age 3.

Physicians diagnose CF using a combination of clinical criteria and diagnostic laboratory testing. Although the sweat test remains the primary diagnostic test for CF, DNA mutation analysis can diagnose over 70 percent of cases, complementing and confirming sweat test results in some instances.

CF exerts its greatest toll on the respiratory and digestive systems, and the severity of respiratory problems often determines quality of life and length

### *Box 1-B—The Human Genome Project*

As the 21st century approaches, Congress and the executive branch have made a commitment to determine the location on the DNA—as has been done for CF—of all other genes in the human body, i.e., to map the human genome. The Human Genome Project is estimated to be a 15-year, $3-billion project. It has been undertaken with the expectation that enhanced knowledge about genetic disorders, increased understanding of gene-environment interactions, and improved genetic diagnoses can advance therapies for the 4,000 or so currently recognized human genetic conditions; a premise supported by the fact that even prior to launching the Human Genome Project, advances in medical genetics have directed the development of new treatment strategies and incrementally improved the management of some genetic conditions.

To address gaps in knowledge about the ethical, legal, and social implications, and perhaps forecast such consequences of this undertaking, the National Institutes of Health (NIH) and the Department of Energy (DOE) each fund an Ethical, Legal, and Social Issues (ELSI) program. Funds for each agency's ELSI effort derive from a set aside of 3 to 5 percent of appropriations for the total genome initiative budget. In fiscal year 1991, DOE's ELSI spending was $1.44 million (3 percent). Fiscal year 1992 spending is targeted at $1.77 million (3 percent). NIH-ELSI spending for fiscal years 1990 and 1991 has been $1,558,913 (2.6 percent) and $4,037,683 (4.9 percent), respectively. For fiscal year 1992, NIH-ELSI aims to spend 5 percent of its human genome appropriation. Several grants supported by NIH/DOE ELSI relate to factors affecting CF carrier screening.

SOURCE: Office of Technology Assessment, 1992.

of survival. Individuals with CF produce thick, sticky mucus. Chronic obstruction and infection of the airways characterize respiratory difficulties and result in lung damage that leads to pulmonary and heart failure. Digestive problems are also common and often predominate over respiratory symptoms early in life. Poor nutrition and impaired growth result because food—particularly fat and protein—is not broken down and absorbed properly.

**Table 1-1—Incidence of Cystic Fibrosis Among Live Births in the United States**

| Population | Incidence (births) |
|---|---|
| Caucasian | 1 in 2,500[a,b,c] |
| Hispanic | 1 in 9,600[d] |
| African American | 1 in 17,000[a,e] to 1 in 19,000[f] |
| Asian American | 1 in 90,000[f] |

[a]T.F. Boat, M.J. Welsh, and A.L. Beaudet, "Cystic Fibrosis," *The Metabolic Basis of Inherited Disease*, C.R. Scriver, A.L. Beaudet, W.S. Sly, et al. (eds.) (New York, NY: McGraw Hill, 1989).
[b]K.B. Hammond, S.H. Abman, R.J. Sokol, et al., "Efficacy of Statewide Neonatal Screening for Cystic Fibrosis by Assay of Trypsinogen Concentrations," *New England Journal of Medicine* 325:769-774, 1991.
[c]W.K. Lemna, G.L. Feldman, B.-S. Kerem, et al., "Mutation Analysis for Heterozygote Detection and the Prenatal Diagnosis of Cystic Fibrosis," *New England Journal of Medicine* 322:291-296, 1990.
[d]S.C. FitzSimmons, remarks at Fifth Annual North American Cystic Fibrosis Conference, Dallas, TX, October 1991.
[e]J.C. Cunningham and L.M. Taussig, *A Guide to Cystic Fibrosis for Parents and Children*, (Bethesda, MD: Cystic Fibrosis Foundation, 1989).
[f]I. MacLusky, F.J. McLaughlin, and H.R. Levinson, "Cystic Fibrosis: Part 1," *Current Problems in Pediatrics*, J.D. Lockhart (ed.) (Chicago, IL: Year Book Medical Publishers, 1985).

SOURCE: Office of Technology Assessment, 1992.

There is no cure for CF. Treatment focuses on managing the respiratory and digestive symptoms to maintain a stable condition and lengthen lifespan. Again, because of CF's varied progression, the regimen and level of therapy depend on the individual. Most therapy involves home treatment (e.g., chest physical therapy to clear mucus from the lungs), outpatient care at one of more than 110 clinics devoted specifically to CF health care, and occasional hospital stays. Today, physicians can look to an ever-expanding array of new pharmaceutical options to manage the care of CF patients; on the horizon are hopes for gene therapy (box 1-C).

Over the last half-century, treatment of CF has evolved so that an illness nearly always fatal in early childhood is now one where life expectancy into adulthood is common. Fifty years ago, most infants born with CF died in the first two years of life. In 1990, median survival was 28 years (figure 1-1)—i.e., of the individuals born with CF in 1962, half were alive in 1990. According to the Cystic Fibrosis Foundation and others, the life expectancy of an infant born with CF in 1992 cannot be estimated, but a few individuals speculate such survival might be 40 years. On the other hand, data from Canada show the steady increase in lifespan since 1940 has plateaued in the last decade. Currently, the median age of an individual with CF in the United States is 12.6 years (figure 1-2).

### *Box 1-C—Cystic Fibrosis Therapies on the Horizon*

In the last several years, scientists have dramatically increased their comprehension of the intricate cascade of processes that ultimately destroy the airways and lead to death in people with CF. With greater knowledge comes targeted strategies to fight the condition. Established CF pulmonary treatments of the past few decades concentrated on fighting infection and clearing airway mucus. Today, new therapies for CF focus on many facets of ameliorating the disease. Some treatments aim to prevent infection and subsequent inflammation altogether. These therapies attempt to intervene at specific junctures in the disease process by decreasing the viscosity of lung secretions, protecting the airway from destruction and preventing infection, or correcting the ionic imbalance.

Two substances—DNase and amiloride—thin CF lung secretions, each through a different mechanism. Both are in clinical trials for approval by the U.S. Food and Drug Administration (FDA). Administration of adenosine triphosphate and uridine triphosphate in conjunction with the diuretic amiloride stimulates choride ion secretion, which is faulty in people with CF; clinical studies also are being carried out for this therapy.

Ironically, the body's natural infection-fighting defense mechanism contributes to the destruction of airways in individuals with CF. Clinical trials are also under way for substances known as antiproteases—including alpha-1-antitrypsin, secretory leukocyte protease inhibitor, and a compound known as ICI 200,880. Antiproteases can protect the airway epithelium from injury mediated by the body's natural bacteria-fighting substances. Finally, although still in the early research stages, recent in vitro evidence demonstrates that cyclic-AMP-stimulating drugs can positively affect chloride balance in some cells from CF patients, suggesting a future avenue for pharmaceutical intervention.

Gene therapy holds the promise of overcoming the condition, perhaps permanently. Unlike treatments that attack symptoms of CF, gene therapy focuses on directly altering DNA to rectify deficits of the disease. In theory, new DNA can be inserted into faulty cells to compensate for the genetic defect. Currently, gene therapy for CF is in the animal experiment stage. Using a crippled virus, the normal human CF gene has been administered directly to the lungs of rats by aerosol spray. Scientists demonstrated this DNA was functional 6 weeks after transfer to the rat lungs—i.e., the genetically engineered DNA was producing normal, human CF gene product. Aerosolized liposomes, fatty capsules that can transport drugs directly into cells, have been used to deliver alpha-1-antitrypsin genes into rabbit lungs, and a similar mechanism might be used to deliver the CF gene to human lungs. Despite significant experimental progress, hurdles remain for gene therapy for CF to be feasible in humans. Long-term safety of the procedure will need to be demonstrated, as will the most appropriate means of transferring the gene and duration of treatment.

SOURCE: Office of Technology Assessment, 1992.

**Figure 1-1—Median Survival of U.S. Cystic Fibrosis Patients Over Time**

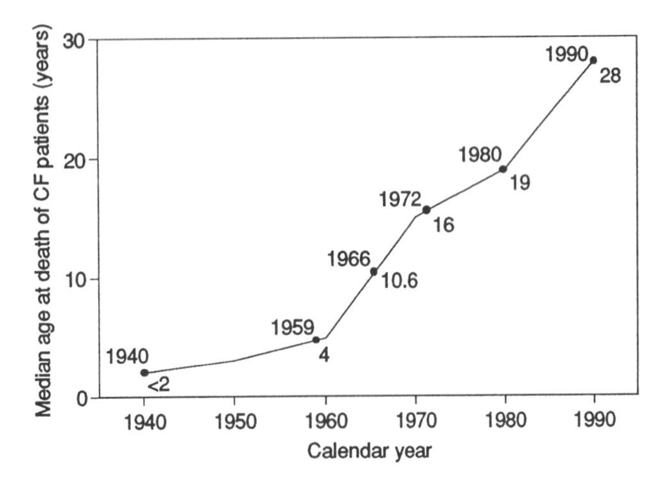


SOURCE: Office of Technology Assessment, 1992, based on S.C. FitzSimmons, "Cystic Fibrosis Patient Registry, 1990: Annual Data Report," Cystic Fibrosis Foundation, Bethesda, MD, January 1992; and S.C. FitzSimmons, Cystic Fibrosis Foundation, Bethesda, MD, personal communication, February 1992.

As with any chronic illness, individuals with CF experience emotional and social strains beyond the physical tolls of the disorder. Children, adolescents, and adults with CF react differently to the condition. For the family of a child with CF, the disease can dominate family activities, particularly if daily therapy is necessary, as is often the case. But while the emotional burden of CF can be difficult, many individuals and their families lead happy, satisfying lives.

## *The Cystic Fibrosis Gene*

CF is a genetic illness transmitted from parents to their children via genetic instructions stored in DNA (figure 1-3). In humans, DNA stores these directions, including those responsible for CF, in genes arrayed on 46 structures called chromosomes (figure 1-4). The gene responsible for CF lies on chromosome 7.

**Figure 1-2—Age Distribution of U.S. Cystic Fibrosis Patients in 1990**

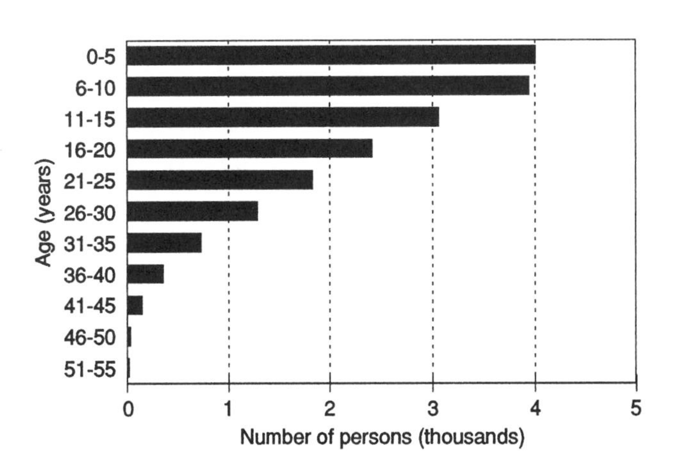


SOURCE: Office of Technology Assessment, 1992, based on S.C. FitzSimmons, "Cystic Fibrosis Patient Registry, 1990: Annual Data Report," Cystic Fibrosis Foundation, Bethesda, MD, January 1992.

**Figure 1-3—The Structure of DNA**

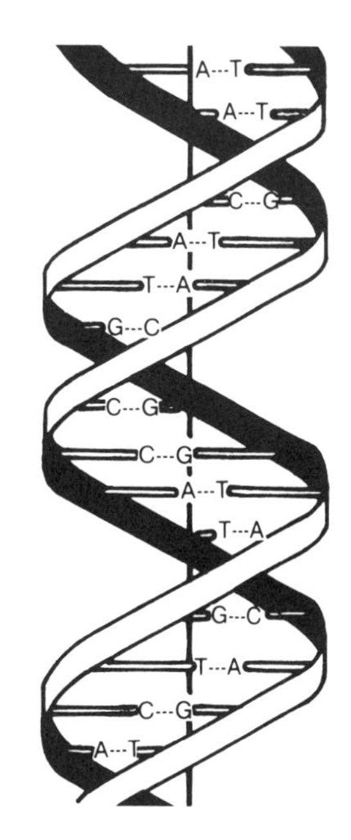


SOURCE: Office of Technology Assessment, 1992.

**Figure 1-4—Human Chromosomes**

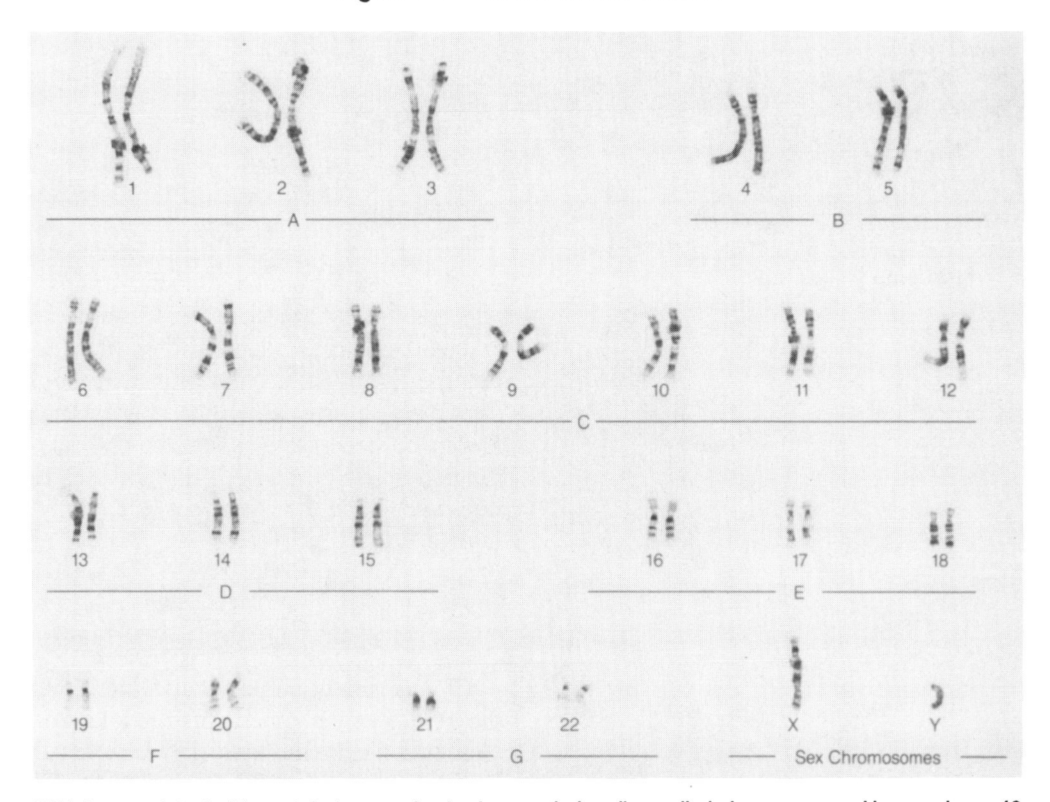


DNA is associated with protein in organized microscopic bundles called chromosomes. Humans have 46 chromosomes: 1 pair of sex chromosomes (two X chromosomes for females; an X and a Y for males) and 22 pairs of autosomes. In 1986, scientists localized the CF gene specifically to chromosome 7.

SOURCE: Vivigen, Inc., Santa Fe, NM, 1992.

Since the 1940s, geneticists have known that CF's pattern of inheritance typifies a recessive condition. For recessive disorders like CF, parents display no symptoms of the disorder, but are asymptomatic carriers. All individuals have two chromosome 7s, but for CF, a carrier mother or father has one chromosome 7 with a CF mutation and one without. The single copy of the nonmutant CF gene in carriers is sufficient to maintain normal physiologic functions. A child is born with CF when he or she inherits the mutant CF gene from each parent—i.e., the child has two chromosome 7s with one CF mutation on each.

The CF gene is distributed over 250,000 contiguous base pairs on chromosome 7 (figure 1-5).

**Figure 1-5—The Cystic Fibrosis Gene**

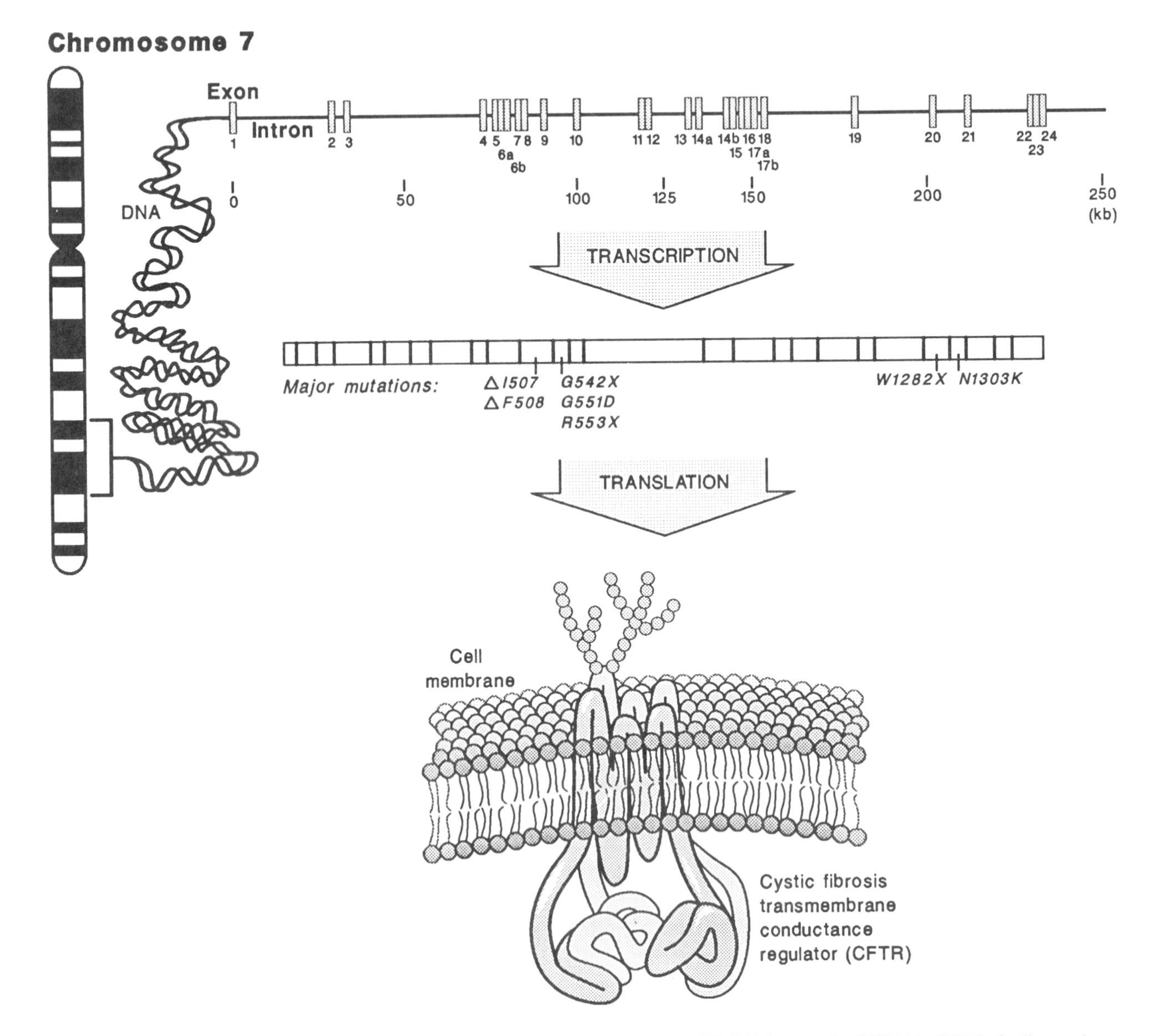


The CF gene is located on the long arm of chromosome 7, where it is spread over 250,000 base pairs (250 kb) of DNA. Coding regions of the DNA, or exons, are separated by noncoding regions, or introns. After the DNA is transcribed into messenger RNA (mRNA) comprised of all 27 exons of the gene, the mRNA is exported from the cell nucleus. Finally, instructions in the mRNA are translated, using special structures in the cell to assemble 1,480 amino acids into the final protein product.

SOURCE: Office of Technology Assessment, 1992, based on M.C. Iannuzzi and F.S. Collins, "Reverse Genetics and Cystic Fibrosis," *American Journal of Respiratory Cellular and Molecular Biology* 2:309-316, 1990.

***Box 1-D—The Gene Product: The Cystic Fibrosis Transmembrane Conductance Regulator***

Cells cannot pump water, but must move fluids across their membranes through a process called osmosis. Osmosis depends largely on ion movement through pores in the membrane (channels) or through transport systems designed to convey ions from one side of the membrane to the other. In individuals with CF, regulation of a particular type of ion transport (chloride; $Cl^-$) is defective.

The product of the CF gene, a protein called the cystic fibrosis transmembrane conductance regulator (CFTR), mediates $Cl^-$ ion flow across membranes. Current evidence suggests that CFTR functions as a channel for $Cl^-$ ions. When the gene carries a ΔF508 or other mutation, it produces a defective CFTR, which in turn disrupts ion flow and results in the physiological effects distinctive of CF (e.g., skin with a salty taste and thick mucus). As the workings of CFTR are clarified, new possibilities for treatment arise.

Conceivably, elucidation of the structure and function of CFTR could facilitate assaying CF carrier status without using DNA analysis. Such assays theoretically could offer an immediate advantage over DNA-based tests. Currently, more than 170 different CF mutations exist, and hence more than 170 assays are necessary to detect them. A functional test could measure the presence of normal or altered CFTR to distinguish unaffected, carrier, or affected individuals. One test might be able to detect the defective CFTR protein no matter which of the 170+ mutations the individual had.

Despite expectations that a functional CFTR test could obviate the need for DNA-based CF carrier tests (and eliminate uncertainty for individuals whose tests are negative), one does not appear imminent. While research to understand CFTR continues to advance rapidly, some of the results appear to cloud, not clarify, the future of a functional test to identify CF carriers. CFTR activity differs depending on the cell type and methods used to measure its activity. In vitro activity also does not correlate with prognosis. Depending on the mutation, a gradient of activity exists; some mutated CFTRs still exhibit activity, while others show none. This variability would make black and white interpretation of a functional assay impossible, and perhaps less informative than DNA analyses.

SOURCE: Office of Technology Assessment, 1992.

Scientists know, however, that not all of these bases get translated into the ultimate CF gene product, called the cystic fibrosis transmembrane conductance regulator (CFTR) (box 1-D). What is also known is CF's pathology stems from a faulty CFTR, and that in most people with CF a three-base pair deletion in each of their CF alleles results in the flawed CFTRs. This three-base pair mutation occurs at position number 508 in the CFTR (abbreviated as delta F508 (ΔF508)). More than 170 additional mutations in the CF gene also lead to faulty CFTRs. Individuals with CF have two of the same, or two different, mutations.

About 70 percent of CF carriers have the ΔF508 mutation.[2] International studies demonstrate ethnic and regional variation in the frequency distribution of this mutation (figure 1-6); as expected, the multicultural nature of the United States reflects this variation. Most of the other 170+ mutations appear in a small fraction of individuals or families, although a few occur at a frequency as great as 1 to 3 percent.

Predicting the precise clinical course of CF—mild versus severe—cannot be done from knowing which mutations are present. Some symptoms (or their lack of severity), however, correlate with particular mutations. Digestive difficulties from pancreatic insufficiency, for example, generally associate with ΔF508.

### *Cystic Fibrosis Mutation Analysis*

With localization of the CF gene, ΔF508, and other CF mutations, it is now possible to directly analyze DNA from any individual for the presence of CF mutations (figure 1-7). Using today's technologies, CF mutation analysis is usually a one-time

[2] Quoted mutation frequencies for ΔF508 and other CF mutations always depend on racial and ethnic background. Throughout this report, OTA presents current expert estimates of appropriate ranges of detection frequencies or sometimes uses a specific figure with qualification (e.g., about 90 percent; approximately 95 percent). OTA adopts such language to avoid restating each time that a frequency depends on racial and ethnic background, not to underemphasize the importance in the distribution variation of CF mutations. In some cases—made clear within the text—a specific frequency is chosen for illustrative or hypothetical purposes.

**Figure 1-6—Occurrence of ΔF508 Mutation in Europe**

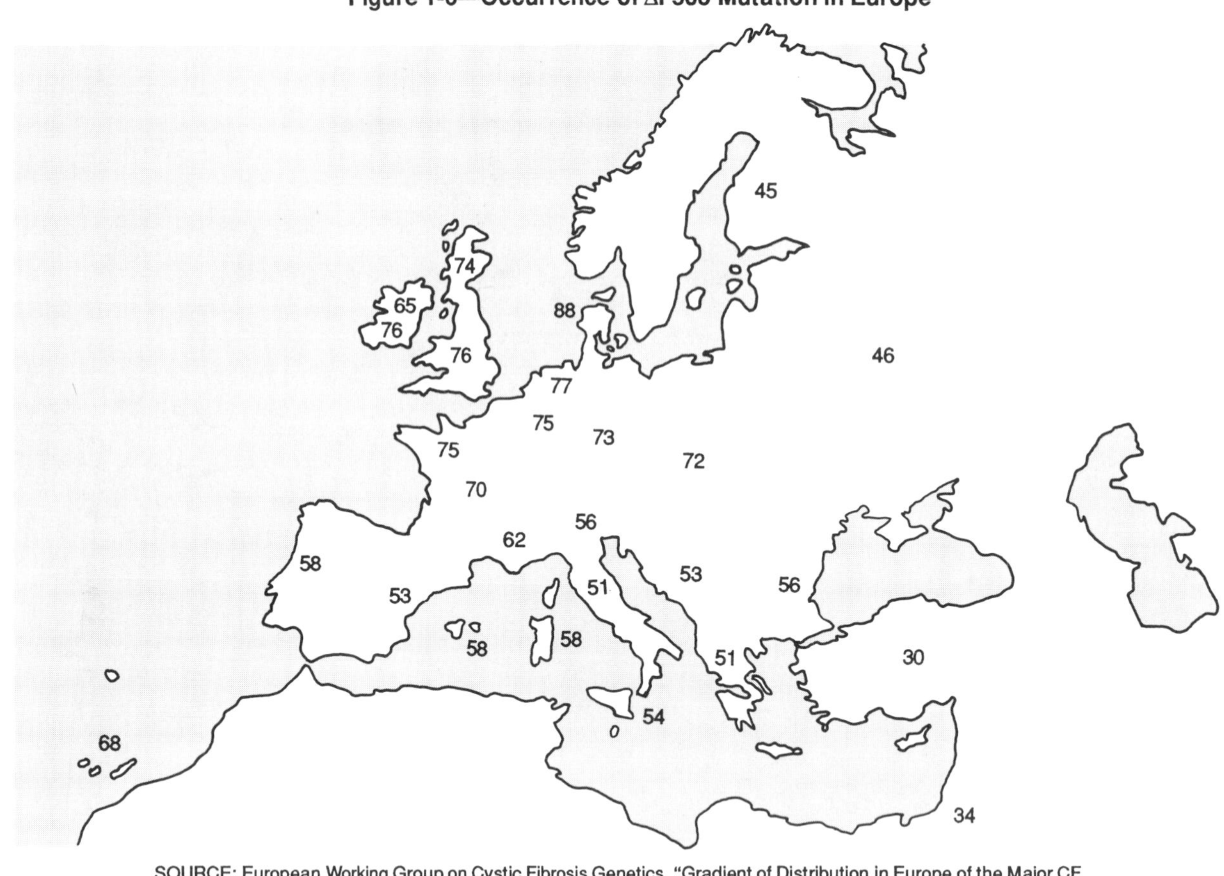


SOURCE: European Working Group on Cystic Fibrosis Genetics, "Gradient of Distribution in Europe of the Major CF Mutation and of Its Associated Haplotype," *Human Genetics* 85:436-445, 1990.

test that can inform an individual whether he or she carries a CF mutation. Carrier *screening* for CF (or CF carrier screening) refers to performing CF mutation analysis on DNA from an individual who has no family history of CF.

Current technology, however, can leave ambiguity, but not because the tests per se are imprecise. Properly performed, DNA-based tests for CF mutations are accurate and specific—meaning if the ΔF508 mutation (or another CF mutation for which the test is run) is present in the individual's genome, the assay detects it more than 99 percent of the time, absent laboratory error. Instead, ambiguity stems from the intrinsic nature of the cause of the disease: Besides ΔF508, more than 170 mutations in the CF gene also cause CF.

In the United States, about 1 in 25 Caucasians carries one CF mutation. Since tests to detect 170+ mutations are impractical, current assays use ΔF508 plus 6 to 12 other CF mutations (ΔF508+6-12) and identify 85 to 90 percent of CF carriers (in Ashkenazic Jews, ΔF508+6 identifies about 95 percent of carriers).[3] Thus, using ΔF508+6-12 means 10 to 15 percent of actual carriers go undetected. In other words, a negative test result does not guarantee that a person is not a carrier.

As mentioned earlier, a child with CF is born only to couples where each partner is a carrier of one CF mutation—though not necessarily the same one for each partner. Such couples are sometimes referred to as carrier couples, or couples who are positive/positive (+/+). For these couples, the chance of having a child with CF is 1 in 4 for each pregnancy. If a couple is positive/negative (+/-)—the father is a carrier, but the mother is not, or vice versa—their offspring can be CF carriers, but cannot have CF.

[3] Again, using ΔF508 alone identifies about 70 percent of CF carriers among American Caucasians of European descent.

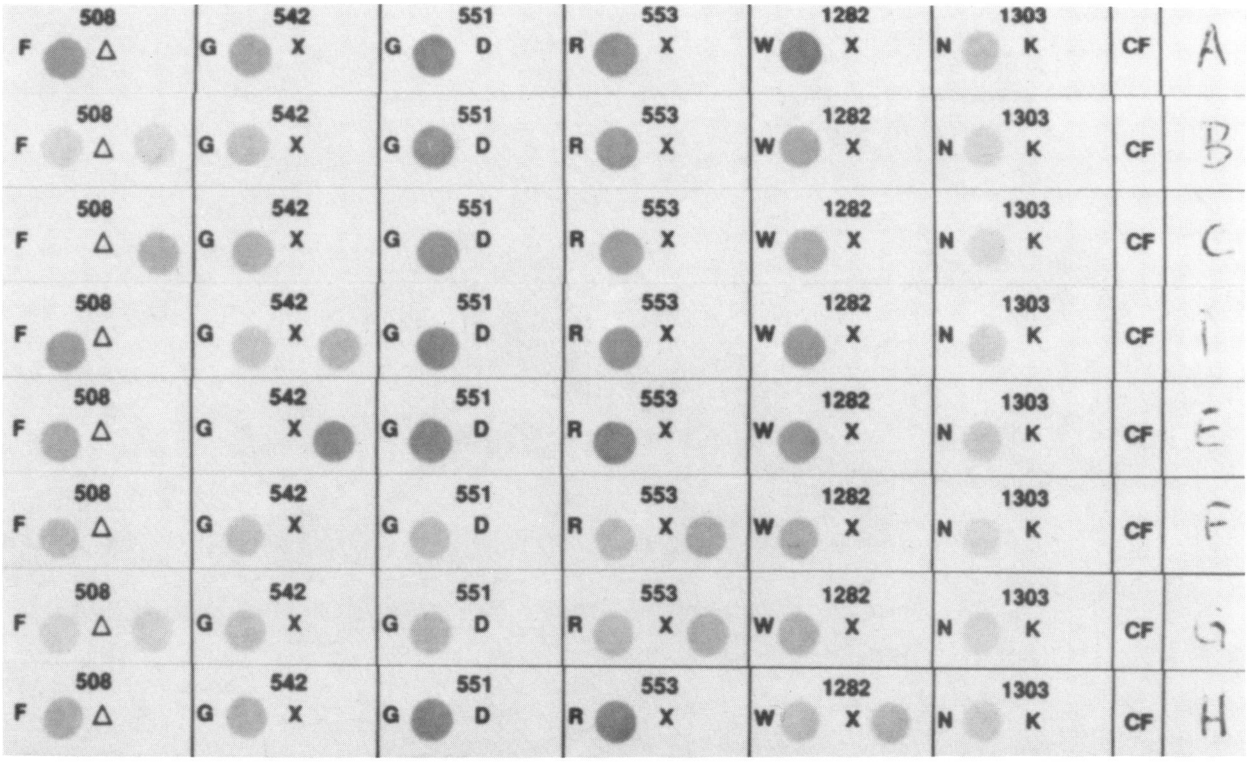

*Photo credit: Roche Molecular Systems, Inc.*

DNA analysis for six common CF mutations. Unique pieces of DNA, called allele specific oligonucleotide probes, are bound to the test strip to detect six common CF mutations; in this photograph, each individual strip runs horizontally. DNA samples from individuals of unknown CF status are obtained, processed, and applied to separate test strips. Here, test strips for eight different individuals are shown (rows A through H). Following hybridization and colorimetric analysis, the patterns of dots on the strips are revealed—and hence the CF status of the individuals.

For each mutation on the strip (ΔF508, G542X, G551D, R553X, W1282X, and N1303K) the left dot, if present, indicates the person has a normal DNA sequence at that part of the CF gene. The right dot, if present, indicates the person has a CF mutation at that site. Individual A, then, has no CF mutations at the six areas of the CF gene analyzed using this test strip, as demonstrated by single dots on the left side for all mutations. In contrast, individuals B,D,F, and H are carriers, as demonstrated by the presence of two dots for one of the CF mutations. Individual C has CF, as demonstrated by a single dot on the right side of the ΔF508 panel; individual E has CF, as demonstrated by the single dot on the right side of the G542X panel. Individual G also has CF, but this person's CF arises from two different mutations—ΔF508 and R553X—as indicated by the pairs of dots in each of these panels.

Using ΔF508+6-12 means that some couples receive test results that indicate one partner is a carrier and one is not, when in fact the negative partner carries one of the rare CF mutations that is not assayed (figure 1-8). Thus, while most couples whose test results are +/- are at zero risk of having a child with CF, some couples with a +/- test result actually are couples whose genetic status is +/+ (but goes undetected) and who are at 1 in 4 risk of a child with CF for each pregnancy. Couples with a +/- test result, then, might misunderstand that their reduced risk of bearing a child with CF is not zero and have a false sense of security about having an unaffected child. If, for example, 100,000 couples experienced a first-time pregnancy, 40 fetuses would be expected to have CF. Prenatal CF mutation analysis with 85 percent sensitivity could detect about 29 fetuses, but 11 would be missed. A few couples who receive a -/-result will also be undetected carrier couples (box 1-E; table 1-2).

## WHY IS CYSTIC FIBROSIS CARRIER SCREENING CONTROVERSIAL?

Prospects of routine CF carrier screening polarize people. Everyone agrees that persons with a family history of CF should have the opportunity to avail themselves of CF mutation analysis, yet controversy swirls around using the same tests in the general

### Figure 1-7—Techniques for DNA Analysis of Cystic Fibrosis Mutations

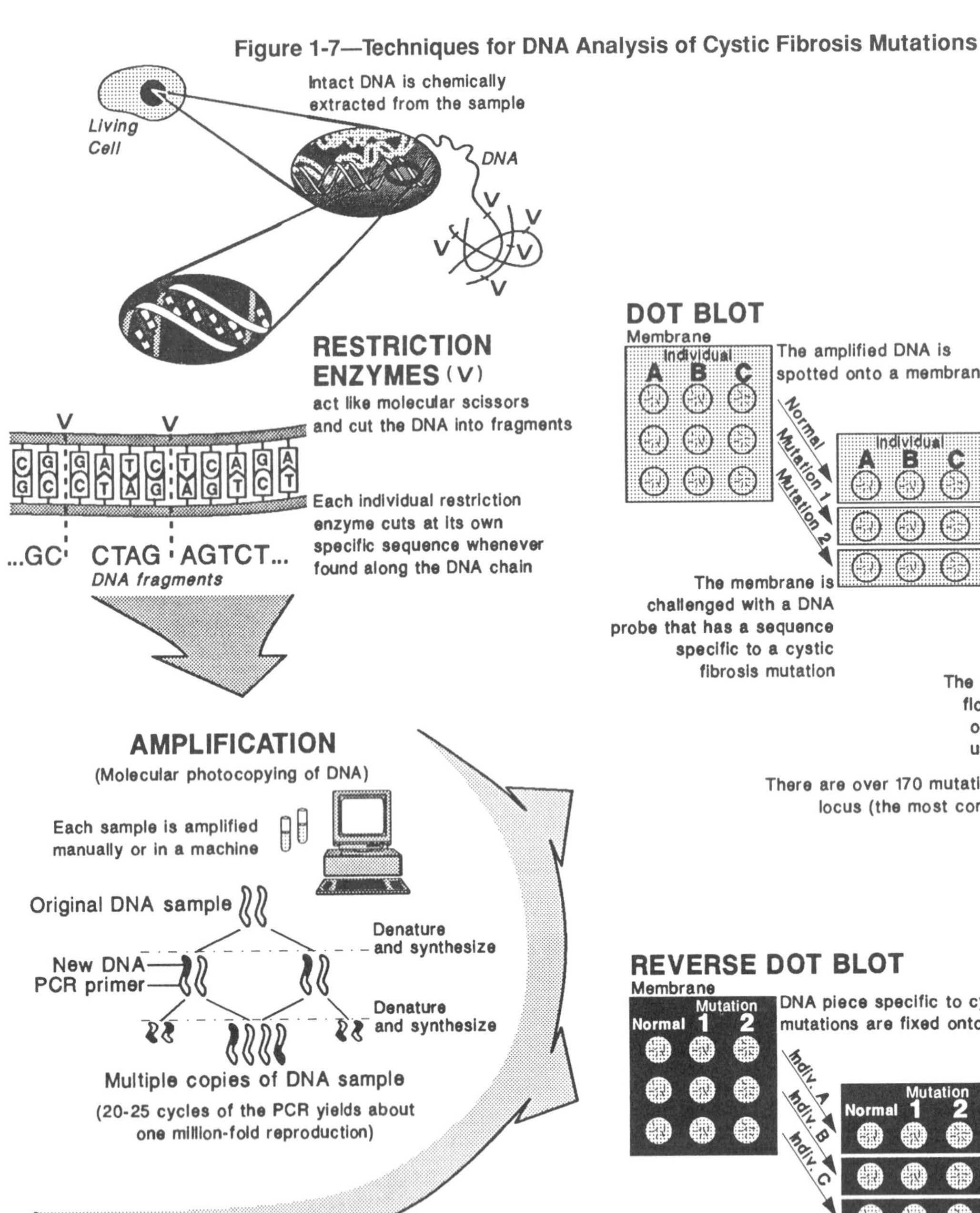


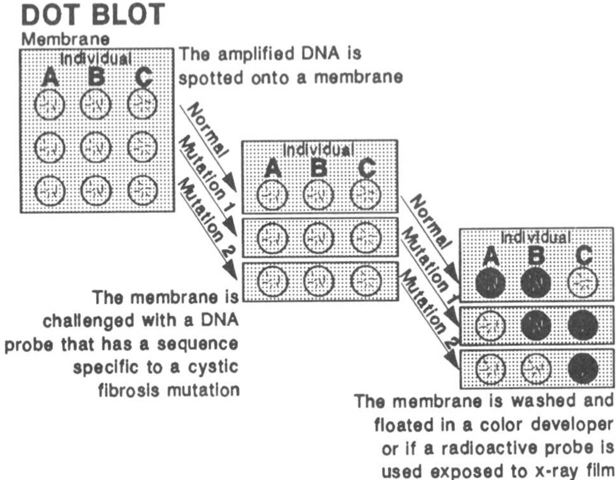

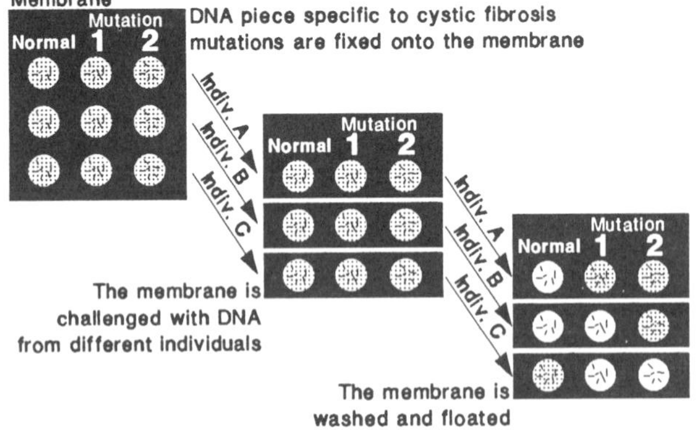


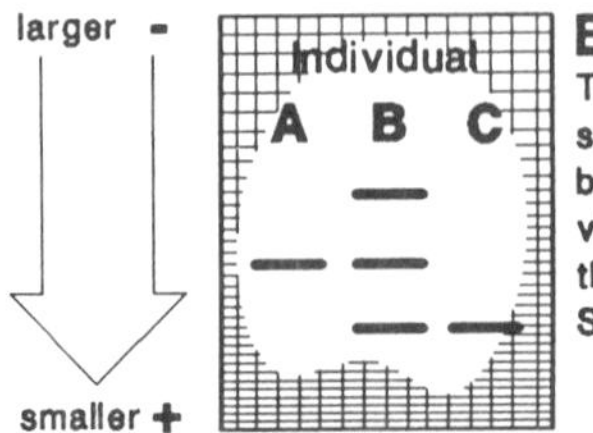


**ELECTROPHORESIS**
The DNA fragments are separated by size into bands in a gel and visualized directly or through a process called Southern blotting

SOURCE: Office of Technology Assessment, 1992.

**Figure 1-8—Cystic Fibrosis Mutation Test Results at 85 Percent Sensitivity**

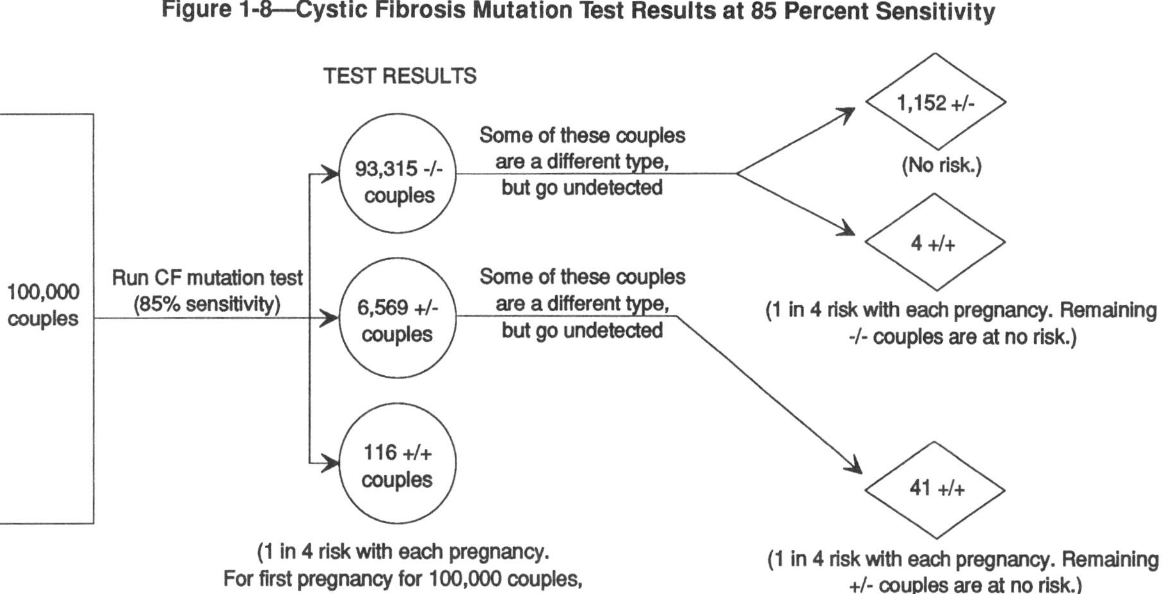


SOURCE: Office of Technology Assessment, 1992.

## *Box 1-E—Cystic Fibrosis Carrier Tests and Detection Sensitivity*

In theory, 4,000 carriers exist among 100,000 random Americans of European descent, because the carrier frequency in this population is about 1 in 25. However, ΔF508+6-12 assays detect about 85 percent of people with CF mutations, so CF carrier screening of this group would identify 3,400 of the 4,000 probable carriers. If the test were 100 percent specific, all 4,000 carriers would be identified.

Similarly, if 100,000 random *couples* were screened, 160 couples would be identified as +/+ (each partner a carrier) if the test were 100 percent sensitive. One-fourth of first-time pregnancies for the 160 +/+ couples would be expected to result in CF-affected fetuses, for a total of 40 expected CF-affected fetuses per 100,000 couples. Instead, at 85 percent sensitivity, about 116 couples will be identified as +/+ and with each pregnancy have a 1 in 4 risk of a child with CF. Results for 93,315 will be -/- (neither identified as a carrier), and about 6,569 couples will have +/- test results (one partner a carrier, the other not identified as a carrier). In fact, approximately 41 of the 6,569 couples with +/- test results are at 1 in 4 risk of bearing a child with CF in each pregnancy, while the remaining 6,528 have no risk—but these two groups cannot be distinguished with an 85 percent test sensitivity (figure 1-8). About 4 of 93,315 couples with -/- test results are also actually at 1 in 4 risk with each pregnancy of having a child with CF.

Thus, of the theoretical 160 +/+ couples, 116 are detectable and 44 are not when the test is 85 percent sensitive. If all 100,000 couples experience a first-time pregnancy, 40 fetuses with CF are expected. With an 85 percent sensitive test, 29 fetuses with CF are detectable via prenatal tests, but 11 will be missed. If the assay elucidates 95 percent of carriers, 144 of 160 couples would be detected. In this case, if all 100,000 couples experience a first-time pregnancy, 36 fetuses with CF could be detected and 4 would be missed.

With a test that detects 85 percent of individuals with CF mutations, a couple whose result is +/- has approximately a 1 in 661 risk of having an affected child with each pregnancy (compared to a general population frequency of about 1 in 2,500). At a detection sensitivity of 95 percent, a couple with a +/- result faces a 1 in 1,964 risk of an newborn with CF with each pregnancy. Detecting a greater proportion of carriers means couples with +/-results can be less anxious about their risk of having a child with CF. Couples who both test negative, while not having zero risk, have a 1 in 109,200 risk of an affected child with each pregnancy at 85 percent test sensitivity.

SOURCE: Office of Technology Assessment, 1992, based on A.L. Beaudet, Howard Hughes Medical Institute, Houston, TX, personal communications March 1992, April 1992; and W.K. Lemna, G.L. Feldman, B. Kerem, et al., "Mutation Analysis for Heterozygote Detection and the Prenatal Diagnosis of Cystic Fibrosis," *New England Journal of Medicine* 322:291-296, 1990.

**Table 1-2—Test Sensitivity and Risk of Child With Cystic Fibrosis**

| Percent mutations detected | Couples at 1 in 4 risk with each pregnancy[a] | | | | Affected fetuses in first pregnancy[a] | | |
|---|---|---|---|---|---|---|---|
| | Actual | +/+ result | +/- result | -/- result | Actual | Detectable | Missed |
| 85 | 160 | 115.6 | 40.8 | 3.6 | 40 | 28.9 | 11.1 |
| 90 | 160 | 129.6 | 28.8 | 1.6 | 40 | 32.4 | 7.6 |
| 95 | 160 | 144.4 | 15.2 | 0.4 | 40 | 36.1 | 3.9 |

[a] per 100,000 couples.

SOURCE: A.L. Beaudet, Howard Hughes Medical Institute, Houston, TX, personal communication, March 1992.

population. What are the elements of the controversy? Can past experiences with other carrier screening initiatives and current research from CF carrier screening pilots resolve some issues?

## *Today's Clinical and Social Tensions*

For years, experts theorized about confronting the potential consequences of increased knowledge of human genetics. In the early 1990s, the CF mutation test moves the debate from the theoretical to the practical. Today, along with clinical tensions surrounding CF carrier screening, are legal, ethical, economic, and political considerations.

No mandatory genetic screening programs of adult populations exist in the United States; OTA finds it highly unlikely that CF carrier screening will set a precedent in this regard. Nevertheless, people disagree about how CF carrier screening of the general population should be conducted.

Proponents of a measured approach to CF carrier screening express concern about several issues that might be raised if use of CF carrier tests becomes routine. Invariably, discussions about CF carrier screening raise concerns about the use of genetic information by insurance companies and become linked to broader social concerns about health care reform in the United States. Related to this are concerns about commercialization of genetic research, i.e., that market pressures will drive widespread use of tests before the potential for discrimination or stigmatization by other individuals or institutions (e.g., employers and insurers) is assessed. Also expressed are questions about the adequacy of quality assurance for DNA diagnostic facilities, personnel, and the tests themselves. Opponents of widespread CF carrier screening also wonder whether the current number of genetic

*Photo credit: Lauren A. Moore*

Approximately 1 in 25 American Caucasians of European descent, 1 in 46 Hispanic Americans, 1 in 60 to 65 African Americans, and 1 in 150 Asian Americans are carriers for CF. About 25 carriers would be expected among this crowd. Current technology would detect 85 to 95 percent of these individuals, depending on their ethnic backgrounds.

specialists can handle a swell of CF carrier screening cases, let alone the cases from tests for other genetic conditions expected to arise from the Human Genome Project. Finally, the extraordinary tensions in the United States about abortion affect discussions about CF carrier testing and screening.

Those who advocate CF carrier tests for use beyond affected families are no less concerned about the issues just raised. Rather, proponents argue that individuals should be routinely informed about the assays so they can decide for themselves whether to be voluntarily screened. They assert that the tests are sensitive enough for current use and will, like most tests, continually improve. These voices believe that failing to inform patients now about the availability of CF carrier assays denies people the opportunity to make personal choices about their reproductive futures, either prospectively—e.g., by avoiding conception, choosing to adopt, or using artificial insemination by donor—or by using prenatal testing to determine whether a fetus is affected.

### *Lessons From Past Carrier Screening Efforts*

Carrier screening is not new to the United States. The 1970s and early 1980s saw a number of genetic screening efforts flourish throughout the country. Federal legislation—chiefly the National Sickle Cell Anemia, Cooley's Anemia, Tay-Sachs, and Genetic Diseases Act (Public Law 94-278; hereinafter the National Genetic Diseases Act) and its predecessors—fueled these programs. Today, what might work for CF carrier screening—and what will not work—can be gleaned from carrier screening for other genetic disorders, even though earlier screening occurred through more centralized efforts. In fact, some argue that creating a defined, federally funded program for CF carrier screening could avoid social concerns, although others assert the contrary.

Frequently considered a successful effort, Tay-Sachs carrier screening was initiated in 1971 at the behest of American Jewish communities. Tay-Sachs disease is a lethal, recessive genetic disorder that primarily affects Jews of Eastern and Central European descent and populations descended from French Canadian ancestors. It involves the central nervous system, resulting in mental retardation and death within the first years of life. Fourteen months of technical preparation, education of medical and religious leaders, and organizational planning preceded massive public education campaigns. Since screening commenced, over one-half million adults have been voluntarily screened; today, it is a part of general medical care.

In contrast, sickle cell programs in the 1970s are generally cited as screening gone wrong. The sickle cell mutation—which like the Tay-Sachs and CF mutations is recessive—affects hemoglobin, the oxygen-carrying molecule in blood. The sickle cell mutation is found predominantly in African Americans and some Mediterranean populations. Most individuals with sickle cell anemia live well into adulthood. Unlike Tay-Sachs screening, much sickle cell screening was mandatory. For the most part, Caucasians designed and implemented programs targeted toward African Americans, leading to proclamations of racist genocide. Even after elimination of most mandatory screening in the late 1970s, actual practice strayed from the stated goals of adequate genetic counseling, public education, and confidentiality of results.

Tay-Sachs carrier screening and sickle cell screening—along with carrier screening for other genetic conditions (e.g., $\alpha$- and $\beta$-thalassemia)—provide perspective for today's discussions about CF carrier screening. Two lessons in particular are clear: Participation should be voluntary and public education is vital. Disagreement exists, however, about the degree to which CF carrier screening can draw on the Tay-Sachs and sickle cell experiences to resolve other considerations (e.g., discrimination). Several factors contribute to questions raised about comparability, including: Today's political climate differs; CF carrier screening has the potential to involve larger numbers of people; and Tay-Sachs and sickle cell screening were implemented, in part, with explicit Government funding in a more programmatic fashion than will be likely for CF carrier screening.

### *Cystic Fibrosis Carrier Screening Pilot Studies*

Opponents of routine CF carrier screening argue that historical perspectives fall short of adequately addressing potential adverse consequences raised by widespread utilization of CF mutation assays, including adequate education and counseling, and prospects for discrimination and stigmatization. They assert that until data are gathered from federally funded pilot projects specific to CF, carrier screening should not be routine. Proponents, on the other hand, argue that sufficient information is available from privately supported CF carrier screening projects, that much historical experience applies, and that any incremental gain that will be gleaned from federally funded studies is insufficient to a priori prevent routine CF carrier screening from proceeding.

### Federally Funded Studies

Despite pleas throughout the genetics community for the Federal Government to fund pilot projects to assess clinical and social considerations raised by the new CF mutation analyses, initial calls for funding of pilots went wanting. In the United Kingdom, the CF Research Trust actively funded and encouraged pilots (box 1-F)—unlike the CF Foundation in the United States, which has focused on investigations to find the CF gene and mutation, but divorces itself from CF carrier screening. Concern about abortion apparently played a major role in the latter policy decision.

After some scrambling, the Ethical, Legal, and Social Issues (ELSI) Program of the National Center for Human Genome Research (NCHGR), National Institutes of Health (NIH), stepped forward to coordinate federally financed pilot studies. In October 1991 (fiscal year 1992), three units of NIH—the National Center for Human Genome Research, the National Institute of Child Health and Human Development, and the National Center for Nursing Research—launched a 3-year research initiative to analyze education and counseling methods related to CF mutation analysis.

Seven research teams, conducting eight studies, received support and will coordinate their efforts (box 1-G). Two of seven clinical studies focus on relatives of individuals with CF (CF carrier testing); the other five focus on the general population. One study involves theoretical modeling. Where appropriate, some features of the research, such as evaluation measures and tools, cost assessment, laboratory quality control procedures, and human subjects protection will be standardized across sites.

### Privately Funded Studies

Prior to the onset of federally sponsored pilot projects, several public and private institutions began to systematically offer CF carrier screening to subsets of the population; pregnant women and their partners, preconceptional adults, teenagers, and fetuses all have been target populations. Most privately funded efforts have been under way since early 1990, and most have collected, or are collecting, data on the incidence of carrier status and mutation frequencies. Some also follow psychosocial issues such as levels of anxiety and retention of information. Most studies can report results, and the various strategies used and different target populations reflect the lack of consensus on the best approach to CF carrier screening (table 1-3).

## WHAT FACTORS WILL AFFECT UTILIZATION?

Initially, routine CF carrier screening will likely occur in the reproductive context; the prenatal population has been the traditional entry point into genetic services for many people. Preconceptional individuals are also a possible population, but for most individuals the first real opportunity for carrier screening takes place post-conception. A focus on pregnant women, however, is not without controversy. Reservations exist about abortion, as do concerns that prenatal testing negatively shapes perceptions of pregnancy, disability, and women. Nevertheless, the primary responsibility for providing CF carrier screening could come to reside with obstetricians, as has occurred with maternal serum alpha-fetoprotein (MSAFP) screening to detect fetuses with neural tube or abdominal wall defects or Down syndrome.

Based on the annual number of births (4.2 million) and spontaneous abortions (an estimated 1.8 million), there are approximately 6 million pregnancies per year for which CF carrier screening might be performed. Twenty-four percent of women giving birth receive no prenatal care until the third trimester, however, so CF carrier screening in the obstetric/prenatal context could initially involve, at most, 10 million[4] men and women per year, depending on who is screened.

For some, the key question still hovering over carrier screening for CF is if, not when. For others, however, the debate has shifted to when. Several institutions already offer CF mutation analysis to individuals, regardless of family history. OTA pro-

[4] This figure does not account for the estimated 2.4 million infertile couples who are trying to conceive and might be interested in CF carrier screening (would increase overall figure). Nor does it estimate the number of Americans not involved in a pregnancy (would increase), the number of individuals involved in more than one conception per year (would decrease), or those who might have been screened during a previous pregnancy (would decrease).

### *Box 1-F—Cystic Fibrosis Carrier Screening in the United Kingdom*

At least five pilot projects exploring the implications of population screening for CF carriers are under way in the United Kingdom; Italy, Denmark, and Austria also have pilot projects. The U.K. pilot projects, begun in 1990, are the most extensive and advanced studies under way. Although health care delivery in the United Kingdom—and the other countries with pilots—differs significantly from that in the United States, some U.K. strategies and results could bear on how routine CF carrier screening might be approached in the United States. Shared concerns include test protocols and techniques, the appropriate target population, psychological aspects (such as anxiety levels for those contemplating CF carrier screening), and the role of primary care providers.

St. Mary's Hospital Medical School in London is evaluating preconception CF carrier screening through three general practice and three family planning clinics. Individuals are approached while they wait for appointments for other reasons, or contact is made by letter prior to a visit (i.e., opportunistic screening). About 66 percent of people approached through general practitioners request screening, and 87 percent of individuals contacted in the family planning clinic seek the test. Participants have been asked how they thought their future reproductive plans might be affected if both they and their partners were found to be carriers. For those with no experience with CF, 38 percent say they would choose not to have children, 78 percent would request prenatal diagnosis should they conceive, and 16 percent would not consider terminating an affected pregnancy. For those who had a relative or knew someone with CF, 45 percent felt they would not have children, 82 percent would seek prenatal diagnosis should they conceive, and 20 percent would not terminate an affected pregnancy. Through mid-1991, St. Mary's had screened about 1,600 individuals at approximately 50 samples a week.

The pilot at St. Bartholomew's Hospital in London, funded entirely through private sources, only offers "couples screening," which solely aims to identify couples at 1 in 4 risk of an affected pregnancy—i.e., couples in which both partners are CF carriers (+/+ couples). Couples are screened and receive their results as a unit—either high or low risk of bearing a child with CF; individual carrier status is not discussed. Even if the geneticist determines that one partner is a carrier, but the other is not (a +/- couple), that couple is informed the same as a couple who both screen negative (a -/- couple): low risk. Because of ethical concerns about concealing information, couples screening is blind. Samples are gathered from each partner, with one randomly screened. If the test is negative, the couple is informed they are at low risk and the second sample goes unscreened. If the sample is positive, the other sample is tested. If the second sample is negative (a +/- couple), the couple is informed they are at low risk, without either being informed that one of them is, in fact, a carrier. Proponents feel this approach is more economical, and believe it reduces the anxiety associated with knowing one's carrier status, since results are reported as a unit. Most observers agree that such a practice would be considered legally and ethically dangerous in the United States.

Funds for three U.K. pilot projects derive from the Cystic Fibrosis Research Trust (CF Trust). They target different populations and seek, in part, to evaluate different parameters. Screening through the University Hospital of Wales in Cardiff is offered opportunistically to adults between 16 and 45 years; prenatal screening will be part of the pilot in 1993. Investigators in Wales evaluated mouthwash, buccal scrapes, and finger pricks as methods for sample collection and concluded mouthwash is the most desireable overall for patient acceptability, successful DNA extraction, and cost. The pilot at Western General Hospital in Edinburgh, Scotland, focuses on prenatal screening, with a long-term goal of preconception screening. The Edinburgh pilot first screens the woman for three mutations; if she is positive, her partner is tested for 15 mutations. Through 1991, over 2,000 samples had been processed, detecting 74 carriers. Guy's Hospital in London offers carrier screening to individuals 18 to 45 years to assess screening through an urban general practice setting. All projects devote considerable effort to examining acceptability, evaluating maternal and paternal anxiety, assessing self-esteem and perceptions of stigma, and developing effective educational material for patients and professionals.

SOURCE: Office of Technology Assessment, 1992.

### Box 1-G—Federally Funded Cystic Fibrosis Carrier Screening Pilot Projects

In October 1991, the National Institutes of Health funded eight clinical assessments of CF carrier testing and screening at seven institutions.

*Children's Hospital Oakland Research Institute, Oakland, CA ($73,196).* Adult siblings of CF patients and their spouses will be interviewed to identify factors motivating or interfering with the pursuit of CF carrier testing in siblings and their partners. In addition to examining interest in testing, this study aims to assess understanding of results, knowledge of medical aspects of CF, and psychological impact following testing.

*Johns Hopkins University, Baltimore, MD ($314,449).* The level of general interest in learning about CF of families and individuals receiving care from a health maintenance organization will be examined. In particular, the study will consider: what factors distinguish those interested in participating in a CF education program from those who are not; examining the characteristics that differentiate people who agree to screening from participants who decide against it, and comparing the responses of individuals identified as CF carriers to those identified as noncarriers, with emphasis on the extent to which these responses are influenced by marital or carrier status of the partner.

*UCLA School of Medicine, Los Angeles, CA ($179,067).* Women of reproductive age and the partners of those who test positive will be screened, including large numbers of Hispanic and Asian Americans, two groups that have not been studied extensively for either their CF mutation frequencies or their response to screening and counseling. Pre- and post-test questionnaires will be used to determine understanding of CF, predictors of consent to screening, and responses to implications of the test results for the various ethnic and socioeconomic subgroups. Strategies for pre- and post-test counseling will be evaluated for effectiveness.

*University of North Carolina, Chapel Hill, NC ($231,916).* Relatives of individuals with CF will receive pretest education, either from a pamphlet in a private physician's office or in a traditional genetic counseling setting. Effectiveness of a precounseling video will be evaluated. Investigators will assess genetic and medical knowledge, psychological status, and selected health behaviors before and after participants receive their test results.

*University of Pennsylvania, Philadelphia, PA ($197,634 and $180,201).* Decision theory and economic techniques will be used to model decisionmaking about CF carrier screening. The study will address: who should be offered screening and the best method; the best course and sequence after results are delivered; rescreening negative individuals as more mutations are identified; and the impact of future treatment on CF carrier screening. Monetary and nonmonetary effects of the alternative strategies raised by these issues will be assessed, as well as the response to screening of groups—i.e., patients, health care providers, and insurers—with varying financial, psychological, and moral perspectives.

A separate clinical study will complement the theoretical work. It will analyze the decisionmaking of couples who are offered CF carrier screening one partner at a time, and whether they choose to have the second partner screened after a negative result for the first. When screening should be offered will be investigated.

*University of Rochester, Rochester, NY ($274,110).* CF mutation analysis will be offered to women of reproductive age to determine what proportion desires it, what proportion that elects screening comprehends test results, and what proportion of partners of screened women elects screening. Anxiety, comprehension, requests for prenatal diagnosis despite low risk, and program costs will be assessed.

*Vanderbilt University, Nashville, TN ($206,513).* The feasibility of a program that incorporates pre- and post-test education for people with negative results, and provides personal counseling to those who test positive, will be evaluated. Written and video materials will be developed. Different settings in which CF carrier screening is offered will be examined, as will factors that affect a couple's decision whether or not to be screened.

SOURCE: Office of Technology Assessment, based on National Center for Human Genome Research, National Institutes of Health, October 1991.

jects approximately 63,000 individuals will be screened for their CF carrier status in 1992—about a 7-fold increase over 1991 (figure 1-9). This rapid upward trend is expected, given the nascent stage of the technology's movement into U.S. medical practice.

Without offering judgment on its appropriateness or inappropriateness, OTA finds that the matter of CF carrier screening in the United States is one of when, not if. Regardless of the number of individuals actually screened, it is clear that, increasingly, patients will be informed about the availability of CF carrier assays and a portion will opt to be screened. What is less clear is the timeframe for physicians to begin routinely informing patients about CF carrier tests. It could be within a year or two, but more likely will be a gradual process over several years. What

## Table 1-3—Privately Funded Cystic Fibrosis Carrier Screening Pilot Projects

| Institution | Target population | Approach | Findings |
|---|---|---|---|
| Baylor College of Medicine (Houston, TX) | Prenatal and preconceptional couples, with and without family history. | Two stages of mutation analysis. Both partners concurrently screened for ΔF508+5. For +/- couples, the negative partner is analyzed for 12 additional mutations at no extra charge. | From 1990-91, 64 at-risk pregnancies detected, of which 14 affected fetuses were diagnosed. Fifty percent of these were electively terminated. No +/-couples requested prenatal fetal diagnosis, no pregnancies were terminated, and clinical evaluation did not indicate undue anxiety.<br>CF carrier screening has been routinely offered ($100 per couple) since September 1991 to all couples of reproductive age who have contact for any reason with Baylor's genetic services. |
| Cornell University Medical College (New York, NY) | Initially, couples with no family history but enrolled in prenatal diagnosis program for other services; currently all couples of reproductive age coming to genetic services, regardless of pregnancy status. | Initially ΔF508; since July 1991 ΔF508+W1282X (at least 30 percent of couples are Ashkenazic Jews). Negative partner in +/-couples is screened for an additional four mutations. | As of March 1992, more than 500 couples screened using a mouth rinse specimen at $100 per couple. About one-third of those offered choose to participate. Followup questionnaires indicate all appear to understand that some at-risk couples will be missed. Virtually all agree screening should be continued, should not be limited to those ethnic groups where detection is highest, nor should be suspended until tests detect more carriers. Primary reason for participation: an interest in learning something relevant to the health of the current pregnancy. Two reasons most often cited by nonparticipants: carrier risk perceived as low or referring physician had not specifically recommended test. |
| Genetics & IVF Institute (Fairfax, VA) | Women undergoing amniocentesis or chorionic villus sampling (CVS), primarily for advanced maternal age, offered concurrent CF mutation analysis. Some had family history. | Initially ΔF508; currently with ΔF508+6. | As of August 1991, 1,327 CVS patients (44 percent) and 370 amniocentesis patients (21 percent) opted for fetal carrier screening. Fifty pregnancies identified as carrier fetuses, 47 to couples with no family history. Twelve couples declined further testing; remaining 38 sought testing for themselves. |
| McGill University (Montreal, Canada) | High school students. | ΔF508 | Conducted in May 1990, 40 percent of about 600 students chose to participate; two carriers were identified. Interviews of these individuals and their families revealed they were positive toward their new knowledge; other family members requested testing.<br>Followup questionnaires revealed participants who were negative were reasonably well-informed about the clinical phenotype and inheritance of CF. Most understood negative test did not rule out carrier status and were satisfied they had participated. |
| Permanente Medical Group, Inc. of Northern California-Integrated Genetics (Framingham, MA)-Vivigen (Santa Fe, NM) | Pregnant women of European Caucasian descent or Hispanic ethnicity. | Woman is screened first for ΔF508+5 mutations, with sequential screening for ΔF508+11 of partner if woman is positive. | As of March 1992, 78 percent of women offered CF mutation analysis have accepted (As enrollees of the Kaiser Permanente health maintenance program, there is no out-of-pocket expense.)<br>Kaiser has developed an informational and educational videotape to test on control and experimental groups, and is using several psychosocial survey instruments to assess individuals' understanding of pathology and genetics of CF, both before and after screening. Once 5,000 individuals have participated, Permanente Medical Group will decide whether, and how, to proceed with CF carrier screening of plan members. |
| Roche Biomedical Laboratories (Research Triangle Park, NC) | Prenatal couples. | Samples collected simultaneously from both partners. Woman's sample screened first for ΔF508+3; if positive, partner's sample screened for ΔF508+3. | Project is nationwide, since prior to initiation in July 1991, a letter of announcement was sent to 100 obstetricians around the country.<br>CF mutation analyses are performed on buccal cell samples (mouth scrape) collected at home. The brushes are placed in color-coded tubes for each sex, and mailed directly to Roche by the individuals. Originally intended to last 6 months, the timeframe has been extended to 1 year, since subscription rate has been less than expected (50 percent as of September 1991). |

SOURCE: Office of Technology Assessment, based on R. Barathur, Specialty Laboratories, Santa Monica, CA, personal communication, September 1991; J.G. Davis, New York Hospital, New York, NY, personal communication, March 1992; S.D. Fernbach, Baylor College of Medicine, Houston, TX, personal communication, January 1992; F. Kaplan, C. Clow, and C.R. Scriver, "Cystic Fibrosis Carrier Screening by DNA Analysis: A Pilot Study of Attitudes Among Participants," American Journal of Human Genetics 49:240-243, 1991; J.D. Schulman, Genetics & IVF Institute, Fairfax, VA, personal communication, December 1991; and D.R. Witt, Kaiser Permanente, San Jose, CA, personal communication, March 1992.

Figure 1-9—Cystic Fibrosis Carrier Screening, 1989-92

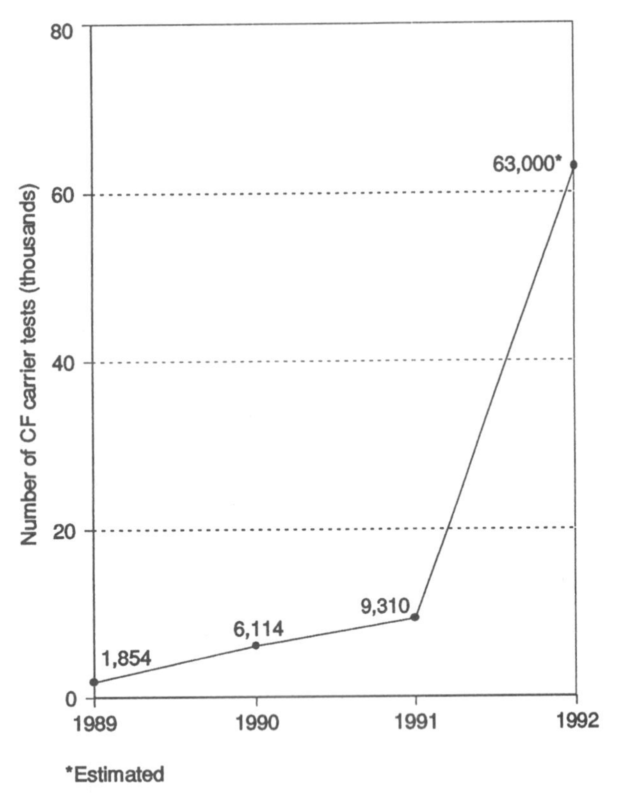


*Estimated

SOURCE: Office of Technology Assessment, 1992.

factors affect—or will affect—routine carrier screening for CF? Eight aspects predominate:

- genetic services delivery and customs of care,
- public education,
- professional capacity,
- financing,
- stigmatization, classification, and discrimination issues,
- quality assurance of clinical laboratories and DNA test kits,
- automation, and
- costs and cost-effectiveness.

Of these issues, all but cost-effectiveness extend beyond CF to global concerns about future tests to assess other genetic risks. This section describes OTA's findings in each of these areas. Presented later is an analysis of what policy issues emerge from these findings and Congress' role in shaping the debate raised by these issues.

## *Genetic Services: Standards of Care and Ensuring Quality*

One broad question expresses a facet of the current clinical controversy: Who serves as gatekeeper of a new technology? The degree to which large numbers of Americans opt to learn their CF carrier status depends first on their interaction with the genetic services system in the country. Utilization of DNA-based CF mutation analysis will depend on the extent to which physicians, genetic counselors, and other health professionals customarily inform individuals about the test's availability. In turn, moving from innovation to standard practice often depends on professional guidelines or statements. Disagreement exists about the applicability of CF carrier tests to individuals without positive family histories, which has led to tensions, with opposite sides questioning the motives of the other. Additionally, consumer acceptance will depend on perceptions that the professional services they receive with screening are of high quality.

### Standards of Care

Should all individuals be informed about tests to identify CF carrier status? Society has no definitive way of determining when physicians should routinely advise people about the availability of tests that could reveal their propensity to have a child with a genetic disorder. Physician practice might be driven by consumer demand, patient autonomy, liability fears, economic self-interest, or a combination of these factors. CF carrier screening presents a classic instance of the perennial problem of appropriately controlling the evolution of practice standards as a new technology becomes available. Thus, deciding the appropriate timing for routinely telling everyone about CF mutation tests is a contentious issue.

Physicians can now offer individuals with no family history of CF a test that can determine, with 85 to 95 percent sensitivity, whether they are CF carriers. With professional opinion in a state of flux—and knowledge of the assay's existence continuing to spread among patients—physicians might wonder whether they are obligated to inform patients of its availability, even before patients ask about it.

Some consumers are interested in genetic tests and CF carrier screening. A 1986 OTA telephone survey of a national probability sample of adult Americans reported that about 9 of 10 approved of making genetic tests available through doctors. Eighty-three percent said they would take a genetic test before having children, if it would tell them whether their children would probably inherit a fatal genetic disease.[5] OTA's 1991 survey of genetic counselors and nurse geneticists found that 18.5 percent of respondents said they were "frequently" or "very frequently" asked by clients about DNA-based CF tests; about 71 percent said the number of inquiries increased from 1989 to 1991. On the other hand, some physicians report that actual willingness to undertake CF carrier screening is currently modest. In part, such reticence stems from the cost of CF mutation analysis, which patients must generally self-pay. It might also arise from a barrier common to many types of medical screening: lack of interest and reluctance to uncover what might be perceived as potentially unpleasant news.

Generally, physicians are obligated to inform patients of the risks and benefits of proposed procedures, so that patients themselves may decide whether to proceed. Where a patient specifically asks about a test, physicians would seem obligated to discuss the test, even if they do not recommend that it be taken. Whether physicians are obligated to query patients about their potential interest in a test the provider views as unwarranted by the patient's circumstances depends on the customary practice of similarly skilled and situated physicians.

Customary practice is often determined by the courts, and courts view statements issued by a relevant professional society as evidence of what a reasonably prudent physician might have done. In mid-1992, after extended discussion, the leadership of the American Society of Human Genetics (ASHG) approved a revised statement that CF mutation analysis "is not recommended" for those without a family history of CF. Some argue that the subtle change in language of the new statement retreats from the absoluteness of a 1990 ASHG statement that stated routine CF carrier screening is "*NOT* yet the standard of care." This view holds that the new statement reflects an evolution of debate within the society—that some believe CF carrier screening *may now be offered* to individuals without a family history of CF, although it might not be the "standard of care." Others argue that ASHG's position is unchanged—that the new statement is tantamount to restating that CF carrier screening *should not be offered* to individuals without a family history of CF. In either case, the statement cannot be interpreted to mean that CF carrier screening *should be offered* to all individuals. The 1990 and 1991 policy statements of professional societies and participants in an NIH workshop stated that CF carrier screening should not be the standard of care.

Today, some physicians take their cues strictly from the early guidelines; the extent to which the 1992 ASHG statement will affect physician practice remains to be seen. Others have concluded that a general population incidence of 1 child with CF per 2,500 births, coupled with the test's imperfect detection sensitivity, makes routinely informing patients about CF mutation analysis unnecessary. Additionally, some physicians might choose not to inform patients of the availability of CF mutation analysis because they judge that the test is too psychologically risky or too expensive to be worth the possible benefits for those without a family history of CF. Still other providers might be unaware of the test or its possible benefits.

Some physicians, however, disagree with existing guidelines and have already chosen to incorporate CF screening into their practices. They believe the assays are sufficiently sensitive for general use, and that even patients with unknown risks of conceiving a child with CF should now have the information to exercise choice in managing their health care. Still other physicians might be offering the assay out of concern that failing to could subject them to charges of medical malpractice if a couple has a child with CF and a court subsequently finds that CF carrier screening had become the standard of care—despite professional statements to the contrary. These practitioners might be concerned by the few cases where courts held that limited adoption of a practice by some professionals is sufficient to call into question the reasonableness of the defendant's practice—regardless of the extent to which that practice was accepted generally by the profession or suggested by professional societies. In fact, with respect to CF carrier screening, customary physician practice might evolve faster than that recommended by physicians' own professional societies, as has occurred for other practices such as amniocentesis.

[5] Survey respondents were not specifically questioned about CF.

## Duties of Care for Genetic Counseling

Once a decision is made to offer information about tests for CF carrier screening—or to provide the assay itself—at least three important issues arise: what constitutes quality genetic counseling, confidentiality of information, and compensation for inadequate counseling or breach of confidentiality.

***Components of Genetic Counseling.*** A genetics professional must understand enough about the patient's health, his or her reproductive plans, and available technologies so that an appropriate family history can be obtained and necessary analyses ordered. Less than this could give patients grounds to complain of a false assurance of safety. More than most aspects of medicine and counseling, genetic counseling involves family issues and family members. For a nonspecialist, it might be enough to recognize the need for a referral.

Having elicited information and obtained test results, the provider must communicate the results in a meaningful way. Translating technically accurate information into understandable information is difficult, but essential. Effective communication also entails recognizing and understanding religious, psychosocial, and ethnocultural issues important to the client and his or her family. People interpret genetic risk information in a highly personal manner and can misperceive, misunderstand, or distort information. For CF carrier screening, an important aspect involves explaining the reproductive risks the client faces and what the condition involves. Perceptions of relative risk significantly affect qualitative decisions. Some consumers could mistake the assay's resolution and perceive that a negative result from use of the latest DNA technology means no risk.

No standard for genetic counseling exists. Some argue in favor of a standard based on what patients would want to know (modeled after informed consent requirements) because there is no fixed professional norm as an alternative, and because adequacy of the information conveyed turns more on the values of the patient being counseled than on professional norms. The prevailing approach in genetic counseling, however, appears to be based on a review of what most professionals do, rather than what an individual patient wants.

***Confidentiality.*** Genetics professionals with information on the carrier status of a patient are legally obligated to keep that information confidential except under a few, specific circumstances. At least 21 States explicitly protect patient information pertaining to medical conditions and treatment; it is also part of the case law in many States without specific statutes. Offending physicians can have their licenses revoked or be subject to other disciplinary action. Patients whose confidential records have been revealed can also bring civil suit against the physician or facility.

Photo credit: Beth Fine

Genetic counseling can help individuals and families understand the implications of positive and negative test outcomes.

Not all genetic information, however, must remain confidential. A provider might wish to reveal genetic information to interested third parties without a patient's permission. Health care professionals are not legally liable or subject to disciplinary action if a valid defense exists for releasing a patient's genetic or other medical information. With CF, the professional might desire to inform a patient's relatives that they also could be at higher than average risk of conceiving a child with CF. If the provider is persuaded that the relatives will not be notified—after a patient has been advised to inform relatives that they too could carry a CF mutation—he or she might believe that breaching confidentiality would be appropriate.

The coming years will see a growing number of situations where health professionals will need to balance confidentiality of patients' genetic information against demands from relatives and other third parties for access to that information. Overall, the risk to the third party from nondisclosure must be balanced against the benefit of maintaining the expected confidentiality of the provider-patient setting. A provider contemplating disclosure to a patient's spouse must weigh the patient's own confidentiality against a spouse's interest in sharing decisions concerning conception, abortion, or preparation for the birth of a child with extraordinary medical needs.

***Compensation for Negligent Genetic Counseling.*** Inadequate genetic counseling can result in a number of outcomes. Patients might forego conception or terminate a pregnancy when correct information would have reassured them. People might choose to conceive children when they otherwise would have practiced contraception, or they might fail to investigate using donor gametes that are free of the genetic trait they wish to avoid. Finally, they might lose the opportunity to choose to terminate a pregnancy.

The birth of a child with a genetic condition could result in malpractice claims of wrongful birth or wrongful life. For wrongful birth claims, most jurisdictions allow compensation for negligent failure to inform or failure to provide correct information in time for parents to either prevent conception or decide about pregnancy termination. With regard to CF, at least one court has ruled that parents may collect the extra medical costs associated with managing the condition. In this case, the couple maintained they would have avoided conceiving a second child had their physicians accurately diagnosed CF in their first child and thus identified each parent as a CF carrier. In wrongful life claims, the child asserts he or she was harmed by the failure to give the parents an opportunity to avoid conception or birth. Most U.S. courts have been reluctant to allow damages because they have been uncomfortable concluding that a child has been harmed by living with severe disabilities when the only alternative is never to have been born.

Practitioners who provide inadequate genetic counseling, including failing to recommend needed tests, might be subject to sanctions—from a reprimand to license revocation—by a regulatory body or a professional society. M.D.-geneticists, as physicians, are formally licensed by States. Ph.D.-geneticists and master's-level genetic counselors are not licensed by States, but until 1992 have been certified (along with physicians) by the American Board of Medical Genetics (ABMG). The continued certification of master's-level counselors by ABMG beyond 1992 is uncertain.

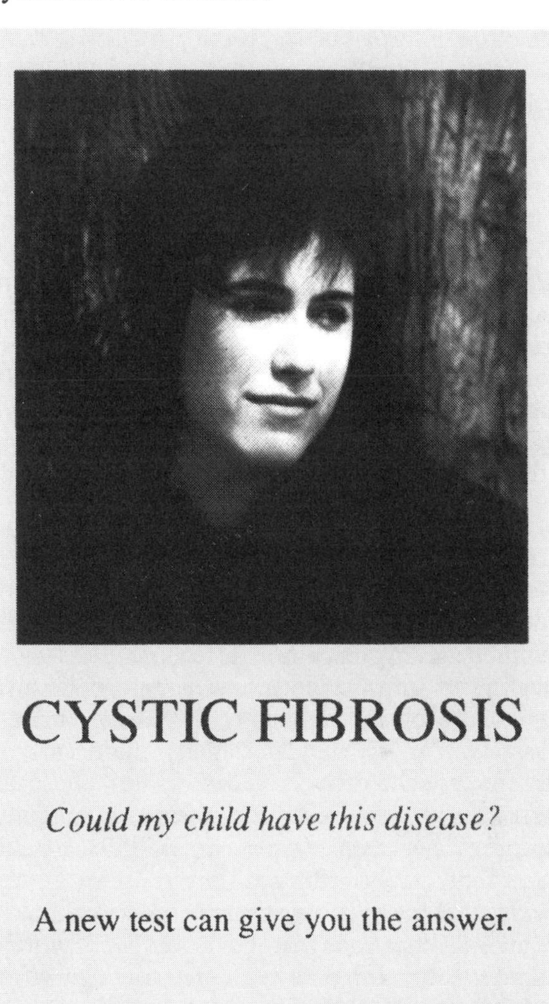


*Photo credit: Peter T. Rowley, University of Rochester School of Medicine*

Educational materials, such as this pamphlet developed at the University of Rochester School of Medicine, Rochester, NY, can be useful for pretest education.

### *Public Education*

Both the way in which a provider communicates information about potential risk to the client (or risk to potential offspring) and the implications of the condition and prognosis influence a client's perception of the information. A person's subjective frame of reference, familiarity with genetics, and ability to understand statistical implications of genetic risks are also important.

Risk perception is always a more important determinant of decisionmaking than actual risk. When confronting the risk of genetic disease in their offspring, and in making reproductive decisions, people tend to place greater weight on their ability to cope with a child with a disability or a fatal disease than on precise numerical risks. One study revealed that regardless of actual risk, parents overwhelmingly see situations as 0 or 100 percent—it will or will not happen—when they believe they cannot cope with the situation.

In addition to subjective factors that influence the interpretation of risk, most individuals have difficulty understanding risk in arithmetic terms, yet comprehending probabilities affects people's understanding of information provided by genetic tests. One study of predominantly Caucasian, middle-class women in Maryland found more than 20 percent thought that "1 out of 1,000" meant 10 percent, and 6 percent of respondents thought it meant greater than 10 percent. A 1991 national survey of public attitudes toward genetic tests reveals that belief in the accuracy of the technology is one of the strongest predictors of favorable attitudes toward genetic tests; that same survey of 1,006 Americans found that less than half were able to answer correctly four of five technical questions regarding genetic tests.

The need for better scientific literacy has been a topic of wide discussion in recent years, and mechanisms to achieve this goal apply equally to genetics education. Increased public education in genetics would benefit individuals' perceptions and understanding about genetic test results—likely reducing time needed for individual counseling.

Public education programs targeted to genetic diseases have been nearly nonexistent since those established under the National Genetic Diseases Act were phased out in 1981. The National Science Foundation (NSF) has supported teacher training programs in genetics for school teachers in Kansas, for example, but no NSF-funded, national effort exists. Teachers who participated in the Kansas program subsequently increased time devoted to genetics instruction at the high-school level by three-fold. Instruction in elementary schools increased 22-fold. More recently, the U.S. Department of Energy (DOE) began funding a 3-year project to prepare 50 selected science teachers per year to become State resource teachers.

Public education can go a long way toward preparing individuals for the decision of whether and when to be screened. Positive and negative experiences with large-scale Tay-Sachs, sickle cell, and α-and β-thalassemia carrier screening programs—in the United States and abroad—demonstrate the value and importance of pretest community education.

### *Professional Training and Education*

Many types of health professionals perform genetic counseling: physicians, Ph.D. clinical geneticists, genetic counselors[6], nurses, and social workers. Critics of widespread CF carrier screening question whether the present genetics counseling system in the United States can handle the swell of cases if CF carrier screening becomes routine.

Currently, about 1,000 master's-level genetic counselors practice in the United States. An additional 100 nurse geneticists provide similar services. The ABMG has certified 630 professionals in genetic counseling, including master's-level genetic counselors, nurses, and M.D. and Ph.D. geneticists. If genetic counseling for CF carrier screening were to fall only to board-certified professionals, the available number of professionals might be short of what is needed. OTA's survey of genetic counselors and nurses in genetics also indicates that respondents believe routine CF carrier screening will strain the present genetic services delivery system. Respondents estimated that, on average, 1 hour would be needed to obtain a three-generational family history and to discuss CF carrier screening and genetic risks.

[6] Again, OTA uses the term "genetic counselor" to specifically describe master's-level individuals certified by the ABMG (or board-eligible) because legal distinctions in licensing and reimbursement for services exist among the different types of professionals who perform genetic counseling.

Skeptics of a personnel shortage assert that counseling about CF carrier assays is likely to take place in the general obstetric/prenatal context, however, and they believe 1 hour exaggerates the amount of time that suffices for all prenatal tests, let alone only CF carrier screening. Furthermore, counseling related to CF carrier screening is likely to extend beyond board-certified individuals to include other physicians and allied health professionals. For example, an unknown number of social workers, psychologists, and other public health professionals perform genetic counseling, often to minority and underserved populations.

Ultimately, the issue of adequate services and professional capacity could turn on the extent to which patients receive genetic services through specialized clinical settings, as they largely do now, versus access through primary care, community health, and public health settings. Overall, OTA cannot conclude whether increased numbers of genetic specialists are necessary—arguments exist pro and con. One finding is clear: Increased genetics education for all health care professionals is desirable. Routine carrier screening for CF—and tests yet to be developed for other genetic conditions—will require adequate training and education of individuals in the broader health care delivery system.

Increasing professional education in genetics will not be an easy task. The average 4-year medical school curriculum includes 21.6 hours of genetics instruction. Fifteen master's-level programs in genetic counseling exist, producing approximately 75 graduates per year. Of 200 U.S. universities that offer graduate nursing degrees, only 4 offer programs providing a master's-level genetics major. Only 9 of nearly 100 accredited social work graduate programs in the United States offer special courses on genetic topics. Few schools of public health offer genetics as part of their curriculum; none requires it.

Federal support for genetic services, education, and training has changed dramatically since 1981. Prior to 1981, genetics programs applied through their State for Federal funds under the National Genetic Diseases Act (Public Law 94-278). With creation of the Maternal and Child Health (MCH) Block Grant (Public Law 97-35), State genetic services now compete with other maternal and child health initiatives (box 1-H). Additionally, Federal spending on demonstration projects for service delivery, training, and education has declined after adjustment for inflation. Training support for master's-level genetic counselors is minimal. The U.S. Department of Health and Human Services (DHHS) provides no financial support for training genetic counselors or for improving genetics education in medical schools. Through support to the Council of Regional Networks for Genetic Services (CORN), DHHS provides funds for some continuing professional genetics education programs for physicians, but not for other genetics professionals.

### *Financing*

Health insurance in the United States is not monolithic. U.S. health care financing, which totaled more than $800 billion in 1991, is a mixture of public and private funds. Federal financing includes Medicare, Medicaid, and the Civilian Health and Medical Program of the Uniformed Services (CHAMPUS). Private funding mechanisms include self-funded plans, commercial health insurance plans, Blue Cross and Blue Shield (BC/BS) plans, health maintenance organizations (HMOs), self-pay, and nonreimbursed institutional funding. State high-risk pools—generally using public and private monies—are also an option in some States for people who cannot obtain private health insurance. Rules and regulations governing each sector vary.[7] Thus, separating how the current financing paradigm might affect CF carrier screening—and vice versa—is difficult.

For the majority of Americans, access to health care, and the health insurance that makes such access possible, is provided through the private sector. Some acquire health insurance on their own through individual policies; 10 to 15 percent of people with health insurance have this type of coverage. Of group policies, about 15 percent have some medical underwriting—i.e., medical and genetic information are used to determine eligibility and premiums for health insurance. A large majority of insured individuals and their family members—163 million of the 214 million with health care coverage—obtain coverage via employer-offered large group policies with no medical underwriting. The employer, in

[7] Benefit packages offered by the different providers vary, as do laws governing them. Except for self-funded company health insurance plans, State laws govern both group and individual private health insurance. Thus, a patchwork of laws and regulations oversees commercial insurers. Laws and regulations for commercial insurers differ from those for BC/BS plans. HMOs are regulated by States, with some Federal guidance.

### *Box 1-H—Genetic Services: Federal-State Partnership*

Funding for genetic services derives from a medley of Federal and State sources, and varies greatly from State to State. During the 1970s, genetic services enjoyed substantial Federal funding, in part through congressional mandate. The Omnibus Budget Reconciliation Act of 1981 (Public Law 97-35), however, led to the consolidation of genetic services funding—along with seven other programs—into the Maternal and Child Health (MCH) Block Grant. Overall, funding for maternal and child health services was cut, and the responsibility for distributing the monies and for providing services was passed to the States, which also had to begin using $3 of State funds for every $4 of Federal money received. Prior to the block grant, no matching funds were required.

Under provisions of the MCH block grant, 85 percent of funds go directly to the States for maternal and child health services. States must decide how to allocate the funds among a number of areas, such as general prenatal care, infant nutritional supplementation, and other maternal and child health needs. MCH funds may be used for health care services, education, and administration. In fiscal year 1990, less than 2 percent of MCH funds were used by States to support genetic services other than newborn screening.

In general, MCH funds account for a small portion of State genetic services. Under terms defined by the block grant, each State decides whether or how much money to designate for genetic services. In 1990, 34 States used MCH funds to support some aspect of general genetic services other than newborn screening, including nonpatient-related activities such as administration and planning. In the majority of States, however, MCH funds accounted for less than 25 percent of fiscal year 1990 funding for genetic services. In fiscal year 1990, MCH funding for genetic services other than newborn screening totalled approximately $8 million; State funding accounted for approximately $22 million.

Fifteen percent of the MCH block grant is administered as direct grants for Special Projects of Regional and National Significance (SPRANS). SPRANS monies are grants for specific projects and are not given to each State. SPRANS provides seed money for demonstration, or pilot, projects in a number of areas. After the demonstration period ends, usually in 3 years, alternative funding must be found.

In fiscal year 1990, genetic services received about 9 percent of all SPRANS funds. When adjusted for inflation, however, constant dollar funding for genetic services under SPRANS has decreased almost every year since the block grant's inception. Moreover, SPRANS support of genetic services has decreased from about 90 percent of the SPRANS genetic services budget in 1981 to approximately 66 percent in 1991. Initially, most of the SPRANS genetic services budget established statewide genetics programs, with each State receiving seed money for at least 4 years. The last State received funding in 1990. Other areas of genetic services delivery receiving SPRANS support include ethnocultural projects to increase utilization of genetic services by underserved populations; psychosocial studies; and support groups for young adults and families. In fiscal year 1990, 16 States used approximately $4 million from SPRANS grants to support demonstration projects in clinical genetic services other than newborn screening. In fiscal year 1990, just over one-third of SPRANS' genetic services budget went to the regional networks and the Council of Regional Networks for Genetic Services (CORN). CORN and the regional networks—comprised of genetic service providers, public health personnel, and consumers—serve as resources for communication and coordinate data collection and quality assurance, but do not provide direct services to patients.

In addition to block grant and SPRANS awards, States also fund genetic services from other sources. In fiscal year 1990, at least 26 States derived $46 million in genetic services funding exclusive of newborn screening from provider in-kind and service charges, third-party reimbursement, grants, contracts, newborn screening fees, health insurance surcharges, and mental health/mental retardation funds. For some States, such funding accounts for most of their genetic services funding. For example, newborn screening fees generated 93 percent of genetic services funding in Colorado and 86 percent in Michigan in fiscal year 1990. Similarly, prenatal screening service fees accounted for more than 83 percent of the genetic services budget in California in fiscal year 1990.

All States, the District of Columbia, and Puerto Rico coordinate genetic services statewide; nearly half experienced a decrease in funding for genetic services from fiscal years 1988 through 1991. Individual State genetic service programs face yearly uncertainty about how much—if any—funding they will receive, which makes planning difficult. As general knowledge and public awareness about genetic diseases continues to emerge out of the Human Genome Project, uncertainty in genetic services funding will be increasingly problematic.

SOURCE: Office of Technology Assessment, 1992 based on E. Duffy, Genetic Services Branch, Maternal and Child Health Bureau, U.S. Department of Health and Human Services, Rockville, MD, personal communication, February 1992; and F.J. Meaney, "CORN Report on Funding of State Genetic Services Programs in the United States, 1990," contract document prepared for the U.S. Congress, Office of Technology Assessment, April 1992.

turn, contracts with a commercial insurer, a BC/BS plan, an HMO, or is self-funded.

Self-funded health insurance plans are group policies that merit specific discussion, since they are creatures of Federal, not State, law. Since enactment of the Employee Retirement Income Security Act of 1974 (ERISA; 29 U.S.C. 1131 et seq.), many companies find self-funding beneficial because their employee benefit plans are not subject to State insurance regulation. With an ERISA plan, the employer directly assumes most or all of the financial liability for the health care expenses of its employees, rather than paying premiums to other third-party payors to assume that risk. Self-funded companies enjoy considerable latitude in designing employee coverage standards. Today, about 53 percent of the employment-based group market is self-funded, and therefore unregulated by the States.

In large measure, the number of people who opt to be screened could hinge on who pays, or will pay, for the cost of CF mutation analyses—the individual or a third-party payor. As mentioned previously, some physicians report that reluctance to undertake CF carrier screening seems to stem from the test's cost. Physicians seeing patients who rely on health insurance to cover part of their expenses usually inform them that their coverage probably precludes reimbursement for CF mutation analysis without a family history of CF,[8] and so if they opt to be screened, they will likely need to self-pay. For laboratories that perform genetic tests, the issue of reimbursement also might be crucial to the ultimate volume of future business in this area.

## Private Sector Reimbursement

Health insurance industry representatives assert that most companies will not pay for tests they consider screening assays. Thus, reimbursement for CF carrier tests in the absence of family history will likely remain on a self-pay basis unless they become part of routine pregnancy care—again, as happened for MSAFP screening.

OTA's 1991 survey of commercial insurers, BC/BS plans, and HMOs[9] confirms these policies for individual contracts or medically underwritten groups. OTA found carrier tests for CF, Tay-Sachs, and sickle cell would not be covered by 12 of 29 commercial insurers offering individual coverage for any reason—screening or family history. No company offering individual insurance or medically underwritten policies would cover CF carrier analysis if a patient requested it, but had no family history. If there is a family history, most companies would pay for carrier tests. Similar results were found for BC/BS plans and HMOs, although a few BC/BS plans and a few HMOs reported they would cover carrier tests performed for screening purposes.

As mentioned earlier, initial carrier screening for CF will likely take place in the context of obstetric/prenatal care. For all three respondent populations, prenatal screening tests for CF generally are not covered without a family history, although more would cover prenatal tests solely at patient request (without family history) than cover general carrier screening. Some respondents covered no prenatal tests.

Respondents were asked to indicate whether they agreed or disagreed with the following scenario:

> Through prior genetic testing, the husband is known to be a carrier for CF. Before having children, the wife seeks genetic testing for CF. The insurance company declines to pay for the testing, since there is no history of CF in her family.

For commercial insurers who write either individual policies or medically underwrite group policies, or both, 21 medical directors (41 percent) agreed strongly or somewhat with this scenario; 28 respondents (47 percent) disagreed somewhat or disagreed strongly. In part, these results reflect OTA's survey

[8] Under the present health care system and current reimbursement policies by insurers, the reality is that the opportunity to be screened depends on the ability to self-pay (except for Medicaid). Thus, questions of access to CF carrier tests and genetic services arise. However, the issue of access to CF carrier screening is no different—and inextricably linked—to the broad issue of health care access in the United States, a topic beyond the scope of this report.

Some contend that until the issue of access is resolved, widespread carrier screening should not proceed. On the other hand, others argue that inequitable access is true for health care in the United States, generally. Supporters of carrier screening for CF question why access to genetic tests and services should be held to a higher standard. In this report, OTA analyzes the issue in the context of today's health care system, but points out that for some opponents of routine CF carrier screening, nonuniversal access is an a priori reason for why CF carrier screening should not proceed.

[9] OTA's survey of health insurers does not measure actual practice, unless otherwise specifically indicated. The information presented here should not be interpreted to represent numbers or percentages of entities who actually have dealt with these issues. Health insurers who write individual policies or medically underwritten groups were asked to speculate how they *would* treat certain conditions or scenarios presented (currently or in the future, depending on the question), not whether they, in fact, *had* made such decisions.

327-046 - 92 - 2 : QL 3

finding that several respondents would not cover any carrier tests, even when medically indicated by a family history. On the other hand, not all respondents who agreed with the scenario represented these companies. These individuals appeared not to understand that the situation was not a case of CF carrier screening, but one of testing to ascertain the couple's risk of conceiving an affected fetus in light of the male's family history.

OTA also found variation in how genetic counseling is covered by commercial insurers, BC/BS plans, and HMOs that offer individual policies or medically underwritten group coverage. OTA's survey of genetic counselors and nurse geneticists confirms these results: Reimbursement for genetic counseling by these professionals is more likely when a family history exists.

Finally, as stated earlier, most people obtain health care coverage through group policies. Determining how these thousands of policies would reimburse for CF carrier screening was not possible for this report. Nevertheless, information gathered informally indicates group policy coverage is unlikely to differ significantly from OTA's survey results—i.e., most policies will not cover CF carrier assays unless there is a family history. The Federal Office of Personnel Management, which oversees Federal employee health benefits, has denied reimbursement for preconception CF carrier screening because it views it as preventive, not therapeutic. On the other hand, one private institute's experience with reimbursement to clients for elective fetal CF carrier screening paints a different picture. In a small survey of clients, 16 of 27 reported they had been reimbursed for their tests. Eleven had been reimbursed fully—by either commercial insurers or BC/BS plans—and five had been partially reimbursed. It is likely that reimbursement occurs more frequently in this population than might be expected from OTA's survey because it occurs in the context of pregnancy management, not preconception.

### Public Sector Reimbursement

Although access to CF carrier tests will largely depend on ability to pay because most private insurance does not cover them—at least to the extent that individual policies reflect group polices—some individuals will be Medicaid eligible. Reimbursement for their assays would be partially covered by this State-Federal partnership. In 1991, OTA surveyed directors of State Medicaid programs and found State to State variation in both the types of genetics and pregnancy-related services covered (table 1-4) and the amounts reimbursed to providers for those services. Some States do not cover certain services at all. For all States and services, the dollars reimbursed fall short of the procedures' actual charges.

**Table 1-4—Medicaid Reimbursement for Genetic Procedures[a]**

| | Covered | Not Covered | Individual consideration | Unknown |
|---|---|---|---|---|
| Amniocentesis | 45 | 0 | 1 | 0 |
| Chorionic villus sampling | 31 | 10 | 4 | 1 |
| Ultrasound | 44 | 0 | 2 | 0 |
| Maternal serum alpha-fetoprotein test | 44 | 0 | 2 | 0 |
| DNA analysis | 26 | 6 | 6 | 8 |
| Chromosome analysis | 41 | 1 | 4 | 0 |
| Genetic counseling | 11[b] | 19 | 2 | 3 |

[a]Based on the responses of 45 States and the District of Columbia to a 1991 OTA survey of Medicaid programs.
[b]Eleven other States cover genetic counseling only as a part of office visits.
SOURCE: Office of Technology Assessment, 1992.

## *Stigmatization, Classification, and Discrimination*

Concern is expressed that CF carrier screening might be sought or offered despite an uncertain potential for discrimination or stigmatization by other individuals or institutions (e.g., employers and insurers). Stigmatization of, or discrimination against, persons with certain diseases is not unique to illnesses with genetic origins. Yet as the number and scope of predictive genetic tests increase, so does concern about how perceptions of and behavior towards carriers (or individuals identified with predispositions) will develop.

### Stigmatization and Carrier Status

While a relationship exists between a characteristic's visibility and the amount of stigma it induces, invisible characteristics (e.g., carrier status) are also stigmatized. Stigmatization of CF carriers will probably focus on the notion that it is irresponsible for people who are at genetic risk to knowingly transmit a condition to their children (box 1-I). A 1990 national survey of Americans reported 39 percent said "every woman who is pregnant *should* be tested to determine if the baby has any serious genetic defects." Twenty-two percent responded

***Box 1-I—Bree Walker Lampley and Preventing Versus Allowing Genetic Disability***

In July 1991, Los Angeles radio talk show host Jane Norris launched a firestorm of controversy when she solicited listener comments on Los Angeles television anchorwoman Bree Walker Lampley's pregnancy. Making her disapproval clear, Norris said:

> We're going to talk about a woman in the news and I mean that literally. She's a very beautiful, very pregnant news anchor, and Bree Walker also has a very disfiguring disease. It's called syndactyly [sic] and the disease is very possibly going to be passed along to the child that she's about to have. And our discussion this evening will be, is that a fair thing to do? Is it fair to pass along a genetically disfiguring disease to your child?

Bree Walker Lampley has ectrodactyly, a genetic condition manifest as the absence of one or more fingers or toes. It is an autosomal dominant disorder; hence her potential offspring have a 50-50 chance of inheriting ectrodactyly. Norris' show highlighted the public tension that exists over attitudes toward preventing genetic disability, illness, and disease.

Some listeners agreed with Norris' opinion against knowingly conceiving a child who would be at 1 in 2 risk of "this deformity—webbed hands. . . ." One caller stated she would "rather not be alive than have a disease like that when it's a 50-50 chance." Other callers compared her comments to racism and eugenic genocide: ". . .this tone of yours that just kind of smacks of eugenics and selective breeding. . . . Are you going to talk in the next hour about whether poor women should have kids?"

The opinions offered illustrate the concern over the potential for discrimination or stigmatization as personal knowledge of one's genetic makeup increases. Shortly after the program aired, one disability rights activist pointed out that the radio show reminded her of her discomfort with the Human Genome Project.

On August 28, 1991, Bree Walker Lampley delivered a healthy baby boy, who has ectrodactyly. In October 1991, arguing that a biased presentation with erroneous information was broadcast, Walker Lampley was joined by her husband, several groups, and other individuals in filing a complaint with the Federal Communications Commission (FCC). Norris and the radio station stand by their right to raise the issue and "have no regrets." The FCC rejected Walker Lampley's complaint in February 1992, and no appeal is planned.

SOURCE: Office of Technology Assessment, 1992, based on *Associated Press*, "FCC Rejects Anchorwoman's Complaint Over Call-In Radio Show," Feb. 14, 1992; J. Mathews, "The Debate Over Her Baby: Bree Walker Lampley Has a Deformity. Some People Think She Shouldn't Have Kids," *Washington Post*, Oct. 20, 1991; and J. Seligmann, "Whose Baby Is It, Anyway?," *Newsweek*, Oct. 28, 1991.

that regardless of what they would want for themselves, "a woman *should* have an abortion if the baby has a serious genetic defect." Nearly 10 percent believed laws should require a woman to have an abortion rather than have the government help pay for the child's care if the parents are poor.

Few empirical studies have examined stigmatization of CF carriers directly, but relevant research funded through the NIH/DOE ELSI Programs of the Human Genome Project is under way. One study in Montreal, Canada, reports carriers generally expressed positive views about their newly determined carrier status (screening for ΔF508 only). Most (68 percent) would want their partner tested, and 60 percent said if the partner were a carrier, it would not affect the relationship. Existing research on genetic carriers and stigmatization, generally for Tay-Sachs or sickle cell, have some bearing on carrier screening for CF—chiefly that public education is crucial to overcoming stigmatization.

How CF—as a condition—is viewed by Americans will affect perceptions and potential reproductive stigma of CF carriers. Of prime importance is a commitment to nondirective genetic counseling to reduce perceived biases so individuals can make informed choices about bearing children with CF. Such a professional commitment coupled with increased public awareness and education about CF carrier screening could reduce potential problems of stigmatization of CF carriers, as well as stigmatization for other disorders as genetic screening evolves through the 1990s and beyond.

### Health Care Coverage Access

One of the most frequently expressed concerns about CF carrier screening specifically, and genetic tests generally, is the effect they will have on health care access and risk classification in the United States. Consumers fear being excluded from health care coverage due to genetic and other factors. Such

Photo credit: American Philosophical Society

Eugenics Building, Kansas Free Fair, 1929.

fears persist despite the fact that most contracts for individual health insurance coverage preclude blanket nonrenewal. Similarly, an insurer cannot raise rates for an individual who has been continuously covered if the person develops a new condition. Of special import to small group policies is that it is legal for an insurer not to renew a group contract, or to renew with a steep premium increase, based on the results of one individual's genetic, or other medical, test. Group policies are rarely guaranteed renewable, and most people in the United States are covered by group policies. Many group policies have preexisting condition clauses that preclude, for some period of time, reimbursement for expenses related to health conditions present on the policy's effective date.

One nationwide survey revealed 3 in 10 Americans say they or someone in their household have stayed in a job they wanted to leave mainly to preserve health care coverage. A 1989 OTA survey of Fortune 500 companies and a random sample of businesses with at least 1,000 employees found 11 percent of respondents assessed the health insurance risk of job applicants on a routine basis; another 25 percent assessed health risks sometimes. Nine percent of these respondents also took into account dependents' potential expenses when considering an individual's application. Forty-two percent of respondents said the health insurance risk of a job applicant reduced the likelihood of an otherwise healthy, able job applicant being hired.

**Figure 1-10—Genetic Conditions as Preexisting Conditions: Health Insurers' Attitudes[a]**

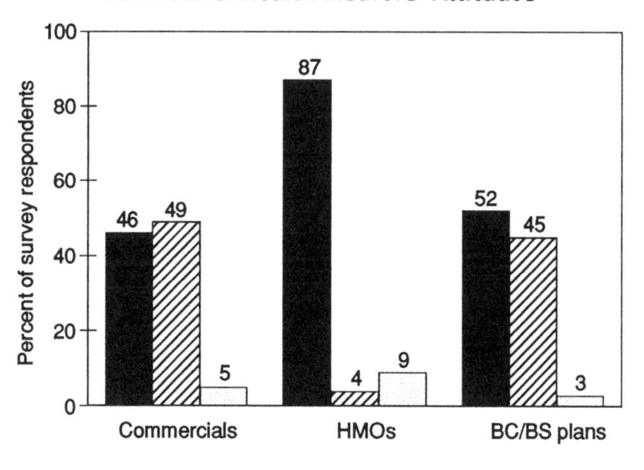


Genetic conditions, such as cystic fibrosis or Huntington disease, are preexisting conditions.

[a]Based on responses to a 1991 OTA survey of commercial insurers, health maintenance organizations, and Blue Cross and Blue Shield plans that offer individual policies or medically underwritten group policies.

SOURCE: Office of Technology Assessment, 1992.

**Figure 1-11—Carrier Status as a Preexisting Condition: Health Insurers' Attitudes[a]**

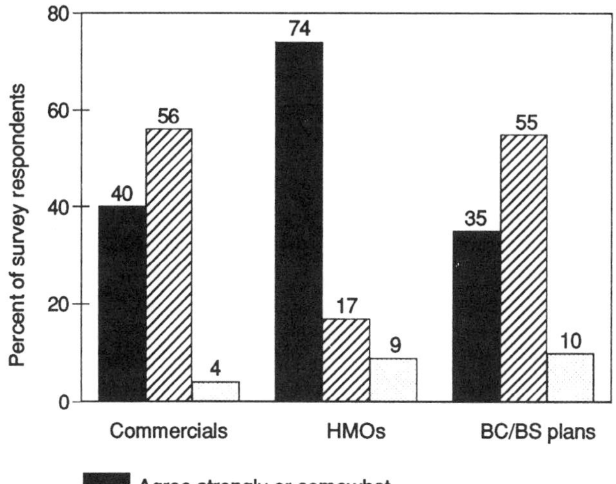


Carrier status for genetic conditions, such as cystic fibrosis or Tay-Sachs, are preexisting conditions.

[a]Based on responses to a 1991 OTA survey of commercial insurers, health maintenance organizations, and Blue Cross and Blue Shield plans that offer individual policies or medically underwritten group policies.

SOURCE: Office of Technology Assessment, 1992.

**Figure 1-12—Genetic Information as Medical Information: Health Insurers' Attitudes[a]**

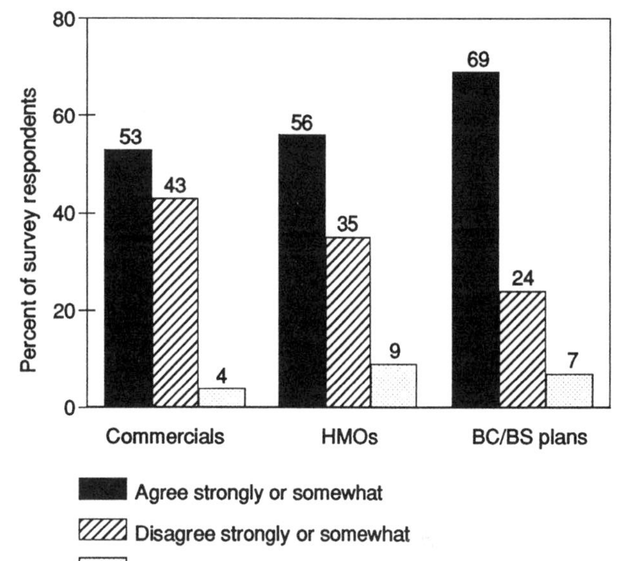


Genetic information is no different than other types of medical information.

[a]Based on responses to a 1991 OTA survey of commercial insurers, health maintenance organizations, and Blue Cross and Blue Shield plans that offer individual policies or medically underwritten group policies.

SOURCE: Office of Technology Assessment, 1992.

OTA found the majority of respondents to its health insurers' survey "agree strongly" or "agree somewhat" that illnesses with genetic bases, such as CF or Huntington disease, are preexisting conditions (figure 1-10). Thus, insurers would exclude reimbursement for such conditions for a period of time if the person could obtain individual or medically underwritten insurance at all. More surprising, since carriers have no symptoms of the disorder, is the finding that respondents, collectively, are nearly evenly split on whether carrier status—e.g., for CF or Tay-Sachs—is a preexisting condition (figure 1-11).

OTA's survey also revealed that genetic information is, for the most part, viewed no differently than other types of medical information (figure 1-12). Personal and family medical histories were the most important factors in determining insurability, according to survey respondents. OTA found medical directors and underwriters felt less strongly about "genetic predisposition to significant conditions" as a facet of insurability than they did about medical history. Of significance to CF carrier screening, a minority of all types of insurers found carrier risk "very important" or "important" to insurability.

Twenty-four percent (7 respondents) of medical directors at commercial insurers writing individual policies said "carrier risk for genetic disease" was "very important" or "important" to insurability; 18 percent (2 respondents) of HMOs responded similarly, as did 8 percent (2 respondents) of BC/BS chief underwriters.

Although an insurer might consider carrier status important to evaluating an application, carrier status does not appear to translate into difficulties for applicants in ultimately obtaining health care coverage from OTA's survey respondents. Ninety-three percent of respondents from commercial insurers and all HMOs offering individual coverage would accept the person with standard rates if the applicant was asymptomatic but had a family history of CF. For BC/BS plans, however, 55 percent would accept at standard rates, 21 percent would accept at the standard rate with an exclusion waiver, and 7 percent would decline to cover the CF carrier. For those who responded they would accept with an exclusion waiver or decline to cover, reluctance to offer standard insurance might stem from not wanting to pay for possible children or from a misunderstanding of the meaning of CF carrier status.

Overall, OTA's survey reveals genetic information is not viewed as a special type of information. In making decisions on insurability and rating based on genetics, what seems important is the particular condition (e.g., CF disease, diabetes, sickle cell anemia), not that the condition is genetically based. The increased availability of genetic information, however, adds to the amount of medical information that insurers can use for underwriting. The availability of this additional information leads to concern that risk assessments will become so accurate on an individual level as to undermine the risk-spreading function of insurance. This, of course, would have profound societal implications.

### Perspectives on the Future Use of Genetic Tests by Health Insurers

Commercial insurers, HMOs, and BC/BS plans already use genetic information in making decisions about individual policies or medically underwritten groups. People seeking either of these types of coverage reveal such information as part of the battery of questions to which applicants respond in personal and family history inquiries. OTA is unaware of any insurer who underwrites individual or medically underwritten groups and requires carrier or presymptomatic tests—e.g., for Huntington or adult polycystic kidney diseases. Even a decade from now, OTA's survey data indicate the vast majority of respondents do not expect to require genetic tests of applicants who have a family history of serious genetic conditions, nor do they anticipate requiring carrier assays even if a family history exists (table 1-5).

Health insurers do not need genetic tests to find out genetic information. It is less expensive to ask a question or request medical records. Thus, whether genetic information is available to health insurers hinges on whether individuals who seek personal policies or are part of medically underwritten groups become aware of their genetic status because of general family history, because they have sought a genetic test because of family history, or because they have been screened in some other context.

OTA's survey reveals health insurers are concerned about the potential for negative financial consequences if genetic information is available to

**Table 1-5—Projected Use of Genetic Information by Insurers in 5 and 10 Years**

| | Respondent | Very likely | Somewhat likely | Somewhat unlikely | Very unlikely | No response[a] |
|---|---|---|---|---|---|---|
| **How likely do you think it is that your company/HMO will in the next 5 years:** | | | | | | |
| Use information derived from genetic tests for underwriting? | *Commercials* | 7 (14%) | 12 (24%) | 16 (31%) | 16 (31%) | 0 ( 0%) |
| | *HMOs* | 1 ( 4%) | 5 (22%) | 9 (26%) | 6 (26%) | 2 ( 9%) |
| | *BC/BS plans* | 3 (10%) | 8 (28%) | 10 (34%) | 6 (21%) | 2 ( 7%) |
| **In the next 10 years:** | | | | | | |
| Use information derived from genetic tests for underwriting? | *Commercials* | 12 (24%) | 20 (39%) | 11 (22%) | 7 (14%) | 1 ( 2%) |
| | *HMOs* | 3 (13%) | 6 (26%) | 8 (35%) | 3 (13%) | 3 (13%) |
| | *BC/BS plans* | 5 (17%) | 13 (45%) | 3 (10%) | 6 (21%) | 2 ( 7%) |

[a] Percentages may not add to 100 due to rounding.

SOURCE: Office of Technology Assessment, 1992.

the consumer, but not them. Thirty-four medical directors (67 percent) from commercial insurers said they "agree strongly" or "agree somewhat" with the statement that "it's fair for insurers to use genetic tests to identify individuals with increased risk of disease." Thirty-eight respondents (74 percent) from commercial insurers agreed strongly or somewhat that "an insurer should have the option of determining how to use genetic information in determining risks."

### Access to Health Insurance After Genetic Tests

Existing information about how genetic test results currently affect individuals' health care coverage is largely anecdotal. One case from the Baylor College of Medicine (Houston, TX) illustrates why concern is expressed about health insurance and genetic screening and testing:

> A couple in their 30s has a 6-year-old son with CF. Prenatal diagnostic studies of the current pregnancy indicate the fetus is affected. The couple decides to continue the pregnancy. The HMO indicated it should have no financial responsibility for the prenatal testing and that the family could be dropped from coverage if the mother did not terminate the pregnancy. The HMO felt this to be appropriate since the parents had requested and utilized prenatal diagnosis ostensibly to avoid a second affected child. After a social worker for the family spoke with the local director of the HMO, the company rapidly reversed its position.

Consumers and patient advocates maintain such situations represent the tip of an iceberg. They assert individuals who avail themselves of genetic tests subsequently have difficulty obtaining or retaining health insurance. Health insurance industry officials argue to the contrary. If the problem was prevalent, they assert, ample court cases could be cited because patients and their attorneys would not be passive recipients of decisions such as that just described.

To explore this issue, OTA asked third parties—nurses in genetics and genetic counselors—for their experiences. In 1991, at least 50 genetic counselors or nurses in clinical practice (14 percent of survey respondents) reported knowledge of 68 instances of patients who experienced difficulty with health insurance due to genetic tests (table 1-6).[10]

It is important to note that most cases described in table 1-6 do not involve recessive disorders and carrier screening for conditions like CF, but involve situations in which genetic test results appear to have been treated the same as adverse test results for nongenetic conditions. Access to health care coverage for CF carriers presumably should not be an issue because CF carriers have no symptoms of the disorder, although OTA's survey of health insurers indicates otherwise in a small fraction of cases. For genetic testing or screening to detect genetic illness (or the potential for illness), however, the possibilities for problems are already unfolding.

The OTA data permit neither extrapolation about the actual number of cases that have occurred in the United States, nor speculation about trends. An estimated 110,600 individuals were seen in 1990 by the genetic counselors and nurses responding to OTA's survey, but OTA did not advise respondents to limit descriptions of clients' insurance difficulties to 1990; it is unlikely that all reported cases occurred in 1990.

### The Americans With Disabilities Act of 1990 and Genetics

In 1990, Congress enacted the Americans With Disabilities Act (ADA; Public Law 101-336), a comprehensive civil rights bill to prohibit discrimination against individuals with disabilities. The ADA encompasses private sector employment, public services, public accommodations, and telecommunications. It does not preempt State or local disability statutes.

Under the ADA, a person with a disability includes someone who has a "record" of or is "regarded" as having a disability, even if no actual incapacity currently exists. A "record" of disability means the person has a history of impairment. This provision protects those who have recovered from a disability that previously impaired their life activities (e.g., people recovered from diseases such as cancer who might still face discrimination based on misunderstanding, prejudice, or irrational fear). Additionally, individuals regarded as having disabilities include those who, with or without an impairment, do not have limitations in their major life functions, yet are treated as if they did have such

[10] OTA does not judge the validity—positively or negatively—of the claims. Some cases might have been settled in favor of the individual because the initial judgment was deemed improper or illegal. Others might have been cases where an applicant attempted to select against an insurer by misrepresenting his or her health history, which would have been resolved against the individual.

**Table 1-6—Case Descriptions of Genetic Testing and Health Insurance Problems[a]**

| |
|---|
| Positive test for adult polycystic kidney disease resulted in canceled policy or increased rate for company of newly diagnosed individual. |
| Positive test for Huntington disease resulted in canceled policy or being denied coverage through a health maintenance organization. |
| Positive test for neurofibromatosis resulted in canceled policy. |
| Positive test for Marfan syndrome resulted in canceled policy. |
| Positive test for Down syndrome resulted in canceled policy or increased rate. |
| Positive test for alpha-1-antitrypsin defined as preexisting condition; therapy related to condition not covered. |
| Positive test for Fabry disease resulted in canceled policy. |
| Woman with balanced translocation excluded from future maternity coverage. |
| Positive Fragile X carrier status and subsequent job change resulted in no coverage. |
| After prenatal diagnosis of hemophilia-affected fetus, coverage denied due to preexisting condition clause. |
| Denied coverage or encountered difficulty retaining coverage after birth of infant with phenylketonuria. |
| Woman diagnosed with Turner's syndrome denied coverage for cardiac status based on karyotype. Normal electrocardiogram failed to satisfy company. |
| Family with previous Meckel-Gruber fetus denied coverage in subsequent applications despite using prenatal diagnosis and therapeutic abortion. |
| Mother tested positive as carrier for severe hemophilia A. Prenatal diagnosis revealed affected boy; not covered as preexisting condition when pregnancy carried to term. |
| After a test revealed that a woman was a balanced translocation carrier, she was initially denied coverage under spouse's insurance because of risk of unbalanced conception. Subsequently overturned. |
| Woman without prior knowledge that she was an obligate carrier for X-linked adrenoleukodystrophy found out she was a carrier. She had two sons, both of whom were healthy, but each at 50 percent risk. Testing was done so they could be put on an experimental diet to prevent problems that can arise from mid- to late childhood or early adulthood. One boy tested positive. The family's private pay policy (Blue Cross/Blue Shield) is attempting to disqualify the family for failing to report the family history under preexisting conditions. |
| After birth of child with CF, unable to insure unaffected siblings or themselves. |

[a]1991 OTA survey of genetic counselors and nurses in genetics. Not all cases, or multiple cases involving same disorder, listed.

SOURCE: Office of Technology Assessment, 1992.

limitations. This provision is particularly important for individuals who are perceived to have stigmatic conditions that are viewed negatively by society.

Examining genetics and the ADA from three broad categories—genetic conditions, genetic predisposition, and carrier status—sheds some light on how the ADA might interface with CF carrier screening and future genetic tests (figure 1-13).

*Genetic Conditions.* Disability is defined only according to the degree of impairment and how severely the disability interferes with life activities, with no distinction between those with genetic origins and those without. A genetic condition that does not cause substantial impairment might not constitute a disability, unless others treat the person as disabled. Thus, significant cosmetic disfigurements (e.g., from burns or neurofibromatosis) could be classified as disabilities if public prejudices act to limit the life opportunities of people who have them. Congress and the courts have long recognized disabilities of primary or partial genetic origin,

**Figure 1-13—Genetics and the Americans With Disabilities Act of 1990**

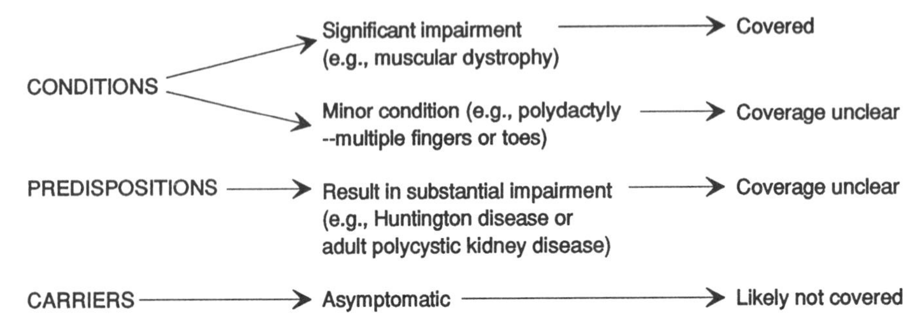


SOURCE: Office of Technology Assessment, 1992.

including Down syndrome, CF, muscular dystrophy, epilepsy, diabetes, and arthritis.

*Genetic Predisposition.* ADA judges disability not just by an objective measure of inability to perform tasks, but also subjectively by the degree to which the public makes the condition disabling through misunderstanding or prejudice. This latter definition might apply to individuals who are asymptomatic but predicted to develop disease in the future—if the public perceives them as having a disability because they might or will get ill. Some argue the ADA's legislative history indicates genetic predisposition might be encompassed. One Congressman stated during the 1990 debate over the conference report that persons who are theoretically at risk "may not be discriminated against simply because they may not be qualified for a job sometime in the future." On the other hand, no further discussion on the issue occurred.

*Carrier Status.* Case law and the ADA's prohibition of discrimination generally hold that employment decisions must be based on reasonable medical judgments that show the disability prevents the individual from meeting legitimate performance criteria. For carriers of recessive conditions such as CF, sickle cell anemia, and Tay-Sachs, there is no disability per se; the ADA appears not to cover carriers. Such individuals are, however, at high risk of having an affected child if their partners also carry the trait and could be misunderstood to be affected by the disease. Discrimination against carriers could arguably constitute discrimination if based on a perception of disability.

*The Equal Employment Opportunity Commission (EEOC) Regulations.* In 1991, EEOC promulgated regulations for implementing the ADA. The regulations do not specifically prohibit discrimination against carriers or persons who are identified presymptomatically for a late-onset genetic condition (e.g., adult polycystic kidney disease or Huntington disease)—despite the fact that the NIH/DOE ELSI Working Group and the NIH/DOE Joint Subcommittee on the Human Genome urged EEOC to clearly protect these individuals. It its interpretive guidance, EEOC notes "the definition of the term 'impairment' does not include characteristic predisposition to illness or disease." From EEOC's perspective, carriers are not encompassed by the ADA's provisions. With respect to individuals diagnosed presymptomatically, EEOC concluded that "such individuals are protected, either when they develop a genetic disease that substantially limits one or more of their major life activities, or when an employer regards them as having a genetic disease that substantially limits one or more of their major life activities."

### The Americans With Disabilities Act and Health Insurance

The ADA also might prohibit discrimination based on an employer's fear of future disability in an applicant's family that would affect the individual's use of health insurance and time away from the job. Nevertheless, the ADA does not speak to this point directly, and so leaves open for future interpretation whether employers may discriminate against carriers who are perceived as more likely to incur extra costs due to illnesses that might occur in their future children. The ADA specifically does not restrict insurers, health care providers, or other benefit plan administrators from carrying out existing underwriting practices based on risk classification. Nor does the ADA make clear whether employers may question individuals about their marital or reproductive plans prior to offering employment or enrollment in an insurance plan. Furthermore, after a person is hired, ERISA-based, self-funded insurance plans can alter benefits to exclude or limit coverage for specific conditions; the ADA does not preempt ERISA.

## *Quality Assurance of Clinical Laboratories and DNA Test Kits*

Quality assurance for CF carrier screening means ensuring the safety and efficacy of the tests themselves, whether they are performed de novo in clinical diagnostic laboratories or via test kits. The quality of the laboratory's performance affects the quality of the counseling services. Ensuring that consumers receive high-quality technical and professional service is the responsibility of providers, under the shared oversight of the Federal Government, State and local governments, private entities (including professional societies), and the courts.

### The Clinical Laboratory Improvement Amendments of 1988

Quality assurance to assess clinical laboratory performance is still in flux, in large measure because 1967 legislation governing regulation of clinical testing facilities was overhauled by Congress in

1988 with enactment of the Clinical Laboratory Improvement Amendments of 1988 (CLIA; Public Law 100-578). CLIA subjects most clinical laboratories to an array of accrediting requirements: qualifications for the laboratory director, standards for the supervision of laboratory testing, qualifications for technical personnel, management requirements, and an acceptable quality control program. CLIA authorizes the Health Care Financing Administration (HCFA) to police an estimated 300,000 to 600,000 physician, hospital, and freestanding laboratories to ensure they adhere to a comprehensive quality assurance program. HCFA may impose sanctions, if necessary.

CLIA clearly encompasses facilities performing DNA-based, clinical diagnostic analyses. But, while it details particular performance standards for several types of clinical diagnostic procedures, CLIA does not specifically address DNA-based tests. This lack of detailed directives for DNA-based diagnostics could be beneficial in the short-term, since the field is rapidly changing.

*State Authorities.* CLIA does not preclude States from regulating and licensing facilities within certain guidelines. After a pilot study, for example, the California State Department of Health Services intends to seek approval for State-specific licensing laws and regulations for DNA and cytogenetic laboratories. Similarly, New York has regulated clinical laboratories since 1964, and has established a genetics quality assurance program that includes requirements for licensing personnel, licensing facilities, laboratory performance standards, and DNA-based proficiency testing. Nevertheless, the principal State role in quality assurance for clinical facilities is licensure and certification of medical and clinical personnel, which are the sole provinces of States.

*The Role of Private Organizations.* While CLIA clearly expands the Federal role in clinical laboratory oversight, the law continues to permit, subject to DHHS approval, the involvement of other parties in regulating laboratory practices. Private organizations, including the Joint Commission on Accreditation of Health Care Organizations, may continue to accredit facilities. Private professional societies will likely have the greatest impact in the area of proficiency testing, one component of accreditation. Efforts by CORN and its regional networks, ASHG, and the College of American Pathologists (CAP) stand at the forefront of developing proficiency tests for DNA-based diagnostics.

In 1989, CAP established a committee to develop appropriate guidelines for all clinical tests involving DNA probes or other molecular biological techniques. The CAP committee has administered two DNA-based proficiency testing pilot programs, although their focus was not genetic disorders. CORN, which receives Federal funding and has been involved in quality assurance of genetics facilities since 1985, sponsored a DNA-based genetic test proficiency pilot of 20 laboratories in 1990. The Southeastern region has a regional proficiency testing program, and will be enlarging its planned second survey into a national test, to be completed in 1992; this effort includes CF mutation analysis. Full proficiency testing for DNA-based genetic diagnostics is planned by 1994. CORN and ASHG have liaisons with the others' efforts, and a joint ASHG/CAP DNA-based proficiency testing pilot for genetic diseases commenced in 1992.

Proficiency testing is widely viewed as a key measure of quality assurance. It can provide a reliable and identifiable benchmark to assess per-

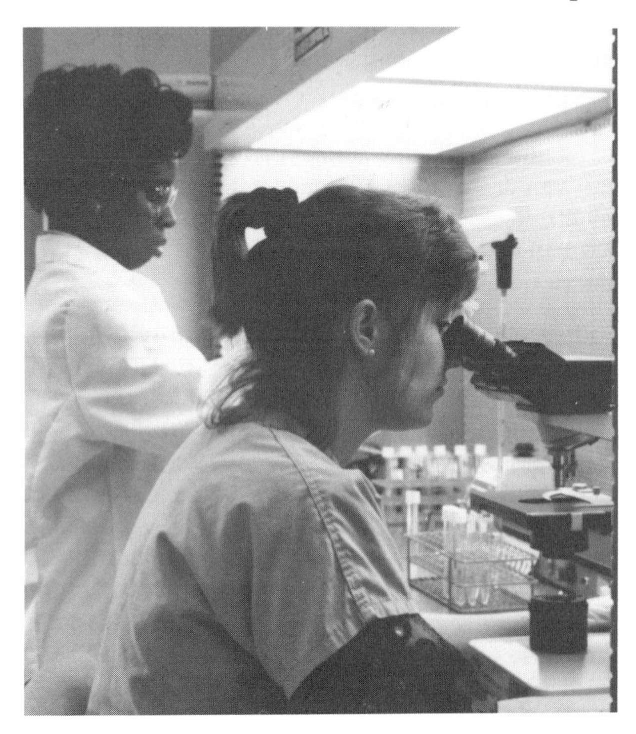

Photo credit: Genetics & IVF Institute

Facilities that perform DNA-based diagnostic tests (e.g., CF mutation analysis) are subject to the Clinical Laboratory Improvements Amendments of 1988.

formance. In the past, professional societies' involvement in proficiency testing to ensure laboratory quality have predominated, and this situation is likely to continue. Cooperation among each of the groups will be essential, as professional-society-based programs could affect proficiency testing for CF mutations (and other DNA tests) long before HCFA proposes proficiency testing rules under CLIA.

### Regulation of DNA Test Kits

Increased use of CF mutation assays for carrier detection will depend, in part, on the development and availability of prepackaged kits. At least two companies—one in the United States and one in the United Kingdom—are testing such kits and anticipate their availability in 1 to 2 years. Before marketing of the kits can occur, however, the U.S. Food and Drug Administration (FDA) must ensure the safety and efficacy of genetic diagnostic test kits, such as those under development for CF mutations. Since genetic diagnostic kits fall within the definition of devices, the extent to which CF mutation kits—or other DNA-based genetic test kits—become available will depend on FDA regulation of devices during development, testing, production, distribution, and use.

FDA's regulatory options range from registering an item's presence in the United States and periodically inspecting facilities to ensure good manufacturing practices, to setting performance and labeling requirements, to premarket review of a device. The agency also may engage in postmarketing surveillance to identify ineffective or dangerous devices. It may ban devices it deems unacceptable. Specific regulation depends on whether FDA classifies the device as Class I, II, or III, with Class III devices receiving the most stringent review.

Since no FDA-approved, DNA-based genetic diagnostic test kit comparable to those being developed for CF carrier analysis exists, it is difficult to predict the ultimate regulatory status of such kits. Preliminary indications are they will be regulated as Class III devices. In response to recent legislation and ongoing congressional concern, FDA appears to be increasing medical device regulation and postmarketing surveillance. If increased FDA scrutiny extends to DNA-based diagnostic test kits, developers can expect more stringent regulation of these products than of previous non-DNA-based genetic test kits. Increased regulation to provide greater assurance of safety and efficacy might, in turn, slow routine CF carrier screening.

*Photo credit: Tony J. Beugelsdijk, Los Alamos National Laboratory*

Automated robotic system used in DNA analysis at Los Alamos National Laboratory.

## *Automation*

The extent to which costs for CF carrier tests decline depends, in part, on automation. Instrumentation will be especially crucial to the development of batteries of tests for multiple genetic disorders. Moreover, compared to most routine clinical tests, current DNA-based CF carrier assays are labor intensive.

Over the past few years, private industry and U.S. national laboratories have developed several instruments that increase the speed and volume of routine DNA diagnostic procedures. Goals for improved instrumentation for DNA analyses stem, in part, from the importance of rapid techniques to the Human Genome Project. Spin-off technologies from DNA mapping and sequencing appear amenable to applications for clinical diagnostics.

Currently, all but one step of what generally constitutes DNA diagnosis is automated or involves instrumentation under development. Most components of DNA analysis, however, are automated as individual units; efforts under way seek to coordinate sequential steps. Some machines are not faster than humans, but they can standardize the procedures and decrease human error.

Clearly, .the crucial steps in DNA-based CF carrier assays are, or can be, automated. Advances in instrumentation indicate that automated, rapid carrier screening for CF—or other genetic conditions—is already technologically feasible. OTA finds the field of DNA automation is advancing at a pace that suggests entirely automated DNA diagnosis can be realized in the next few years.

## *Costs and Cost-Effectiveness*

Perhaps the least examined facet of CF carrier screening is cost. Data for parts of OTA's analysis were often lacking and assumptions had to be made. Unlike the seven preceding factors, which in many cases will generically affect utilization of DNA-based tests for disorders other than CF, findings that pertain to cost-effectiveness do not extend beyond CF carrier screening—although the approach used in this report could be applied to screening with other genetic tests.

While economic analyses can inform decisions surrounding resource allocation and access to genetic screening, they have limits. In the context of public policy and genetics, the 1983 President's Commission report on genetic screening articulates solid guidance about the benefits and limits of cost-effectiveness and cost-benefit analyses: These analytical approaches are tools to be used within an overall policy framework, not solely as a method of making or avoiding judgment. There is no intimation in OTA's analysis that something that saves or costs money is more or less desirable from a welfare standpoint.

### Cost of Cystic Fibrosis

The cost of any illness is the answer to the hypothetical question: If the disease disappeared and everything else held constant, how many more dollars would be available to the economy? Many elements are needed to answer this question, but broadly speaking they fall into two categories: information about direct medical costs associated with CF and nonmedical direct costs related to the disease (i.e., family caregiving time).

Direct medical expenses for CF include costs of hospitalization, outpatient care, physical therapy, and drugs. These costs are not the same for everyone with the disease (table 1-7). Clinical symptoms of CF vary widely, although broad divisions in its severity can be drawn. Some individuals require only one inpatient visit every 2 years or so; others

**Table 1-7—Annual Cost of Medical Care for Cystic Fibrosis Patients**

| Treatment | Mild[a] | Moderate | Severe |
|---|---|---|---|
| Acute treatment | | | |
| Antibiotics | $ 2,000[b] | $6,000 | $12,000 |
| IV supplies | 300 | 500 | 900 |
| Hospitalization | 3,500 | 14,000 | 28,000 |
| Miscellaneous | 100 | 200 | 400 |
| Total cost acute | 5,900 | 20,700 | 41,300 |
| Chronic management | | | |
| Visits to CF Center | 600 | 800 | 1,200 |
| Medications | 2,000 | 3,000 | 4,000 |
| Total cost chronic | 2,600 | 3,800 | 5,200 |
| Total cost acute and chronic treatment | 8,500 | 24,500 | 46,500 |

[a]There is another category, "submild," whose illness requires infrequent hospitalization—less than once per year. Based on existing data, about 40 percent of patients are submild and 40 percent are mild (about one hospital episode per year). Approximately 13 percent of all individuals with CF had two or three hospitalizations per year; this group represents the moderate portion. Finally, about 6 percent of all patients had four or more hospitalizations per year and comprise the severe patient group.

No data exist on the average expenses of the submild group, but a reasonable assumption might be their expenses are about twice the average medical care cost of the average American under 65 years of age, or $2,000. In fact, costs might be slightly higher; actual costs for one submild case (parents providing physical therapy and no hospitalizations in 9 years) were approximately $4,700 in 1990; the cost of drugs alone was $1,900. Nevertheless, the OTA analysis errs on the conservative side and uses $2,000 in determining the average medical care cost of an individual with CF.

[b]All values in 1989 dollars.

SOURCE: Wilkerson Group, Inc., *Annual Cost of Care for Cystic Fibrosis Patients* (New York, NY: Wilkerson Group, Inc., 1991).

have problems so severe as to require four or more hospitalizations per year. Similar variation exists for other medical expenses. Overall, taking these several factors into account, average annual medical expenses for CF patients are estimated at $10,000. Assuming a median life expectancy in 1990 of 28 years the present value of lifetime medical expenses is approximately $146,430 (1990 dollars using a 5 percent discount rate).

The main nonmedical direct cost associated with CF is parental time beyond the time required for a child without the illness. CF centers estimate that parents often must spend 2 hours per day on therapy for a child with CF. In addition, parents lose time from work when the person falls ill. Time is also spent on physician and clinic visits. OTA uses an estimate of 938 hours per year of extra caregiving to a person with CF, which is generally provided by family members. Assuming an estimated domestic/nursing wage of $10 per hour, the present value of CF-related lifetime nonmedical direct costs is $139,744 (1990 dollars using a 5 percent discount rate).

**Table 1-8—Costs for Cystic Fibrosis Carrier Tests At Selected Facilities**

| Institution | Price per sample |
|---|---|
| Baylor College of Medicine | $ 55 or 200 |
| Boston University | 170 |
| Collaborative Research, Inc. | 173 |
| Cornell University Medical Center | 75 |
| GeneScreen | 165 |
| Genetics & IVF Institute | 225 |
| Hahnemann University | 225 |
| Hospital of the University of Pennsylvania | 150 |
| Integrated Genetics | 150 |
| Johns Hopkins University Hospital | 270 |
| Mayo Medical Laboratories | 200 |
| St. Vincent's Medical Center | 150 |
| University of Minnesota | 136 |
| University of North Carolina | 150 |
| Vivigen, Inc. | 200 to 220 |

SOURCES: Office of Technology Assessment, 1992, and M.V. Pauly, "Cost-Effectiveness of Screening for Cystic Fibrosis," contract document prepared for the U.S. Congress, Office of Technology Assessment, August 1991.

## Cost of Cystic Fibrosis Mutation Analysis

Since CF is the most common, life-shortening, recessive disorder among Caucasians in the United States, commercial interest in the test is high. Currently, at least six commercial companies perform DNA-based CF mutation analyses, as do at least 40 university and hospital laboratories. Table 1-8 presents data on test charges for several private and public facilities; the average price per sample is about $170. With increased volume of tests and automation, however, many predict the cost per CF mutation assay will decrease. OTA uses a cost per test of $100 because the analysis focuses on the potential future of large-scale CF carrier screening and presumes economies of scale will apply.

Indirectly related to cost-effectiveness, but directly related to how much CF mutation analysis will cost in the future, is the issue of patents, licensing, and royalty fees for genetic diagnostics. A patent is pending for the CF gene, for example. Similarly, royalty licenses must be paid for the process—the polymerase chain reaction, or PCR—by which CF mutation analysis is performed. Thus, royalty licensing fees will be reflected in costs of the tests to consumers. Currently, debate is increasing on the issue of intellectual property protection and the Human Genome Project. A resolution of this controversy, if any, will affect costs of DNA-based diagnostic tests and hence cost-effectiveness of screening for genetic disorders.

## Costs and Cost-Effectiveness of Carrier Screening for Cystic Fibrosis

Data about the cost of screening large numbers of individuals for CF carrier status do not exist. In estimating the cost of carrier screening for CF, OTA included costs of the CF mutation analyses, chorionic villus sampling for fetal testing, and costs for pretest education and post-test counseling. Taken together, these costs were analyzed in the context of several scenarios for preconception screening of women (and possibly their partners) and prenatal screening of pregnant women (and if necessary their partners and the fetus).

Regardless of the strategy or scale, CF carrier mutation analysis provides information to an individual about his or her likelihood of having a child with CF should the partner also be a carrier. Hence, at its core, a cost-effectiveness analysis of CF carrier screening involves assumptions about reproductive behavior. A base case was established for the following six variables:

- 80 percent of women elect screening,
- 85 percent sensitivity of the CF mutation assay,
- 8.4 percent of +/+ couples are infertile,
- 10 percent of +/+ fertile couples choose not to conceive,
- 90 percent of +/+ fertile couples conceive, and 100 percent use prenatal testing, and
- 100 percent of CF-affected pregnancies detected are terminated.

As alternatives, other assumptions were made for several additional scenarios by varying the factors in turn (or combination) to yield a series of cost-effectiveness estimates. In evaluating costs and savings, changes in behavior were considered only for +/+ couples, and costs and savings were calculated for a hypothetical population of 100,000 eligible women (or couples). The economic costs include costs associated with CF carrier screening. The economic savings include avoiding the direct medical and nonmedical costs associated with having a child with CF. The base case and all scenarios were then compared to costs in the absence of screening.

One scenario, for example, assumed 50 percent of women chose to participate, another assumed all individuals elected screening. Another screened the woman and man simultaneously, rather than screening the man only when the woman was positive.

Photo credit: Robyn Nishimi

Others used 50 percent as the frequency of affected pregnancies terminated. Overall, whether CF carrier screening can be paid for on a population basis through savings accrued by avoiding CF-related medical and caregiving costs depends on the assumptions used—including how many children people will have, average CF medical costs, and average time and cost devoted to caring for a child with CF, as well as variations in reproductive behaviors, costs of CF mutation analyses, and screening participation rates.

Eight of 14 scenarios examined by OTA result in a net cost over no screening. Under six cases, however, CF carrier screening is cost-effective, but most of these scenarios involve 100 percent participation, test sensitivity, or selective termination—all unlikely to be realized in the near term, if ever. Nevertheless, CF carrier screening can save money compared to no screening even under less absolute circumstances. The balance between net savings versus net cost in nearly all scenarios is fine. How many individuals participate in screening is relatively unimportant to cost-effectiveness, but it is clear the frequency of affected pregnancies terminated and the assay's price will ultimately affect this balance.

## WHAT IS THE ROLE OF CONGRESS?

Speculation about the impact of a CF carrier test on individuals and society has existed for years. Today, that speculation is being transformed into reality. In this report, OTA identifies eight factors affecting implementation of CF carrier screening. From the analysis of these factors, OTA concludes that Congress could play a role in six broad policy areas:[11]

- genetics education and the public,
- personnel,
- genetics and discrimination,
- clinical laboratory and medical device regulation,
- instrumentation, and
- integration of DNA assays into clinical practice.

### *Genetics Education and the Public*

For people to make informed decisions about whether CF mutation assays would be useful to them, they must understand what CF is, know what carrier status means, and have some understanding of the probabilistic nature of genetic tests. Beyond comprehending technical information, the public should also appreciate the positive and negative social implications that could adhere. Better public education would also mean fewer total counseling hours would be needed.

Mechanisms by which Congress can generally improve science and education in the United States were assessed in a separate OTA report.[12] Federal efforts specifically targeted to educating the public about human genetics are diffuse, but do exist. If Congress determines that increased genetics educa-

[11] Congress also plays a role in an additional policy issue raised by CF carrier screening and the development of other genetic tests—i.e., health care access. As mentioned, however, access to CF carrier tests, and services related to them, is no different—and inextricably linked—to the broad issue of health care reform in the United States, a topic beyond the scope of this report.

[12] U.S. Congress, Office of Technology Assessment, *Educating Scientists and Engineers: Grade School to Grad School*, OTA-SET-377 (Washington, DC: U.S. Government Printing Office, June 1988).

tion is a priority, it could urge interagency coordination and/or appropriate increased funds. In particular, Congress could exploit three general avenues to increase public education about genetics: school-based science education, patient education, and widespread public appeal.

Existing agencies and programs have some efforts related to public education in genetics, each serving different purposes. These efforts can serve as the foundation for new initiatives. The National Institutes of Health/U.S. Department of Energy's Ethical, Legal, and Social Issues Programs of the Human Genome Project, for example, have awarded grants that target each of the avenues just described, including curriculum development, science teacher education, evaluation of improved means to deliver genetic information to patients, and a mass media production that will be available through public television. If Congress concludes that ELSI Programs should increase their attention to public education, it could direct them to seek and award a greater number of grants focused on this issue. In doing so, Congress could direct that a greater proportion of such awards be made with existing funds, at the expense of other areas. Or, Congress could direct that more than the expected 5 percent set aside from the fiscal year 1993 Human Genome Project appropriation be devoted to the ELSI Programs—at the expense of the scientific and technical components—and that the increased funds be allocated to public education grants. Finally, Congress could increase the ELSI Programs' funding specifically for public education.

The National Science Foundation serves as the lead Federal agency for science education, particularly teacher education and training. Thus, with respect to specifically enhancing public knowledge through school-based science education, Congress could encourage NSF—directly through appropriations or indirectly through oversight—to increase attention to education in human genetics. Currently, supplemental genetics education for K-12 teachers is piecemeal; NSF has funded a few projects to train high school and grade school teachers about genetics, but no nationwide effort exists.

The DHHS National Center for Education in Maternal and Child Health serves as the Federal repository for a wide range of materials related to human clinical genetics—ranging from genetics training manuals for social workers to patient information pamphlets for a number of genetic diseases; it once served as an active clearinghouse to disseminate information about genetics nationwide. Due to budgetary constraints, the center now functions more as a passive resource to provide information on request, rather than performing aggressive outreach. Through oversight, Congress might judge that the lost function of the center should be reinstated, but it would need to recognize that increased funds would be necessary to achieve this goal.

### *Personnel*

Several types of health care professionals perform genetic counseling—master's level genetic counselors, physicians, Ph.D.-level clinical geneticists, nurses, and social workers. No coordinated Federal training and education framework exists to serve all. The Federal Government provides financial support for education and training of certain health personnel through Title VII and Title VIII of the Public Health Service Act. Title VII provides education support to the fields of medicine, osteopathy, dentistry, veterinary medicine, optometry, podiatry, public health, and graduate programs in health administration. It does so through grants and contracts to institutions, and through loans to individuals. Title VIII focuses primarily on advanced training of nurses. The MCH block grant also supports some genetics-related training and education.

If Congress determines that training of additional genetics personnel—beyond those practicing or in the pipeline—is essential to maintain quality care, it could enact legislation that amends Title VII or Title VIII to include master's-level genetic counseling programs. It could also encourage increased genetics education for the other health professions encompassed by these acts. Grantees and contractors that receive Title VII or Title VIII funds, for example, might be required to increase genetics-related curriculum for all health professionals. Congress could also increase appropriations under the MCH block grant, or stipulate that States receiving MCH funds earmark a designated level of State funds to education, training, or both.

Genetics education for those already practicing is as important as genetics training and education for new health professionals. In part, the issue of adequate services and professional capacity depends on whether patients continue to receive genetic services through specialized clinical settings, as most do now, versus access through primary care, community health, and public health settings. If the nonspecialized clinical route becomes more common, it will require that existing genetic specialists provide adequate genetics education to other practitioners in the U.S. health care system. Congress could focus on two executive branch entities to accelerate this provider-to-provider knowledge transfer. First, it could continue to encourage the NIH/DOE ELSI Programs of the Human Genome Project to fund grants for this purpose. Second, Congress could enhance, through increased appropriations, professional training and continuing education efforts under the MCH block grant.

### *Genetics and Discrimination*

Concern about discrimination arises from new capabilities to assess genetic information. This concern currently focuses on the Americans With Disabilities Act and subsequent rulemaking by the U.S. Equal Employment Opportunity Commission. First, as enacted, the ADA left open the question of whether genetic predisposition to illness or carrier status were covered as protected classes. In its final rule, EEOC rejected the premise that genetic predisposition or carrier status are covered under the ADA for employment purposes. Because some debate exists as to the intent of Congress in this area, Congress could revisit the issue to clarify its intentions with respect to genetic and disability discrimination under ADA. Many opine that litigation will ultimately define the scope of the ADA.

Second, ADA is silent on whether employers may discriminate—for the purposes of hiring—against individuals (e.g., CF carriers) who are perceived as more likely to incur extra costs due to illnesses that could occur in their future children. An OTA survey of Fortune 500 companies and companies with 1,000 or more employees revealed that 9 percent of employers surveyed account for dependents' potential expenses when considering an individual's application. If Congress determines the potential health insurance costs of an applicant's dependent should not be considered in hiring decisions, it could signal its intent through legislation.

Finally, concerns about discrimination in insurance coverage and repercussions on health care access arise in the era of new genetic tests, but insurance regulation in the United States is largely a matter for the States. Nevertheless, one aspect of health insurance relates to both the ADA and Federal law regarding employee benefits (i.e., the Employee Retirement Income Security Act of 1974). The number of individuals receiving health care coverage via ERISA-based, self-funded plans is increasing. Under ERISA, which preempts State insurance law, any self-funded company can cap, modify, or eliminate employees' health care benefits for a particular condition at any time, as long as the company complies with the notice requirements in the plan agreement. Such conditions are in no way limited to genetic illnesses. Congress could prohibit such actions, if it deems it necessary, by amending ERISA, the ADA, or both.

### *Clinical Laboratory and Medical Device Regulation*

Congress has a long legislative history in regulating clinical laboratories and medical devices. In the past 4 years, Congress has moved twice—the Clinical Laboratory Improvement Amendments of 1988 and the Safe Medical Devices Act of 1990 (SMDA)—to address perceived deficiencies in each area. Absent additional action by Congress, the regulatory framework for clinical laboratories and medical devices will evolve from these two statutes. Currently, the regulatory status for both is in flux, as executive branch agencies only now are developing specific rules and regulations.

If Congress believes the new DNA-based genetic diagnostics require clinical laboratory quality assurance considerations beyond the 1988 legislation, it could amend CLIA to specify criteria for DNA assays. On the other hand, the field of clinical DNA diagnostics is changing rapidly. Congress might prefer to maintain the Health Care Financing Administration's flexibility in adapting to these changes. In that case, Congress could monitor HCFA's approach to DNA analyses through its oversight of HCFA's implementation of CLIA, generally.

With respect to medical devices, no FDA-approved DNA test kit for CF mutation analysis exists, although kits are being tested with companies' expectation of their availability in 1 to 2 years. Congress can amend SMDA if it believes DNA test kits constitute so novel a device that SMDA's provisions for premarket evaluation and postmarket surveillance do not suffice. Evaluating FDA's regulation of DNA diagnostics in the absence of a product could prove difficult, however, and so Congress might prefer to take no action at this time.

### *Instrumentation*

The ability to test quickly and accurately will be crucial to inexpensive CF carrier screening. It will be even more important if panels of genetic assays for an array of disorders are to be developed. Currently, all but one step of techniques used in DNA diagnostic analysis are automated, but there is little integration of the components. If Congress determines that the goal of quick, accurate batteries of DNA tests is important, it could make such integration a Federal research priority under the Human Genome Project by designating that certain levels of appropriations be targeted to tailoring instrumentation and automation to DNA diagnostics. Currently, the Human Genome Project serves as the primary funding locus for developing instrumentation to automate DNA analysis—chiefly through appropriations to U.S. national laboratories.

### *DNA Assays and Clinical Practice*

In today's social, economic, and legal climate, OTA believes that, as a practical matter, a federally funded or controlled program for *population-based* CF carrier screening is not on the horizon. In the 1990s, CF mutation analysis could become routine, but not likely as part of a unified, national program. If Congress determines in the distant future that a programmatic public health model for CF carrier screening or other genetic conditions is necessary, it can look to the National Genetic Diseases Act to craft a population-based program. In 1992, the issue at hand is: How, and to what extent, will CF carrier tests—and other genetic tests in the pipeline—integrate into contemporary medical practice?

Many perspectives on how CF carrier screening should be implemented exist, including a socially regulated program, a free market model, and a focus on patient autonomy and choice. Those who support a regulated framework in the fashion of a public health model (e.g., newborn genetic screening) believe public health's historical use of institutional mechanisms and social approaches is appropriate and necessary for quality assurance and consumer protection. Others take a dim view of a regulated model for CF carrier screening because they believe that consumers are best served by having CF carrier tests available through general medical practice and by providing them the opportunity to choose and manage their own health care. They argue that formal, government-sponsored structure translates to regulated medicine, which they oppose, because it can interfere with patient care.

No definitive way exists to determine when providers should routinely inform people about the availability of genetic tests, and in some respects, Congress has less a role to play in this policy issue than in the preceding five. Nevertheless, Congress can influence when and how genetic tests are integrated in two specific ways.

First, 2 years lapsed between identification of the CF gene and its mutations and the initiation of federally sponsored pilot studies to assess routine CF carrier screening. Before other DNA-based tests come on-line, Congress could encourage the genetic services delivery and genetic research agencies of the executive branch to coordinate efforts to develop an institutional means to ensure evaluation of genetic tests through federally sponsored consensus conferences, workshops, and pilot projects (if necessary) prior to their being incorporated into routine medical care. In doing so, concerns raised that CF carrier screening is being rushed into practice might be assuaged if future tests receive federally led, timely evaluation. On the other hand, critics of Federal intervention will continue to argue that federally sanctioned efforts will slow access to tests and information that some consumers would find desirable.

Second, once a test becomes fully integrated into clinical practice, Congress can direct the Agency for Health Care Policy and Research to examine whether practice guidelines for CF carrier screening, or other genetic tests, are appropriate. Supporters of practice guidelines believe they offer the potential to decrease malpractice claims, control health care costs, improve quality, and generally influence the use of a technology. Detractors argue such guidelines differ little from professional statements, will increase malpractice claims, and suggest regulated medicine.

## PROSPECTS FOR THE FUTURE

Leaving aside the precise timing of routine CF carrier screening, it is clear the number of DNA-based tests for genetic disorders and predispositions will increase rapidly over the next decade, almost certainly by an order of magnitude. OTA considers it likely that the time available, if any, for debate and

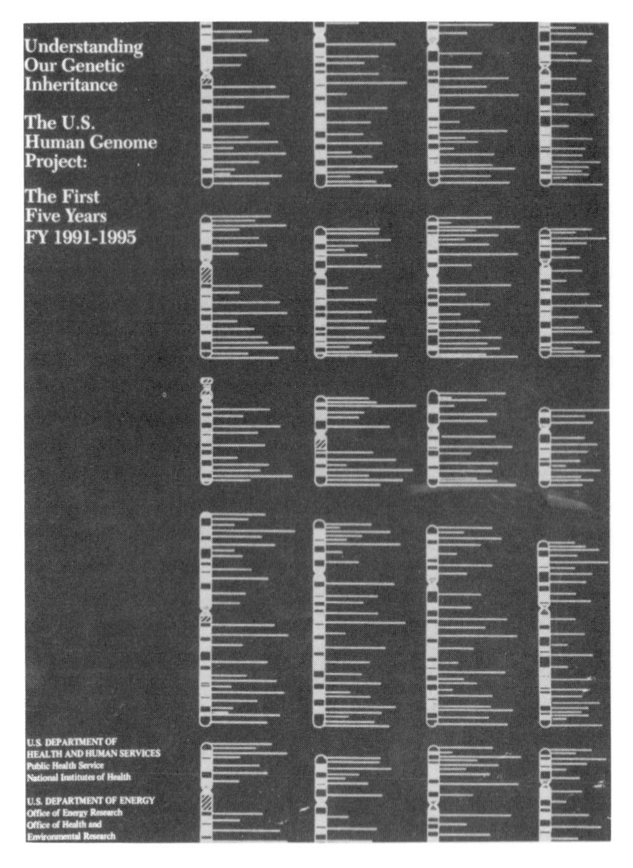


*Photo credit: Office of Technology Assessment*

The U.S. Human Genome Project, jointly funded by the U.S. Department of Health and Human Services and the U.S. Department of Energy is estimated to be a 15-year, $3 billion project. As the project continues to unfold, Congress will likely face policy issues stemming from both the discovery of new information and applications of the information.

discussion on dissemination and use of new genetic tests will be compressed as pressure to use them rises. Given this scenario, some of the policy questions raised in this report extend beyond implications for CF carrier screening.

On one hand, CF carrier screening can be used to construct a paradigm that describes a set of policy issues for genetic tests to come. Access to health care merits specific mention because it is repeatedly raised as a concern tied to the increasing availability of genetic information—i.e., will the new knowledge elucidated through the Human Genome Project positively or negatively affect how Americans obtain or retain health care coverage? Certain additional themes will apply: ensuring clinical laboratory competence, quality assurance of the tests, maintaining high-quality service delivery, promoting public education, supporting provider training, and safeguarding against discrimination and stigmatization. Of course, as American policies and politics change—or remain the same—the approaches to address these issues might differ.

Another generic issue, but one likely to ignite controversy with each new test, is the pace at which the assay should be integrated into general medical practice. Early use of CF mutation analysis is in the obstetric and prenatal context, and this trend will likely continue. As such, it serves as a good model to examine the broader consequences of genetic screening when this context is the chief avenue of a test's introduction. But experience with CF carrier screening is less applicable for tests that detect adult, late-onset genetic disorders (e.g., Huntington disease or familial breast cancer) or tests that predict genetic predisposition to multifactorial conditions (e.g., coronary artery disease, and, again, breast cancer). This issue—how customs of care evolve—could decline as broad categories of predictive genetic tests develop. It might not, however, because every disease—and how people perceive each—differs.

One consideration for the future not fully explored in this report is indirectly related to cost-effectiveness, but directly related to how much CF mutation analysis—and other diagnostic genetic tests—will cost in the future. At issue are patents, licensing, and royalty fees for both products (e.g., the CF gene, for which a patent is pending) and processes (e.g., PCR, for which Roche Molecular Systems holds the patent) that are important to DNA-based diagnostics. Although automation appears likely to lower costs of DNA diagnostics, intellectual property protection, the impact of which cannot be fully assessed, to some extent might counter lower prices realized by new instrumentation. Issues surrounding intellectual property, scientific exchange, commercial development, and the Human Genome Project have existed since that project's outset. They continue to loom and might need congressional attention if they become pressing. Witness, for example, the new debate surrounding patenting certain DNA sequences.

Certain factors related to CF carrier screening will be less germane to analyzing the implications of other emerging tests that assess genetic risks. In particular, cost-effectiveness is a case-by-case matter. Likewise, the issue of making automation a priority through Federal funding for instrumentation research and development presumably will dissipate.

Finally, fundamental to consideration of CF carrier screening is the issue of genetic counseling and abortion. Prenatal screening will probably comprise the largest portion of CF carrier assays, at least initially. Thus, as with prenatal tests generally, the extraordinary friction about abortion in this country is inevitably linked to the implications of CF carrier testing and screening. But as knowledge from the Human Genome Project accumulates, so will the number and definitiveness of genetic tests, and so presumably the social, ethical, and political tension. Some tests will be more likely than others to have prenatal applications, but as long as utilization of the new assays by pregnant women is possible, some will opt for abortion.

While not explicitly overturning *Roe* v. *Wade*, the 1992 U.S. Supreme Court decision in *Planned Parenthood of Southeastern Pennsylvania* v. *Casey* means women's access to legal abortions now turns largely on State law. The decision appears to affirm that women may choose to terminate pregnancies prior to fetal viability, but States may make this more difficult than it has been prior to the ruling. The court's ruling in the Pennsylvania case indicates States may enact laws related to information delivery, waiting periods, services provision, and restrictions on public financing or use of public facilities, as long as such laws do not present a substantial obstacle to a woman's choice. If Congress believes States should be preempted from enacting such laws, it could pass Federal legislation prohibiting State restrictions in any of these areas.

As well, the 1991 U.S. Supreme Court decision in *Rust* v. *Sullivan* upheld Federal regulations stating that patients at clinics receiving certain Federal funds (i.e., from Title X of the Public Health Service Act) may not receive information about the option of terminating a pregnancy at risk for a child with a genetic disorder. In March 1992, an executive order modified the original regulation and stated that such information may be provided by a physician, although the legal standing of that order is in question. The vast majority of practitioners providing services in such clinics—nurses and genetic counselors—still may not inform patients of this option. Congress came close to rescinding the entire restriction when a majority of Members of Congress voted to

*Photo credit: Chip Moore*

The U.S. Supreme Court

overturn the regulation in 1991. If Congress believes nonphysician health care professionals should be allowed to counsel patients about abortion following diagnosis of fetal abnormalities, it could reexamine the issue and enact an exception for counseling related to genetic conditions or overturn the regulation entirely.

Nearly 10 years ago, the President's Commission for the Study of Ethical Problems in Medicine and Biomedical and Behavioral Research concluded the fundamental value of CF carrier screening lies in its potential for providing people with information they consider beneficial for autonomous reproductive decisionmaking. CF carrier screening, however, is not just about a person's future reproductive choices. CF carrier screening represents the first of many DNA-based tests to come and raises many issues. Policy decisions made about it will reverberate far beyond this specific case.

Chapter 2

# Introduction

# Contents



### *Boxes*



### *Figures*



### *Tables*

# Chapter 2
# Introduction

People want—expect—perfectly healthy babies. When a child is born with a genetic condition, parents suffer anxiety, endure anguish, and experience guilt: "This baby is sick because of us."

This report is about one of these inherited conditions: cystic fibrosis (CF). CF is a life-shortening disorder. It is a genetic condition—i.e., one that follows a clear pattern of inheritance in families—and is the most common, lethal recessive disorder in American Caucasians of European descent. Each year in the United States, about 1 in 2,500 babies is born with CF (10,35,47)—i.e., about 1,700 to 2,000 babies with CF are born annually (25). Approximately 1 in 9,600 Hispanic, 1 in 17,000 (9) to 19,000 (50) African American and 1 in 90,000 (50) Asian American newborns have CF.

Medicine has long recognized the consequences of CF (table 2-1) on several organ systems, particularly the lungs and pancreas. Only recently, however, have scientists pinpointed the most common change, or mutation, in the genetic material—DNA—that accounts for the majority of CF cases (44,66,68). Because CF is a recessive trait, a child with CF must receive two mutant CF genes, one inherited from each parent, who are CF "carriers," but who do not have the disorder (figure 2-1). Thus, while approximately 30,000 people in the United States have CF, as many as 8 million people could be carriers of one CF mutation. What are the implications of informing this latter pool of individuals—or a subset of those of reproductive age and younger—about tests that reveal CF carrier status?

## TERMINOLOGY

Human genetics, like all scientific disciplines, is rife with jargon, and subtle distinctions in language can matter a great deal. People, reports, or institutions rarely define terms of art in precisely the same manner. To avoid confusion, OTA uses several terms as follows.

OTA defines ***genetic testing*** as the use of specific assays to determine the genetic status of individuals already suspected to be at high risk for a particular inherited condition. While any individual can be

**Table 2-1—History of Cystic Fibrosis: Selected Highlights**

| | |
|---|---|
| 1650 | Literature refers to now characteristic CF pancreatic and lung symptoms association with salty skin and early death. |
| 1705 | A book of folk philosophy states that a salty taste means a child is bewitched. |
| 1857 | *The Almanac of Children's Songs and Games*, Switzerland, quotes from Middle Ages: "Woe is the child who tastes salty from a kiss on the brow, for he is hexed, and soon must die." |
| 1938 | First reported description of disease, calling it "cystic fibrosis of the pancreas." |
| 1946 | Antibiotics found effective for treating CF-related lung infection. |
| 1946 | Inheritance pattern—autosomal recessive—suggested. |
| 1953 | Sweat abnormality in CF first described. |
| 1955 | First review of use of pancreatic enzymes to treat CF. |
| 1959 | Safe and accurate way to diagnose CF, "sweat testing," reported. |
| 1960 to present | Accelerated improvement in survival. |
| 1968 | Mechanism underlying CF-related male infertility demonstrated. |
| 1981 to 1983 | Basis for sweat abnormality (i.e., electrolyte transport problems) described. |
| 1986 | CF gene localized to chromosome 7. |
| 1989 | CF gene and its most common mutation identified. |
| 1990 | CF mutation assays available from selected genetic laboratories, companies, and medical centers. |
| 1990 | CF mutation corrected in laboratory cells. |
| 1991 | Functions of CF gene described. |

SOURCE: Office of Technology Assessment, 1992, based on L.M. Taussig, *Cystic Fibrosis* (New York, NY: Thieme-Stratton, Inc., 1984).

**Figure 2-1—Inheritance of Cystic Fibrosis**

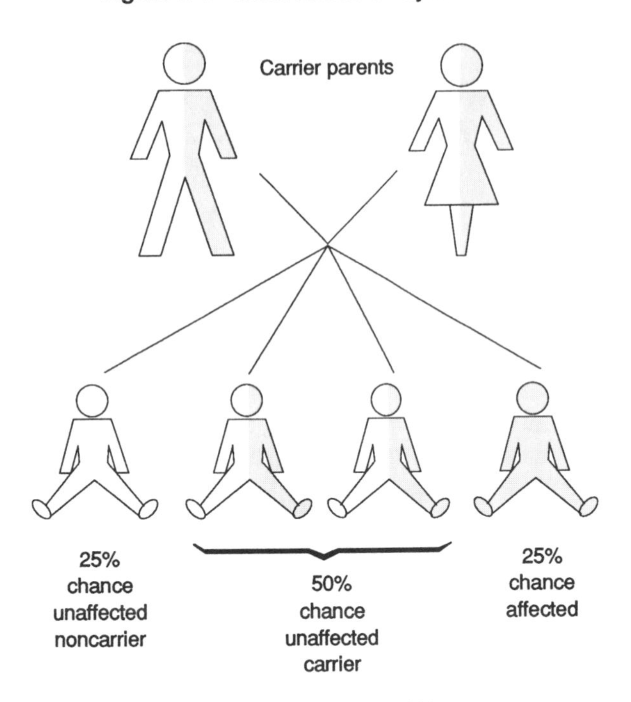


SOURCE: Office of Technology Assessment, 1992.

considered "at high risk" for a particular unknown trait, and hence be "tested," "at high risk" here denotes the presence of a family history or clinical symptoms. The terms ***genetic test***, ***genetic assay***, and ***genetic analysis*** are used interchangeably to mean the actual laboratory examination of samples.

***Genetic screening*** usually uses the same assays employed for genetic testing, but is distinguished from genetic testing by its target population. OTA uses the term "screening" selectively. In this report, it refers to analyzing samples from individuals without a family history of the disorder, groups of these individuals, or populations. ***Carrier screening for CF*** (or ***CF carrier screening***), then, involves performing tests on persons for whom no family history of the disorder exists to determine whether they have one normal and one aberrant copy of the CF gene, but not the disorder (which results from having two aberrant CF genes).[1]

Many individuals are CF carriers but do not have a positive family history. In fact, 4 of 5 babies born with CF each year—as many as 1,600—are cases where there was no known family history for CF. The difference between *testing* and *screening* is illuminated by considering a person contemplating procreation. He or she could inquire about the availability of an assay to determine the probability that he or she could have a child affected with CF. If there are no relatives with the disorder, the individual could be informed that a test would provide information about his or her genetic status for CF. The person could then elect to be *screened* to determine whether or not he or she is a carrier for CF. If, however, there is a family history of the disease, a practitioner would ideally inform the individual and his or her partner about CF carrier assays and they might choose to be *tested* to determine if they are both carriers.

***Genetic counseling*** is a clinical service that includes providing an individual (and sometimes his or her family) with information about heritable conditions and their risks. When centered around genetic testing and screening, it involves both education and psychological counseling to convey information about the ramifications of possible test outcomes, prepare the client for possible positive or negative analyses, and discuss actual test results. Many types of health professionals perform genetic counseling. OTA reserves the term ***genetic counselor*** for master's-level individuals certified (or board-eligible) by the American Board of Medical Genetics to clarify the discussions of the legal distinctions in licensing and third-party reimbursement among the different types of practitioners. But, OTA uses the term ***genetic counseling*** more generically to refer to the educational and informational process that is performed by genetic specialists, including physicians, Ph.D. clinical geneticists, genetic counselors, nurses, and social workers.

OTA avoids using the term "program" in discussing CF carrier screening in the United States. For many, the term connotes a formal public health effort led or sanctioned by Federal, State, or local governments. In analyzing CF carrier screening, OTA's premise is that large numbers of Americans might be screened for their CF carrier status. OTA remains neutral on whether the assays will be a component of a fixed, regulated scheme or another facet of general medical practice.

[1] In contrast, OTA uses the term *CF screening* (or *screening for CF*) to mean screening individuals to diagnose the presence or absence of the actual disorder, in the absence of medical indications of the disease or a family history of CF. Such screening usually involves newborns (ch. 3), but is rarely done for CF except in Colorado and Wisconsin. CF testing of infants is common if a family history of the condition exists.

### *Box 2-A—Eugenics At the Turn of the Century*

Eugenics refers to processes or policies to either discourage or prevent reproduction by members of society with "undesirable" heritable traits or to encourage or require procreation by individuals who have "desirable" genetic characteristics. Put more broadly, it involves efforts that interfere with individuals' reproductive choices in order to attain a "societal" goal. Drawing on roots developed by Francis Galton, a cousin of Charles Darwin, in England in the late 1800s, eugenics movements flourished in the United States at the turn of the century.

Compulsory sterilization laws were an outgrowth of the U.S. eugenics movement. The Model Eugenics Act, from which many States drafted their laws in the early 1900s, targeted institutionalized tuberculosis patients, people who were blind or deaf, and chronic alcoholics among those who should be sterilized. In 1927, the U.S. Supreme Court upheld the Commonwealth of Virginia's sterilization law, Justice Oliver Wendell Holmes writing:

> We have seen more than once that the public welfare may call upon the best citizens for their lives. It would be strange if it could not call upon those who already sap the strength of the State for these lesser sacrifices, often not felt to be such by those concerned, in order to prevent our being swamped with incompetence. . . . Three generations of imbeciles are enough (16).

Despite the fact that compulsory sterilizations continued into the 1970s, the eugenics movement per se waned in the United States during the 1930s, largely from distaste with Hitler's embrace of eugenics. Wariness over past abuses of genetic information led to the emphasis on nondirective genetic counseling in clinical practice. Nevertheless, the legacy of eugenics—though by and large renounced—continues to color perceptions about large-scale genetic screening, and thus to subtly influence decisions surrounding human genetics and public policy.

SOURCE: Office of Technology Assessment, 1992, based on N.A. Holtzman, *Proceed With Caution: Predicting Genetic Risks in the Recombinant DNA Era* (Baltimore, MD: The Johns Hopkins University Press, 1989); D.J. Kevles, *In the Name of Eugenics* (Berkeley, CA: University of California Press, 1985); K.M. Ludmerer, *Genetics and American Society: A Historical Appraisal* (Baltimore, MD: The Johns Hopkins University Press, 1972); P. Reilly, *Genetics, Law, and Social Policy* (Cambridge, MA: Harvard University Press, 1977); and P. Reilly, *The Surgical Solution* (Baltimore, MD: The Johns Hopkins University Press, 1991).

## RECENT HISTORY OF HUMAN GENETICS AND PUBLIC POLICY

The science of human genetics is embedded in this country's consciousness, and has manifested itself—overtly and covertly—in public policies throughout U.S. history (box 2-A). Race and skin color, for example, are genetically influenced, and have played a direct role in official and unofficial decisionmaking. In some respects, identifying carriers of CF mutations—invisible genetic characteristics—is just another twist in the history of genetics and U.S. public policy, but one that has implications for the majority population in this Nation.

To provide background and perspective for today's debate about CF carrier screening, this section briefly describes watershed events in U.S. politics and human genetics. Not intended to be comprehensive, it focuses on a few, discrete events in the 20th century that place the questions raised by CF carrier screening in context and help frame the issues and options addressed by this report. The impact of broader U.S. laws, such as Title VII of the Civil Rights Act (42 U.S.C. 2000e) and the Americans With Disabilities Act (Public Law 101-336; 42 U.S.C. 12101), are discussed elsewhere in the report.

### *U.S. Law and Genetic Disease*

Most U.S. legislation related to genetic disease is State law covering newborn screening (2,63,78,89). During the 1970s, however, Congress enacted three measures involving carrier screening for several genetic conditions (Public Laws 92-294, 92-414, and 94-278). Today, most State newborn screening laws (and the programs and practices established by them) operate, for the most part, unchallenged. In contrast, the Federal Government's role in public health and genetics has changed historically.

In the late 1960s and early 1970s, sickle cell anemia received prominent attention as a health concern. The African American community felt that sickle cell anemia was a neglected condition, with little Federal research funding directed toward it. As the debate progressed, Federal interest, along with State interest, developed. President Nixon made an appeal for an effort to combat sickle cell anemia in his 1971 health address to Congress (39), and the following year he signed into law the National

Sickle Cell Anemia Control Act (Public Law 92-294). While the act focused on detecting sickle cell anemia, the mechanics of the test also identified carriers for sickle cell. Later that year, Congress moved a second time to enact legislation directed at another genetic disease, β-thalassemia, with the National Cooley's Anemia Control Act (Public Law 92-414).

Both programs represented a significant expansion of Federal support for nonresearch genetic initiatives. Federal programs supported only State efforts with voluntary participation, a measure designed to defuse ongoing controversy over mandatory, coercive screening. And although the statutes' intent was to reduce stigmatization of and discrimination against carriers, these practices continued unabated (64).

In 1976, Congress amended the sickle cell legislation, broadening it to the National Sickle Cell Anemia, Cooley's Anemia, Tay-Sachs, and Genetic Diseases Act (Public Law 94-278; hereinafter the National Genetic Diseases Act). In doing so, it expanded both the scope and authority of activities, as well as the range of genetic disorders for which Federal grants and contracts were awarded. The legislation emphasized voluntary participation and the use of proper guidelines for confidentiality of results; it also stressed that genetic counseling for all participants should be available—goals that experts agree are desirable for CF carrier screening (18,54). In 1978, Congress reauthorized the program, which continued to provide funding for basic and applied research, training, screening, counseling, and information and education programs (Public Law 95-626).

In 1981, the role of the Federal Government in genetic services, education, and training dramatically altered (61). Authorization for programs operated under the National Genetic Diseases Act was replaced by the Maternal and Child Health Block Grant (Public Law 97-35). No longer were Federal funds for genetic services, professional training, and public education programs guaranteed: The majority of fund allocation decisions have since been left to individual States. Programs for genetic services, research, and professional training now compete with other maternal and child health services (box 2-B). And while many States have responded with State or regional programs, the reduced Federal role led a presidentially appointed commission to voice concern about the adequacy and effectiveness of genetic services, education, and training (61).

While difficult to quantify, decreased Federal attention to genetic services, training, and education might have left the country less than well prepared to handle the rapid integration of molecular genetics research into clinical practice. From the late 1970s to the present, basic research in genetics has enjoyed generous Federal sponsorship and returned the dividend of increased knowledge about many genetic conditions. In contrast, Federal funds for projects relating to genetic services show a steady decline since 1981. These genetic services provide the link to translate basic research developments into clinical practice.

A void in Federal funding for genetic services might have exacerbated at least one issue raised by the prospect of routine CF carrier screening: the inadequacy of training-related monies to ensure sufficient genetic counseling services. Similarly, decreased Federal spending for genetic services likely contributed to the initial scrambling to fund pilot studies for CF carrier screening (17,67). In fact, it was left to the National Center for Human Genome Research (NCHGR), National Institutes of Health (NIH)—a research, not service, agency—to step forward and coordinate clinical assessments of genetic services for CF carrier screening (90).

In October 1991, NCHGR funded six 3-year clinical assessment studies to examine education and counseling issues related to CF carrier screening. The National Institute of Child Health and Human Development and the National Center for Nursing Research also funded one project each (53). The Cystic Fibrosis Foundation, which took a lead role in funding investigations to find the CF gene and its mutations, declines to participate in any decisions about pilot projects for CF carrier screening, saying its mission is not prevention, but improving treatment and finding a cure (67).

### *The 1983 President's Commission Report*

In 1980, Congress created the President's Commission for the Study of Ethical Problems in Medicine and Biomedical and Behavioral Research (Public Law 95-622; 42 U.S.C. 300). Among the topics Congress mandated that the Commission examine was the ethical, social, and legal implications of genetic screening, counseling, and educa-

### *Box 2-B—Genetic Services: Federal-State Partnership*

Funding for genetic services derives from a medley of Federal and State sources, and varies greatly from State to State. During the 1970s, genetic services enjoyed substantial Federal funding, in part through congressional mandate. The Omnibus Budget Reconciliation Act of 1981 (Public Law 97-35), however, led to the consolidation of genetic services funding—along with seven other programs—into the Maternal and Child Health (MCH) Block Grant. Overall, funding for maternal and child health services was cut, and the responsibility for distributing the monies and for providing services was passed to the States, which also had to begin using $3 in State funds for every $4 of Federal money received. Prior to the block grant, no matching funds were required.

Under provisions of the MCH block grant, 85 percent of funds go directly to the States for maternal and child health services. States must decide how to allocate the funds among a number of areas, such as general prenatal care, infant nutritional supplementation, and other maternal and child health needs. MCH funds may be used for health care services, education, and administration. In fiscal year 1990, less than 2 percent of MCH funds were used by States to support genetic services other than newborn screening.

In general, MCH funds account for a small portion of State genetic services. Under terms defined by the block grant, each State decides whether or how much money to designate for genetic services. In 1990, 34 States used MCH funds to support some aspect of general genetic services other than newborn screening, including nonpatient-related activities such as administration and planning. In the majority of States, however, MCH funds accounted for less than 25 percent of fiscal year 1990 funding for genetic services (51). In fiscal year 1990, MCH funding for genetic services other than newborn screening totaled approximately $8 million; State funding accounted for approximately $22 million (51).

Fifteen percent of the MCH block grant is administered as direct grants for Special Projects of Regional and National Significance (SPRANS). SPRANS monies are grants for specific projects and are not given to each State. SPRANS provides seed money for demonstration, or pilot, projects in a number of areas. After the demonstration period ends, usually in 3 years, alternative funding must be found.

In fiscal year 1990, genetic services received about 9 percent of all SPRANS funds. When adjusted for inflation, however, constant dollar funding for genetic services under SPRANS has decreased almost every year since the block grant's inception. Moreover, SPRANS support of genetic services has decreased from about 90 percent of the SPRANS genetic services budget in 1981 to about 66 percent in 1991. Initially, most of the SPRANS genetic services budget established statewide genetics programs, with each State receiving seed money for at least 4 years. The last State received funding in 1990 (27). Other areas of genetic services delivery receiving SPRANS support include ethnocultural projects to increase utilization of genetic services by underserved populations; psychosocial studies; and support groups for young adults and families. In fiscal year 1990, 16 States used approximately $4 million from SPRANS grants to support demonstration projects in clinical genetic services other than newborn screening (51). In fiscal year 1990, just over one-third of SPRANS' genetic services budget went to the regional networks and the Council of Regional Networks for Genetic Services (CORN) (27). CORN and the regional networks—comprised of genetic service providers, public health personnel, and consumers—serve as resources for communication and coordinate data collection and quality assurance, but do not provide direct services to patients.

In addition to block grant and SPRANS awards, States also fund genetic services from other sources. In fiscal year 1990, at least 26 States derived $46 million in genetic services funding exclusive of newborn screening from provider in-kind and service charges, third-party reimbursement, grants, contracts, newborn screening fees, health insurance surcharges, and mental health/mental retardation funds. For some States, such funding accounts for most of their genetic services funding. For example, newborn screening fees generated 93 percent of genetic services funding in Colorado and 86 percent in Michigan in fiscal year 1990. Similarly, prenatal screening service fees accounted for more than 83 percent of the genetic services budget in California in fiscal year 1990 (51).

All States, the District of Columbia, and Puerto Rico coordinate genetic services statewide; nearly half experienced a decrease in funding for genetic services from fiscal years 1988 through 1991 (51). Individual State genetic service programs face yearly uncertainty about how much—if any—funding they will receive, which makes planning difficult. As general knowledge and public awareness about genetic diseases continues to emerge out of the Human Genome Project, uncertainty in genetic services funding will be increasingly problematic.

SOURCE: Office of Technology Assessment, 1992.

### *Box 2-C—The 1975 National Research Council Report, "Genetic Screening: Programs, Principles, and Research"*

In response to a letter from the American Society of Human Genetics, the National Research Council (NRC) of the National Academy of Sciences convened a committee in 1972 specifically to analyze neonatal screening for phenylketonuria and generally to assess the relation between genetics and preventive medicine. In particular, the committee was charged with addressing the questions: To what degree has genetics played a part in thinking about and practice of disease prevention? How should this relationship be fostered and extended?

Key recommendations of the committee were that participation be left to the discretion of the person tested and that information obtained as a result of a test not be made available to others except with the consent of the patient. The committee also advised that professionals responsible for screening programs be aware of and regularly assess potentially damaging effects of screening, including invasion of privacy, breach of confidentiality, civil rights violations, and psychological effects from being labeled a genetic carrier. Principles described in the report still underlay genetic screening and testing today.

The NRC report was not initiated by the Federal Government, but it was supported with Federal funds from the National Science Foundation. It made a critical impact in shaping future discussions, such as the 1983 President's Commission report, *Screening and Counseling for Genetic Conditions: The Ethical, Social, and Legal Implications of Genetic Screening, Counseling, and Education Programs.*

SOURCE: Office of Technology Assessment, 1992, based on Committee for the Study of Inborn Errors of Metabolism, National Research Council, *Genetic Screening: Programs, Principles, and Research* (Washington, DC: National Academy of Sciences, 1975).

tion programs. In 1983, the Commission published the results of its deliberations (61).

In carefully weighing the advantages and disadvantages of applications of advances in medical genetics, the Commission found, on the whole, that these advances have greatly enhanced health and well-being (62). Drawing on the literature (55) (box 2-C) and public hearings, the Commission reached 15 conclusions, including recommendations about the confidentiality of genetic information and mandatory versus voluntary screening (61).

The Commission's report on genetic screening is noteworthy for its examination of past experience with screening programs (e.g., for Tay-Sachs disease, sickle cell anemia, and phenylketonuria) and its prescience in using CF carrier screening as a specific case study. The Commission's analysis explored ethical aspects of genetic screening in anticipation of issues it predicted would be raised by large-scale carrier screening for CF. It concluded that the fundamental value of CF carrier screening lies in its potential for providing people with information they consider beneficial for autonomous reproductive decisionmaking (61,62). Nine years ago, the Commission identified some of the same controversies being discussed today.

### *The Human Genome Project*

As the 21st century approaches, Congress and the executive branch have made a commitment to support scientific efforts to determine the location on the DNA of all genes in the human body (as has been done for CF)—in short, to map the human genome. The Human Genome Project is estimated to be a 15-year, $3 billion project. It has been undertaken with the expectation that enhanced knowledge about genetic disorders, increased understanding of gene-environment interactions, and improved genetic diagnoses can advance therapies for the 4,000 or so currently recognized genetic conditions; a premise supported by the fact that even prior to the Human Genome Project, advances in medical genetics have guided the development of new treatment strategies and incrementally improved the management of some genetic conditions over the years (22,23).

In many respects, the Human Genome Project served as the catalyst for congressional interest in this OTA assessment. Despite scientific and technological promises of the project, fears have been raised about how information gained from it—such as identification of CF mutations—will be used (37,52,56,57,80). These concerns will involve policy decisions for Congress.

To address gaps in knowledge and perhaps forecast the social consequences of the Human Genome Project, NIH and the Department of Energy (DOE) each fund an Ethical, Legal, and Social Issues (ELSI) Program. Funds for each agency's ELSI effort derive from 3 to 5 percent of appropriations set aside from the total genome initiative budget. In fiscal year 1991, DOE's ELSI spending was $1.44 million (3 percent). Fiscal year 1992 spending is targeted at $1.77 million (3 percent) (26). NIH-ELSI spending for fiscal years 1990 and 1991 was $1.56 million (2.6 percent) and $4.04 million (4.6 percent), respectively. For fiscal year 1992, NIH-ELSI aims to spend 5 percent of the NCHGR appropriation ($4.98 million) (45). Table 2-2 lists the types of efforts that have been funded to date by the ELSI program.

**Table 2-2—Research Grants Funded by the Ethical, Legal, and Social Issues Program, National Institutes of Health and U.S. Department of Energy (May 1991)**

| Source | Description |
|---|---|
| DOE......... | Project to prepare 50 selected science teachers per year for 3 years to become State resource teachers in human genetics. Workshops will also update and expand curriculum materials. |
| DOE......... | Project to examine legal protections of confidentiality of genetic information and to study the availability of and need to collect genetic data to plan public health service programs. |
| DOE......... | Study to assess the significance of discrimination directed against individuals and family members because of real or perceived differences in their genetic constitution. |
| DOE......... | Project to survey ethical attitudes toward the medical applications of genetic information and to conduct a legal study of confidentiality of genetic data. |
| DOE......... | Report examining the current funding mechanisms in the biological and biomedical sciences of major Federal agencies and private organizations to determine the impact of funding on the ability to recruit and retain young investigators. |
| DOE......... | Eight-part television series, "The Secret of Life." |
| DOE......... | Curriculum development module and instructional activities, "Mapping and Sequencing the Human Genome: Science, Ethics, and Public Policy," for first-year high school biology students. |
| DOE......... | Conference and laboratory workshop for nonscientists drawn from four groups: public policymakers, civic leaders, program officers of health-related foundations, and science journalists. |
| DOE......... | Conference: "Justice and the Human Genome Project." |
| NIH.......... | Study, including public education and participation, to determine the impact of the Human Genome Project on women and to identify ways of avoiding or reducing potential gender injustice. |
| NIH.......... | Historical analysis of the relevance of eugenics to genomics for the specific case of cancer theory and policy. |
| NIH.......... | Study to examine the ethical and legal implications of genetic information on understanding health, normality, and disease causation. |
| NIH.......... | Project to develop a human molecular genetics curriculum module for honors, main-stream, and low-achieving high school students and adults in a continuing education program. (Cofunded with the NIH Center for Research Resources.) |
| NIH.......... | Series of projects to update and inspire secondary school science teachers in genome technologies and their implications, including newsletter for educators, "hands on" demonstrations to the public and at schools, and workshops. |
| NIH.......... | National survey of public knowledge and perceptions of genetic testing and the Human Genome Project. (Cofunded with the National Science Foundation (NSF).) |
| NIH.......... | Survey of physicians' and master's-level genetic counselors' knowledge of and attitudes toward genetic testing. Survey and interview of commercial interests in and impact on human genetics research. |
| NIH.......... | U.S.-Canadian survey of geneticists', genetic counselors', and genetic clinic patients' views on a variety of situations in genetics that pose ethical dilemmas. A separate grant involves a survey of geneticists from 34 additional nations about the same situations. |
| NIH.......... | Sociological study exploring the meaning of human genetics in popular culture (e.g., fiction, film, news accounts) to understand lay interpretations of genetic concepts. |
| NIH.......... | Comparison of feminist, medical, and bioethical analyses of impact of genetic testing on parent-child relationships. (Cofunded with NSF.) |
| NIH.......... | Study of the concept of genetic susceptibility and the basis and limits of privacy of genetic information about individuals. |
| NIH.......... | Ethnographic study of the impact of genome research on the social organization of biological science. |
| NIH.......... | Report on professional standards for forensic DNA typing. (Cofunded with NSF, Federal Bureau of Investigation.) |
| NIH.......... | DNA sequencing of mitochondrial DNA to define the technical and statistical limits of this approach to human identification applications (e.g., identifying victims of accidents, natural disasters, and wars; reuniting separated families; investigating claims of identity; and aiding criminal investigations). |

(Continued on next page)

Table 2-2—Research Grants Funded by the Ethical, Legal, and Social Issues Program, National Institutes of Health and U.S. Department of Energy (May 1991)—(Continued)

| Source | Description |
|---|---|
| NIH | Study of the impact of genetic testing and counseling on medicine and the doctor-patient relationship. |
| NIH | Paradigm analysis to develop a comprehensive and systematic framework to resolve ethical issues raised by genomic information in clinical genetics. |
| NIH | Study of the historical and social impact of amniocentesis. |
| NIH | Study examining historical case studies to examine the potential risks of stigmatization associated with genetic testing, screening, and diagnosis. |
| NIH | Interdisciplinary study of the implications for insurance of increasing information from the Human Genome Project. |
| NIH | Study of the impact of genetics on access to health care. |
| NIH | Historical analysis of the impact of the genetics of human leukocyte antigens on criminology and the genetics of race. |
| NIH | Training manual and communication materials to train genetic counselors to, in turn, conduct courses for primary care providers. |
| NIH | Intensive short course for scientists and bioethicists on the ethical, legal, and social implications of the Human Genome Project. |
| NIH | Education workshop series for State legislators and other State officials. |
| NIH | Public lecture series on the ethical, legal, and social implications of the Human Genome Project. |
| NIH | Forum for genetic disease support groups on the Human Genome Project and its ethical, legal, and social implications. |
| NIH | Eight CF pilot screening projects (six by National Center for Human Genome Research, Ethical, Legal, and Social Issues Program, and one each by the National Center for Nursing Research and National Institute of Child Health and Human Development.) |
| NIH | Conferences: "Strategies for Documentation of Research on the Human Genome," "Human Genome Workshop: Ethics, Law, and Social Policy," "Legal and Ethical Issues Raised by the Human Genome Project," "A Legal Research Agenda for the Human Genome Initiative," "The Genetic Prism: Understanding Health and Responsibility," "Ethics, Values, Professional Responsibilities," "Biotechnology and the Diagnosis of Genetic Disease," "Testing for Germ Line p53 Mutations in Cancer Families," "Human Genome Research in an Interdependent World," "Ethical and Legal Implications of Genetic Testing," "Computers, Freedom, and Privacy," "The Human Genome Project: A Choices and Challenges Forum," "A Conference on Human Genome Research Implications," and "Genetic Factors in Crime: Findings, Uses, and Implications." |
| NIH/DOE | Conference: "Genetics, Religion, and Ethics." |
| NIH/DOE | Project to examine issues of privacy, stigma, and discrimination, particularly as they relate to culturally diverse groups—both those who have and have not used genetic services. |
| NIH/DOE | Study investigating newborn genetic screening programs and policies governing State-sponsored genetic screening. Minority populations' access to and use of genetic services will be examined, including the nature of services available to rural populations. |
| NIH/DOE | Television documentary, "The Future of Medicine." |
| NIH/DOE | Report addressing a variety of issues presented by the rapid proliferation of genetic tests capable of predicting future disease. |

SOURCE: Office of Technology Assessment, 1992.

# THE INTERESTED PARTIES: PRESSURES FOR AND AGAINST SCREENING

Why is carrier screening for CF a controversy? Experts agree that persons with a family history of CF should have the opportunity to avail themselves of the new, DNA-based tests. No one espouses mandatory screening. Who opposes voluntary screening, and on what grounds? Who supports CF carrier screening, and why? Do past experiences with large-scale genetic screening (e.g., maternal serum alpha-fetoprotein, sickle cell, or Tay-Sachs) provide a framework to answer questions raised by routine CF carrier screening? What is the role of genetics in public health (box 2-D)?

Many parties have a stake in resolving questions raised by our increased understanding of human genetic disease, including CF. These stakeholders include consumers, health care providers, and commercial ventures. Also weighing in on the evaluation of the technical, legal, ethical, and economic considerations are experts and professional societies in each of these fields. This section briefly describes the tensions that have arisen and identifies areas

### *Box 2-D—Genetic Screening and the Practice of Public Health*

In some respects, friction over routine carrier screening for CF reflects different notions of public health and its interaction with genetics. What is public health, today? How do genetic testing and screening for CF fit in its practice? Do they fit at all? Does the evolving practice of clinical genetics challenge many common assumptions about the limitations on, and aims of, public health authorities?

Public health is a dynamic field, and its history records struggles over the limits of its mandate. Public health attempts to prevent disease, prolong life, promote physical health through sanitation of the environment, control contagious infections, educate individuals and whole populations about health, and organize medical services for the early diagnosis and preventive treatment of disease. Since it can involve social machinery to ensure maintenance of health (59,69), such institutional mechanisms might sooner or later violate—or be perceived to violate—private beliefs, private property, or the prerogatives of other institutions (73). Today, some public health initiatives, such as quarantine or immunization, are mandatory. Compulsory components, however, are only a narrow slice of what constitutes the practice of public health. There is nothing inherently coercive or mandatory about public health per se: Witness, for example, public education about drug abuse, sanitation, or voluntary cholesterol or blood pressure screening.

Debates surrounding public health issues, such as the spread of infectious disease, often involve an adversarial model focused largely on balancing individual rights against community rights, on the assumption that the two are in conflict (1,60). For public health issues like genetic testing and screening, however, individual interests might be in harmony with public interests, and thus cooperative models of individual and governmental action (3,34,59) could be more appropriate. On the one hand, who better to make the choice of whether to conceive a child with a genetic disorder than the individuals who will both gain from and provide support to the child. At the same time, as the Human Genome Project project continues to identify genetic risks that everyone faces in procreation, genetic diagnosis and counseling becomes an aspect of personal health for the entire community—and hence perhaps governmental action.

Nevertheless, disputes about the role of public health practices in genetics arise and often adopt polemic tones. The balance between individual freedom, individual responsibility, and government responsibility for health is especially delicate in areas such as carrier screening for CF. If examined from the view of public health measures to control disease, CF and other genetic illnesses are fundamentally different from infectious disease. Unlike familiar public health measures such as vaccination or sanitation policies, CF carrier screening does nothing to protect individuals from the causes of disease, nor does it directly improve personal health. CF carrier screening conveys information about future scenarios—i.e., the potential of CF occurring in one's offspring, not oneself. Viewed negatively from this perspective, some maintain that public health and genetics equate with eugenic motives. Still others take a dim view of a public health role for CF carrier screening, not because of eugenic overtones, but because they believe that consumers are best served by having CF carrier tests available through general medical practice. They argue that formal effort translates to regulated medicine, which they oppose.

Balanced against these perspectives, however, are beliefs of others that public health currently centers on identifying, educating, and counseling individuals and populations about achieving good health. From this perspective, genetic screening falls squarely beneath the public health rubric, which should play an important and appropriate role in CF carrier identification. These voices argue that there is nothing inherently eugenic about the role of public health in genetics. To the contrary, many believe public health's historical tradition with institutional mechanisms and social approaches is appropriate and necessary for quality assurance and consumer protection.

It is easy to see how a formal CF carrier screening policy could be perceived as a form of eugenics if it were assumed that all persons found to be carriers would, or should, act to prevent the birth of a child with a genetic condition. Thus, while some maintain that such is not the case and that the public health goal met by routine CF carrier screening is to provide information and options, others assert that early diagnosis or reducing incidence of genetic illness on a population basis is also an implicit goal. In any case, whether information about carrier status affects the incidence of CF ultimately depends on how individuals use information provided by screening, and reducing incidence of the disorder might not be a goal per se of carrier screening, but could be a consequence.

SOURCE: Office of Technology Assessment, 1992.

**Figure 2-2—Chromosome 7 and the Cystic Fibrosis Gene**

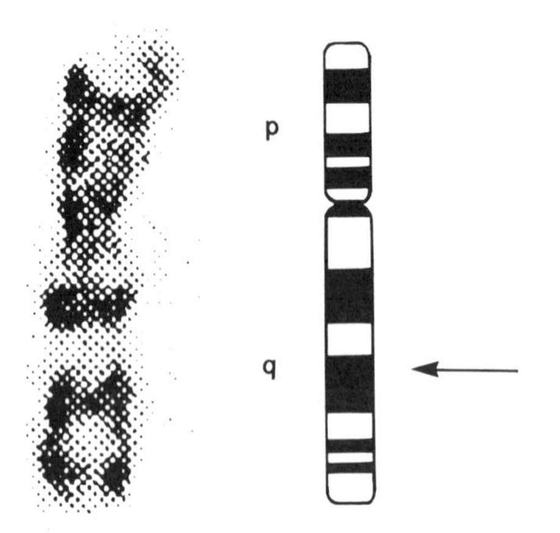


In humans, DNA is associated with protein, in bundles called chromosomes. Each chromosome contains many genes, but only the chromosomes—which can be ordered in pairs by their size and shape—are microscopically identifiable. Humans have 46 chromosomes: 1 pair of sex chromosomes (two X chromosomes for females; an X and a Y for males) and 22 pairs of autosomes. In 1986, scientists discovered that the CF gene was on chromosome 7. **Left:** Chromosome 7, as visualized by light microscopy. **Right:** Schematic of chromosome 7; arrow denotes location of CF gene on the long arm of the chromosome.

SOURCE: Office of Technology Assessment, 1992.

about which concern has been expressed. Subsequent chapters elaborate on and analyze these issues.

## *Scientific and Clinical Tensions*

Elucidation of the location of the CF gene and the most common mutation leading to the condition—commonly referred to as delta F508 (ΔF508) (figure 2-2) (44,66,68)—has been quickly followed by a widely available, direct-DNA assay for carrier testing and screening. Using today's technology, it is usually a one-time test that can inform an individual whether he or she carries a CF mutation and could thus pass it to his or her offspring (who would be affected if it also received a CF mutation from the other parent). In theory, carrier screening for CF could encompass 100 to 125 million Americans of reproductive age, but will probably involve significantly fewer numbers.

Routine CF carrier screening will likely integrate into medicine in the reproductive context first—chiefly obstetric/prenatal, but also preconceptional. A focus on pregnant women, however, is not without controversy (13,20,48,49). Some have concerns about abortion, and some have reservations that prenatal testing negatively shapes perceptions of pregnancy, disability, and women (48,49). Nevertheless, based on the annual number of births (4.2 million) (31,88) and spontaneous abortions (an estimated 1.8 million) (31), there are approximately 6.0 million pregnancies per year for which CF carrier screening might be performed. Twenty-four percent of women giving birth receive no prenatal care until the third trimester (88), however, so CF carrier screening in the obstetric/prenatal context would involve, at most, 10 million[2] men and women per year, depending on who is screened. Followup carrier screening that focused on relatives of people identified as carriers initially, rather than mass screening, also significantly reduces the number who theoretically must be screened to identify a majority of carriers (24).

The current test, however, leaves ambiguity when results are negative. About 1 in 25 Caucasians carry a CF mutation, but the ΔF508 test identifies only 70 to 80 percent of actual CF carriers[3] in this population, depending on a person's ethnicity (30,47). More than 170 additional genetic alterations in the gene also cause CF—i.e., a person with CF can have the same mutation on his or her chromosomes or two different mutations. Assays using ΔF508 plus 6 to 12 other CF mutations (ΔF508+6-12) identify 85 to 90 percent of CF carriers, depending on the population being screened (21,58). (In Ashkenazic Jews, ΔF508+6 identifies nearly 95 percent of carriers (71).) Thus, a negative test result does not guarantee that a person is not a carrier. He or she could carry one of the rare CF mutations that was not assayed and hence still be a carrier. For a test that detects 85 percent of carriers, about 1 in 165 individuals who test negative using ΔF508+6-12 will have an undetected mutation; at 90 percent sensitivity, 1 in 246 individuals who test negative will be a carrier (47).

[2] This figure does not account for the estimated 2.4 million infertile couples who are trying to conceive and might be interested in CF carrier screening (would increase overall figure). Nor does it estimate the number of men and women not involved in a pregnancy (would increase), the number of individuals involved in more than one conception per year (would decrease), or those who might have been screened during a previous pregnancy (would decrease).

[3] It should be emphasized that the ΔF508 DNA-based test is not 70 to 80 percent accurate. Evidence indicates that the test per se is specific, and that DNA tests yield accuracy greater than 99 percent (11,46). That is, if the ΔF508 mutation is present in the individual's genome, the test detects it, absent laboratory error. Like all diagnostic tests, a certain number of false positive or false negatives can arise during the course of testing. Quality control and quality assurance, discussed in chapter 5, are designed to reduce this number to a small figure.

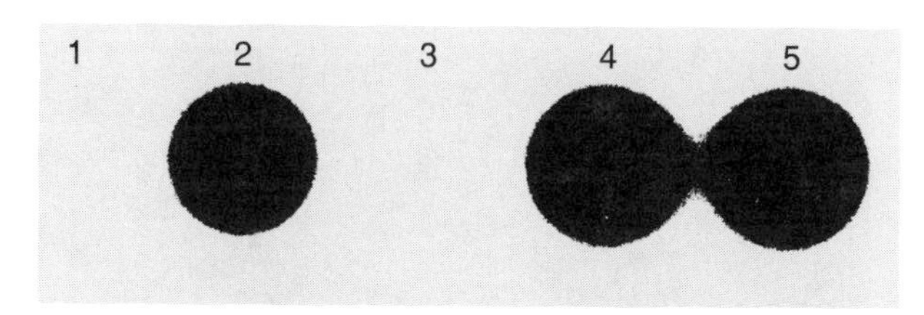


*Photo credit: IG Laboratories, Inc.*

DNA analysis for the most common mutation responsible for CF, ΔF508. A dot indicates the individual has a ΔF508 mutation. The absence of a dot means the person does not have a ΔF508 mutation, but he or she could carry one of the other 170+ CF mutations.

Couples where each partner is a carrier are sometimes referred to as carrier couples, or couples who are positive/positive (+/+). For these couples, the chance of having a child with CF is 1 in 4 for each pregnancy. If the CF test detected 100 percent of mutations, a couple in which one partner is positive and one negative (+/-) would not be at risk of bearing a child with CF. Tests to detect 170+ mutations are impractical, however, and even if they were feasible, not all CF mutations have been identified. Using ΔF508+6-12 means that for +/-couples, the negative partner could carry one of the rare mutations that the assay is not structured to detect. Couples where one partner is a carrier and the other's result is negative might misunderstand that their reduced risk is not zero risk.

For example, if 100,000 random couples were screened, 160 couples would be identified as +/+ if the test were 100 percent sensitive; one-fourth of first-time pregnanices for these 160 couples (i.e., 40) would be expected to result in CF-affected fetuses. Instead, at 85 percent sensitivity, about 116 couples will be identified as +/+ and with each pregnancy have a 1 in 4 risk of a child with CF. Results for 93,315 will be -/-, and about 6,569 couples will have +/- results. In fact, approximately 41 of the 6,569 couples with +/- results are at 1 in 4 risk of bearing a child with CF in each pregnancy, while the remaining 6,528 have no risk—but these two groups cannot be distinguished with an 85 percent test sensitivity (6,47).

About 4 of the 93,315 couples with -/- test results are also actually at 1 in 4 risk with each pregnancy of having a child with CF. Thus, of the theoretical 160 +/+ couples, 116 are dectable and 44 are not when the test is 85 percent sensitive. In other words, if all 100,000 couples experience a first-time pregnancy, 40 fetuses with CF are expected. But with an

**Table 2-3—Test Sensitivity and Risk of Child With Cystic Fibrosis**

| Percent mutations detected | Couples at 1 in 4 risk with each pregnancy[a] | | | | Affected fetuses in first pregnancy | | |
|---|---|---|---|---|---|---|---|
| | Actual | +/+[b] | +/-[b] | -/-[b] | Actual | Detectable | Missed |
| 85 | 160 | 115.6 | 40.8 | 3.6 | 40 | 28.9 | 11.1 |
| 90 | 160 | 129.6 | 28.8 | 1.6 | 40 | 32.4 | 7.6 |
| 95 | 160 | 144.4 | 15.2 | 0.4 | 40 | 36.1 | 3.9 |

[a] per 100,000 couples.
[b] Test results.

SOURCE: A.L. Beaudet, Howard Hughes Medical Institute, Houston, TX, personal communication, March 1992.

85 percent sensitive test, 29 are detectable and 11 missed. If the assay elucidates 95 percent of carriers, 144 of 160 couples would be detected. In this case, if all 100,000 couples experience a first-time pregnancy, only 4 couples at 1 in 4 risk of having a child with CF would be missed (table 2-3) (4,6).

With a test that detects 85 percent of CF carriers, a couple with a +/- result has approximately a 1 in 661 risk of having an affected child with each pregnancy (compared to a general population frequency of about 1 in 2,500). At a 95 percent detection rate, a couple whose result is +/- faces a 1 in 1,964 risk of an newborn with CF with each pregnancy (47). When the test detects a greater proportion of mutations, +/- couples can be told with greater confidence that their risk of having a child with CF is more remote; hence they might be less anxious about uncertainty. Couples who both test negative, while not having zero risk, would have a 1 in 109,200 risk of an affected child with each pregnancy (85 percent sensitivity) (47).

Some scientists, clinicians, and organizations argue that even achieving detection levels of 90 to 95 percent is insufficient to justify routine CF carrier screening—that other requirements must be met (4,7,12,18,29,32,41,54,93). They assert that CF mutation tests are appropriate only for testing individuals or families with a known history of CF or in pilot projects. Another view holds that individuals should not be advised about CF carrier screening, but for those who actively seek it and who receive sufficient education and counseling, screening is acceptable (42). Others, while also advocating pilot studies, believe the current state-of-the-art is sufficient for the test to now be offered routinely to persons of unknown risk during the course of general or obstetric/prenatal care (5,12,14,33,65,70). The latter proponents argue that consumers should be informed about the test and be given an opportunity

327-046 - 92 - 3 : QL 3

to choose whether to take it or not. Related to the issue of informing individuals about CF carrier assays is concern on the part of some physicians that withholding information about their availability leaves them vulnerable to malpractice suits.

### *Social Pressures*

Science is so much a part of society that it is no longer useful, or helpful, to consider its impact in isolation (36). While CF carrier screening is first a question of science, it is also a question of personal values (28). Not surprisingly, then, pressures for and against CF carrier screening do not center solely on scientific issues. Intertwined are matters of law, ethics, and economics. Compelling arguments that assess, weigh, and consider these factors are being made for and against routine CF carrier screening. This section briefly touches on the social pressures involved; the ensuing chapters analyze them in greater detail.

For some questions, the debate is highly charged, emotional, and divisive—e.g., prenatal testing and the option of abortion. The extraordinary tensions in the United States about abortion affect, to a certain extent, the analysis of the implications of CF carrier testing and screening. A couple where both partners are positive for ΔF508+6-12 can undergo prenatal testing to determine whether the fetus will have CF. The Cystic Fibrosis Foundation, for example, divorces itself from the CF carrier screening debate, and the abortion issue apparently played a major role in this policy (67). Nevertheless, although abortion tinges the debate, reproductive aspects of CF carrier screening encompass broader choices, including avoiding conception, seeking adoption, or choosing artificial insemination by donor.

Another concern expressed by opponents of CF carrier screening is that market pressures will drive widespread use of tests before the potential for discrimination or stigmatization by other individuals or institutions (e.g., employers and insurers) is assessed (8,15,93). This view contends that commercialization and advertising will lead some to opt for screening without fully realizing the implications of, for example, insurance considerations. On the other hand, patient demand is a major element of market forces. Thus, some point out that commercialization of genetic tests is not the factor responsible for increased interest in genetic assays, but rather that commercialization is the response to public demand.

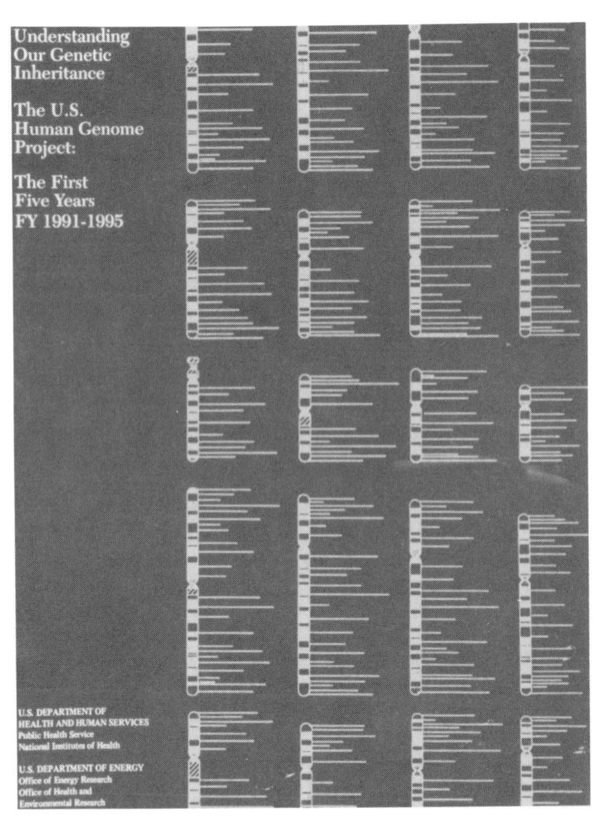


*Photo credit: Office of Technology Assessment*

Five-year plan for the U.S. Human Genome Project, jointly funded by the U.S. Department of Health and Human Services and the U.S. Department of Energy.

Because the price to consumers for CF tests averages about $170 per test, opponents also raise questions about costs. While some clients can afford out-of-pocket payments for CF carrier assays, issues of access arise for those who cannot pay but wish to use the tests. Moreover, even with less expensive tests, CF assays, like all diagnostic tests, are subject to limitations defined by laboratory quality control and quality assurance. Thus, what standards should prevail? How should quality be monitored? Finally, opponents of widespread CF carrier screening ask: How can the limited genetic services delivery system in the United States handle the swell of CF carrier screening cases, let alone cases of other genetic conditions arising from increased knowledge from the Human Genome Project? These voices express concern on both quantitative and qualitative fronts: that inadequate numbers of personnel exist (93) and that optimal methods for educating and counseling related to CF carrier screening need definition for those personnel who are available (40).

Those who advocate CF carrier tests for use beyond affected families are no less concerned about the issues just raised. Rather, proponents argue on other legal and ethical grounds that screening should move forward and individuals routinely informed about the assays so they can voluntarily choose to avail themselves of the tests (12,33,65,70). They assert that the tests are sensitive enough for current use and will, like most tests, continually improve. Since 80 percent of babies born with CF are to couples with no previous family history (42), these voices believe that failing to inform patients now about the availability of CF carrier assays denies people the opportunity to make personal choices about their reproductive futures, either prospectively—e.g., by avoiding conception, choosing to adopt, or using artificial insemination by donor—or by using prenatal testing to determine whether a fetus is affected.

**Figure 2-3—Trends in the Number of Samples Screened for Cystic Fibrosis Carrier Status, 1989-91**

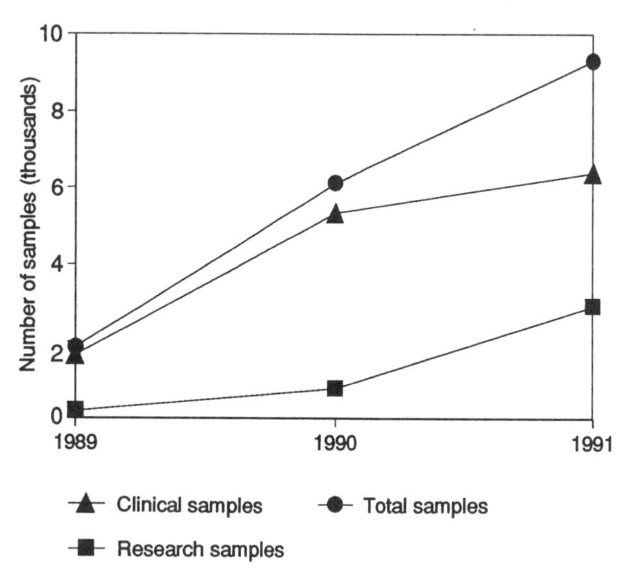


SOURCE: Office of Technology Assessment, 1992.

# THE OTA ASSESSMENT

For years, scientists, clinicians, lawyers, ethicists, and policymakers theorized about the potential consequences that increased knowledge of human genetics would bring. In the early 1990s, CF mutation tests move the debate from the theoretical to the practical. With this report, OTA assesses both the current technical capability of direct, DNA-based tests to detect mutations in the CF gene and what this capability means for individuals and society.

For some, the key question hovering over routine carrier screening for CF is if, not when. For others, the debate has shifted to when, not if. Without making judgment on its appropriateness or inappropriateness, OTA finds that the matter of CF carrier screening in the United States is one of when, not if. The expansion in the number of tests for CF carrier status will likely continue (figure 2-3); OTA estimates that as many as 63,000 individuals[4] could be screened for their CF carrier status in 1992. A rapid upward trend is not entirely unexpected, however, given the nascent stage of the technology's movement into U.S. medical practice. What is unclear is the extent to which its integration will be sustained.

Regardless of the number of individuals actually screened, it is clear that, increasingly, patients will be informed about the availability of CF carrier assays and a proportion will opt to be screened. Nevertheless, the timeframe for physicians to begin routinely informing patients about CF carrier tests is uncertain. It could be within a year or two, but more likely will be a gradual process over several years—time enough, perhaps, for policymakers to address the issues raised by this report.

Leaving the precise timing of CF carrier screening aside, the number of DNA-based tests for genetic disorders and predispositions unquestionably will increase rapidly over the next decade, almost certainly by an order of magnitude. OTA considers it likely that the time available, if any, for debate and discussion on dissemination and use of new genetic tests will be compressed, as pressure to use them rises. Given this scenario, some of the policy questions raised in this report extend beyond implications for CF carrier screening.

[4] This number is based on a canvas of 41 facilities performing CF mutation analysis (30 responding) and tests performed in federally and privately funded pilot studies. It underestimates the number of individuals who will be informed about CF carrier screening, since: not all who are informed agree to screening, standards of care will evolve, and not all facilities responded.

## *OTA Surveys*

In collecting information for this assessment, OTA found specific details were needed to answer questions about two areas covered by the report:

- What are the attitudes of genetic counselors and nurses in genetics toward CF carrier testing and screening? To date, what have been their experiences with CF mutation analyses? What are their current caseloads and what changes do they expect with routine CF carrier screening? Have their patients had difficulties with health insurance coverage due to results from genetic tests?
- What are the attitudes and policies of health insurers toward genetic testing and screening for CF carriers? Do these differ by provider type, i.e., commercial health insurers, Blue Cross/Blue Shield (BC/BS) plans, and health maintenance organizations (HMOs)? How does genetic information, including information from genetic assays, affect underwriting? What role do they envision for genetic tests in their future business practices?

OTA addressed these questions by conducting mail surveys. First, OTA surveyed the June 1991 members of the National Society of Genetic Counselors and the International Society of Nurses in Genetics. OTA focused on genetic counselors and nurses in genetics to avoid duplication with, and to compare its results to, other surveys of medical and clinical geneticists (38,91). A separate background paper describes this survey's approach and presents data OTA collected that are not directly related to CF carrier screening (86).

Second, to address questions related to practices and attitudes toward genetic information for individual health insurance policies or medically underwritten groups, OTA sent tailored questionnaires to three survey populations: medical directors of commercial insurers; medical directors and chief underwriters of all BC/BS plans; and medical directors at the 50 largest HMOs, largest HMO within a State, or largest by HMO model type. This report summarizes these data and examines their implications for the policy issues surrounding carrier screening for CF. As with the results from the genetic counselors/nurses in genetics survey, a separate OTA background paper describes this survey's methods and results in greater detail (87).

**Table 2-4—Public Attitudes About Making Genetic Tests Available Through Physicians**

Question: If there were genetic tests that would tell a person whether they or their children would be likely to have serious or fatal genetic diseases, would you approve or disapprove of making those tests available through a physician?

| | Percent[a] |
|---|---|
| Approve | 89 |
| Disapprove | 9 |
| Not sure | 2 |

[a] Percentages are presented as weighted sample estimates. The unweighted base from which the sampling variance can be calculated is 1,273.

SOURCE: Office of Technology Assessment, *New Developments in Biotechnology: Public Perceptions of Biotechnology*, OTA-BP-BA-45 (Washington, DC: U.S. Government Printing Office, May 1987).

Paralleling the paucity of information about counselors' and insurers' attitudes toward genetic testing and screening is our lack of knowledge about how consumers view these practices today. A new OTA survey of Americans' attitudes toward genetic testing and screening was not feasible for this report, however. Nor was a comparative analysis possible of the views of the general population versus CF-affected individuals or families. Other studies, however, have surveyed certain aspects of consumer attitudes toward prenatal diagnosis of CF (43,92) and neonatal and carrier screening for CF (19).

A 1986 OTA telephone survey (77) of a national probability sample of adult Americans reported that about 9 of 10 Americans approved of making genetic testing available through doctors (table 2-4). Furthermore, 83 percent of respondents reported they would take a genetic test before having children if such a test would tell them whether their children would probably inherit a fatal genetic disease (table 2-5). Survey respondents were not, however, specifically questioned about CF.

An independent survey in 1990 queried Americans about their attitudes toward genetic tests in a different manner, but found overall public opinion toward them was favorable. Sixty-six percent of respondents believed "genetic screening will do more good than harm." Even when informed as part of a question that treatment was impossible for most serious genetic conditions despite the availability of prenatal diagnosis, 69 percent said they would want prenatal testing if they (or their partner) were pregnant (72).

**Table 2-5—Consumer Attitudes Toward Genetic Tests**

Question: If genetic tests become available that would indicate whether or not it was likely that your children would inherit a fatal genetic disease, would you personally take such a test before having children or not?

| | Percent[a] |
|---|---|
| Would take test | 83 |
| Would not take test | 15 |
| Not sure | 3 |

[a] Percentages are presented as weighted sample estimates. The unweighted base from which the sampling variance can be calculated is 1,273. Percentages do not add to 100 due to rounding.

SOURCE: Office of Technology Assessment, *New Developments in Biotechnology: Public Perceptions of Biotechnology*, OTA-BP-BA-45 (Washington, DC: U.S. Government Printing Office, May 1987).

## *Scope and Organization of This Report*

As mentioned earlier, the primary focus of this report is the implications of routine carrier screening for CF. Secondarily, the report analyzes the appropriateness of using CF as a model for policy decisions raised by tests for other genetic conditions: To what extent is there an algorithm that describes the policy implications of the broad array of current and potential genetic tests? Where possible, the report analyzes how experiences with CF carrier screening can be used to construct a generic set of policy issues. Conversely, where concerns and possible solutions for CF carrier screening are inappropriate or less relevant, the report identifies these areas.

To provide a perspective on CF, medical information about the disease—its diagnosis, its therapy, and its prognosis—is presented in chapter 3. To set the stage for the legal, economic, social, and policy analyses of CF carrier screening, chapter 4 describes the genetics of CF: It covers the technical basis for—and limitations of—DNA tests for CF mutations. Chapters 5 through 9 analyze five key aspects of CF mutation analysis: quality assurance, education and counseling facets, financing, social and legal dimensions of discrimination issues, and costs and cost-effectiveness. CF carrier screening programs in the United Kingdom are described in chapter 10, which also analyzes if lessons learned from these efforts can aid decisionmaking in the United States. Appendixes A and B describe the international epidemiology of CF mutations and case studies of other carrier screening efforts, respectively.

This report does not present an ethical analysis per se of the implications of routine CF carrier screening because the fundamental principles identified in the 1983 President's Commission report remain unchanged (61,62). And although the boundary of this report encompasses carrier screening for CF, previous OTA reports analyze other issues related to new genetic technologies, including: genetic monitoring and screening in the workplace, the implications of the Human Genome Project, the commercial development of tests for human genetic disorders, human gene therapy, forensic uses of DNA tests, and technologies to detect heritable mutations (74-77, 79,80,83-85). Finally, detailed analyses of allied issues, such as safety and efficacy of amniocentesis, prenatal care and pregnancy management (78), termination of pregnancy, and assisted conception (81,82) are beyond the scope of this report.

# CHAPTER 2 REFERENCES


1. Adelman, C.S., "The Constitutionality of Mandatory Genetic Screening Statutes," *Case Western Reserve Law Review* 31:897-948, 1981.
2. Andrews, L.B., *State Laws and Regulations Governing Newborn Screening* (Chicago, IL: American Bar Foundation, 1985).
3. Bayer, R., *Private Acts, Social Consequences: AIDS and the Politics of Public Health* (New York, NY: The Free Press, 1989).
4. Beaudet, A.L., "Invited Editorial: Carrier Screening for Cystic Fibrosis," *American Journal of Human Genetics* 47:603-605, 1990.
5. Beaudet, A.L., Howard Hughes Medical Institute, Houston, TX, remarks at the 8th International Congress of Human Genetics, Washington, DC, October 1991.
6. Beaudet, A.L., Howard Hughes Medical Institute, Houston, TX, personal communication, February 1992.
7. Biesecker, L., Bowles-Biesecker, B., Collins, F., et al., "General Population Screening for Cystic Fibrosis Is Premature," *American Journal of Human Genetics* 50:438-439, 1992.
8. Billings, P.R., "Mutation Analysis in Cystic Fibrosis," *New England Journal of Medicine* 323:62, 1990.
9. Boat, T.F., "Cystic Fibrosis," *Textbook of Respiratory Medicine*, J.F. Murray and J.A. Nadel (eds.) (Philadelphia, PA: W.B. Saunders, 1988).
10. Boat, T.F., Welsh, M.J., and Beaudet, A.L., "Cystic Fibrosis," *The Metabolic Basis of Inherited Disease*, C.R. Scriver, A.L. Beaudet, W.S. Sly, et al. (eds.) (New York, NY: McGraw Hill, 1989).

11. Boehm, C.D., and Kazazian, H.H., Jr., "Prenatal Diagnosis by DNA Analysis," *The Unborn Patient: Prenatal Diagnosis and Treatment*, 2nd ed., M.R. Harrison, M.S. Golbus, and R.A. Filly (eds.) (Philadelphia, PA: W.B. Saunders Co., 1991).
12. Booth, W., "Genetic Screening for Cystic Fibrosis Provokes Anxious Debate," *Washington Post*, Aug. 10, 1991, p. A3.
13. Bowles-Biesecker, B., University of Michigan Medical Center, Ann Arbor, MI, personal communication, December 1991.
14. Brock, D., "Population Screening for Cystic Fibrosis," *American Journal of Human Genetics* 47:164-165, 1990.
15. Brown, D., "Individual 'Genetic Privacy' Seen as Threatened," *Washington Post*, Oct. 20, 1991.
16. *Buck* v. *Bell*, 274 U.S. 200 (1927).
17. Caskey, C.T., Howard Hughes Medical Institute, Houston, TX, personal communication and remarks at "The Impact of Human Molecular Genetics on Society," Banbury Center, Cold Spring Harbor, NY, November 1990.
18. Caskey, C.T., Kaback, M.M., and Beaudet, A.L., "The American Society of Human Genetics Statement on Cystic Fibrosis Screening," *American Journal of Human Genetics* 46:393, 1990.
19. Cobb, E., Holloway, S., Elton, R., et al., "What Do Young People Think About Screening for Cystic Fibrosis?," *Journal of Medical Genetics* 28:322-324, 1991.
20. Collins, F.S., Howard Hughes Medical Institute, Ann Arbor, MI, personal communication, January 1992.
21. Collins, F.S., "Cystic Fibrosis: Molecular Biology and Therapeutic Implications," *Science* 256:774-779, 1992.
22. Cook-Deegan, R.M., *Gene Quest: Science, Politics, and the Human Genome Project—A Prepublication Draft for Scholars* (Washington, DC: Georgetown University National Reference Center for Bioethics Literature, 1991).
23. Cook-Deegan, R.M., "The Human Genome Project: The Formation of Federal Policies in the United States, 1986-1990," *Biomedical Politics*, K.E. Hanna (ed.) (Washington, DC: National Academy Press, 1991).
24. Cox, T.K., and Chakravarti, A., "Detection of Cystic Fibrosis Gene Carriers: Comparison of Two Screening Strategies by Simulations," *American Journal of Human Genetics* 49(Supp.):327, 1991.
25. Cunningham, J.C., and Taussig, L.M., *A Guide to Cystic Fibrosis for Parents and Children* (Bethesda, MD: Cystic Fibrosis Foundation, 1989).
26. Drell, D., Human Genome Program, Office of Health and Environmental Research, U.S. Department of Energy, Germantown, MD, personal communication, November 1991.
27. Duffy, E., Genetic Services Branch, Maternal and Child Health Bureau, U.S. Department of Health and Human Services, Rockville, MD, personal communication, February 1992.
28. Eddy, D.M., "How To Think About Screening," *Common Screening Tests*, D.M. Eddy (ed.) (Philadelphia, PA: American College of Physicians, 1991).
29. Elias, S., Annas, G.J., and Simpson, J.L., "Carrier Screening for Cystic Fibrosis: Implications for Obstetric and Gynecologic Practice," *American Journal of Obstetrics and Gynecology* 164:1077-1083, 1991.
30. Estivill, X., Casals, T., Mortal, N., et al., "ΔF508 Gene Deletion in Cystic Fibrosis in Southern Europe," *Lancet* 2:1404, 1989.
31. Forrest, J.D., Gold, R.B., and Kenney, A-M., *The Need, Availability and Financing of Reproductive Health Services* (New York, NY: The Alan Guttmacher Institute, 1989).
32. Gilbert, F., "Is Population Screening for Cystic Fibrosis Appropriate Now?," *American Journal of Human Genetics* 46:394-395, 1990.
33. Gilbert, F., Cornell University Medical College, "Cystic Fibrosis Carrier Screening in the General Population as We Enter 1992," personal communication, January 1992.
34. Gostin, L.O., Curran, W.J., and Clark, M., "The Case Against Compulsory Case Finding in Controlling AIDS—Testing, Screening and Reporting," *American Journal of Law and Medicine* 12:7-53, 1987.
35. Hammond, K.B., Abman, S.H., Sokol, R.J., et al., "Efficacy of Statewide Neonatal Screening for Cystic Fibrosis by Assay of Trypsinogen Concentrations," *New England Journal of Medicine* 325:769-774, 1991.
36. Hanna, K.E., "Conclusions," *Biomedical Politics*, K.E. Hanna (ed.) (Washington, DC: National Academy Press, 1991).
37. Holtzman, N.A., *Proceed With Caution: Predicting Genetic Risks in the Recombinant DNA Era* (Baltimore, MD: The Johns Hopkins University Press, 1989).
38. Holtzman, N.A., Johns Hopkins University, Baltimore, MD, remarks before the Ethical, Legal, and Social Issues Committee of the National Center for Human Genome Research, National Institutes of Health, Arlington, VA, January 1991.
39. Hubbard, R., and Henifen, M.S., "Genetic Screening of Prospective Parents and of Workers: Some Scientific and Social Issues," *International Journal of Health Services* 15:231-251, 1985.
40. Juengst, E.T., National Center for Human Genome Research, National Institutes of Health, personal communication, December 1991.
41. Kaback, M., "Should Cystic Fibrosis Testing Be Available to the General Population?," *Perspectives in Genetic Counseling* 12(1):5, 1990.

42. Kaback, M., University of California, San Diego, San Diego, CA, remarks at "Cystic Fibrosis Counseling: A Two-Day Workshop for Genetic Counselors," Sarah Lawrence College, Bronxville, NY, January 1992.
43. Kaback, M., Zippin, D., Boyd, P., et al., "Attitudes Toward Prenatal Diagnosis of Cystic Fibrosis Among Parents of Affected Children," *Cystic Fibrosis: Horizons. Proceedings of the 9th International Cystic Fibrosis Congress*, D. Lawson (ed.) (New York, NY: John Wiley & Sons Inc., 1984).
44. Kerem, B.-S., Rommens, J.M., Buchanan, J.A., et al., "Identification of the Cystic Fibrosis Gene: Genetic Analysis," *Science* 245:1073-1080, 1989.
45. Langfelder, E.J., National Center for Human Genome Research, National Institutes of Health, personal communication, November 1991.
46. Lebo, R.V., Cunningham, G., Simons, M.J., et al., "Defining DNA Diagnostic Tests Appropriate for Standard of Clinical Care," *American Journal of Human Genetics* 47:583-590, 1990.
47. Lemna, W.K., Feldman, G.L., Kerem, B.-S., et al., "Mutation Analysis for Heterozygote Detection and the Prenatal Diagnosis of Cystic Fibrosis," *New England Journal of Medicine* 322:291-296, 1990.
48. Lippman, A., "Prenatal Genetic Testing and Screening: Constructing Needs and Reinforcing Inequities," *American Journal of Law and Medicine* 17:15-50, 1991.
49. Lippman, A., "Mother Matters: A Fresh Look at Prenatal Diagnosis and the New Genetic Technologies," *Reproductive Genetic Testing: Impact on Women*, proceedings of a conference, Nov. 21, 1991.
50. MacLusky, I., McLaughlin, F.J., and Levinson, H.R., "Cystic Fibrosis: Part 1," *Current Problems in Pediatrics*, J.D. Lockhart (ed.) (Chicago, IL: Year Book Medical Publishers, 1985).
51. Meaney, F.J., "CORN Report on Funding of State Genetic Services Programs in the United States, 1990," contract document prepared for the U.S. Congress, Office of Technology Assessment, April 1992.
52. Murray, T.H., "Ethical Issues in Human Genome Research," *FASEB Journal* 5:55-60, 1991.
53. National Center for Human Genome Research, National Institutes of Health, "NIH Collaboration Launches Research on Education and Counseling Related to Genetic Tests," *Human Genome Project Progress*, Oct. 8, 1991.
54. National Institutes of Health, Workshop on Population Screening for the Cystic Fibrosis Gene, "Statement From the National Institutes of Health Workshop on Population Screening for the Cystic Fibrosis Gene," *New England Journal of Medicine* 323:70-71, 1990.
55. National Research Council, Committee for the Study of Inborn Errors of Metabolism, *Genetic Screening: Programs, Principles, and Research* (Washington, DC: National Academy of Sciences, 1975).
56. National Research Council, Committee on Mapping and Sequencing the Human Genome, *Mapping and Sequencing the Human Genome* (Washington, DC: National Academy Press, 1988).
57. Nelkin, D., and Tancredi, L., *Dangerous Diagnostics: The Social Power of Biological Information* (New York, NY: Basic Books, 1989).
58. Ng, I.S.L., Pace, R., Richard, M.V., et al., "Methods for Analysis of Multiple Cystic Fibrosis Mutations," *Human Genetics* 87:613-617, 1991.
59. Parmet, W.E., "AIDS and Quarantine: The Revival of an Archaic Doctrine," *Hofstra Law Review* 14:137-162, 1985.
60. Parmet, W.E., "Legal Rights and Communicable Disease: AIDS, the Police Power, and Individual Liberty," *Journal of Health Politics, Policy, and Law* 14:741-771, 1989.
61. President's Commission for the Study of Ethical Problems in Medicine and Biomedical and Behavioral Research, *Screening and Counseling for Genetic Conditions: The Ethical, Social, and Legal Implications of Genetic Screening, Counseling, and Education Programs* (Washington, DC: U.S. Government Printing Office, 1983).
62. President's Commission for the Study of Ethical Problems in Medicine and Biomedical and Behavioral Research, *Summing Up: The Ethical and Legal Problems in Medicine and Biomedical and Behavioral Research* (Washington, DC: U.S. Government Printing Office, 1983).
63. Reilly, P., *Genetics, Law, and Social Policy* (Cambridge, MA: Harvard University Press, 1977).
64. Reilly, P., "Government Support of Genetic Services," *Social Biology* 25:23-32, 1978.
65. Reilly, P., "Advantages of Genetic Testing Outweigh Arguments Against Widespread Screening," *The Scientist*, Jan. 21, 1991.
66. Riordan, J.R., Rommens, J.M., Kerem, B.-S., et al., "Identification of the Cystic Fibrosis Gene: Cloning and Characterization of Complementary DNA," *Science* 245:1066-1072, 1989.
67. Roberts, L., "Cystic Fibrosis Pilot Projects Go Begging," *Science* 250:1076-1077, 1990.
68. Rommens, J.M., Iannuzzi, M.C., Kerem, B.-S., et al., "Identification of the Cystic Fibrosis Gene: Chromosome Walking and Jumping," *Science* 245:1059-1065, 1989.
69. Rosenkrantz, B., *Public Health and the State: Changing Views in Massachusetts, 1842-1936* (Cambridge, MA: Harvard University Press, 1972).

70. Schulman, J.D., Maddalena, A., Black, S.H., et al., "Screening for Cystic Fibrosis Carriers," *American Journal of Human Genetics* 47:740, 1990.
71. Shoshani, T., Augarten, A., Gazit, E., et al., "Association of a Nonsense Mutation (W1282X), the Most Common Mutation in Ashkenazi Jewish Cystic Fibrosis Patients in Israel, With Presentation of Severe Disease," *American Journal of Human Genetics* 50:222-228, 1992.
72. Singer, E., "Public Attitudes Towards Genetic Testing," *Population Research and Policy Review* 10:235-255, 1991.
73. Starr, P., *The Social Transformation of American Medicine* (New York, NY: Basic Books, 1982).
74. U.S. Congress, Office of Technology Assessment, *The Role of Genetic Testing in the Prevention of Occupational Disease*, OTA-BA-194 (Washington, DC: U.S. Government Printing Office, April 1983).
75. U.S. Congress, Office of Technology Assessment, *Human Gene Therapy*, OTA-BP-BA-32 (Washington, DC: U.S. Government Printing Office, December 1984).
76. U.S. Congress, Office of Technology Assessment, *Technologies for Detecting Heritable Mutations in Human Beings*, OTA-H-298 (Washington, DC: U.S. Government Printing Office, September 1986).
77. U.S. Congress, Office of Technology Assessment, *New Developments in Biotechnology: Public Perceptions of Biotechnology*, OTA-BP-BA-45 (Washington, DC: U.S. Government Printing Office, May 1987).
78. U.S. Congress, Office of Technology Assessment, *Healthy Children: Investing in the Future*, OTA-H-345 (Washington, DC: U.S. Government Printing Office, February 1988).
79. U.S. Congress, Office of Technology Assessment, *New Developments in Biotechnology: The Commercial Development of Tests for Human Genetic Disorders*, staff paper, February 1988.
80. U.S. Congress, Office of Technology Assessment, *Mapping Our Genes—The Genome Projects: How Big, How Fast?*, OTA-BA-373 (Washington, DC: U.S. Government Printing Office, April 1988).
81. U.S. Congress, Office of Technology Assessment, *Infertility: Medical and Social Choices*, OTA-BA-358 (Washington, DC: U.S. Government Printing Office, May 1988).
82. U.S. Congress, Office of Technology Assessment, *Artificial Insemination: Practice in the United States: Summary of a 1987 Survey*, OTA-BP-BA-48 (Washington, DC: U.S. Government Printing Office, August 1988).
83. U.S. Congress, Office of Technology Assessment, *Genetic Witness: Forensic Uses of DNA Tests*, OTA-BA-438 (Washington, DC: U.S. Government Printing Office, July 1990).
84. U.S. Congress, Office of Technology Assessment, *Genetic Monitoring and Screening in the Workplace*, OTA-BA-455 (Washington, DC: U.S. Government Printing Office, October 1990).
85. U.S. Congress, Office of Technology Assessment, *Medical Monitoring and Screening in the Workplace—Results of a Survey*, OTA-BP-BA-67 (Washington, DC: U.S. Government Printing Office, October 1991).
86. U.S. Congress, Office of Technology Assessment, *Cystic Fibrosis and DNA Tests: Policies, Practices, and Attitudes of Genetic Counselors—Results of a Survey*, OTA-BP-BA-97 (Washington, DC: U.S. Government Printing Office, forthcoming 1992).
87. U.S. Congress, Office of Technology Assessment, *Cystic Fibrosis and DNA Tests: Policies, Practices, and Attitudes of Health Insurers—Results of a Survey*, OTA-BP-BA-98 (Washington, DC: U.S. Government Printing Office, forthcoming 1992).
88. U.S. Department of Health and Human Services, *Healthy People 2000: National Health Promotion and Disease Prevention Objectives, Conference Edition* (Washington, DC: U.S. Government Printing Office, 1990).
89. U.S. Department of Health and Human Services, Health Services Administration, Genetic Services Branch, *State Laws and Regulations on Genetic Disorders* (Washington, DC: U.S. Department of Health and Human Services, 1980).
90. U.S. Department of Health and Human Services, National Institutes of Health, National Center for Human Genome Research, RFA #HG-91-01, *NIH Guide to Grants and Contracts* 20(14), Apr. 5, 1991.
91. Wertz, D.C., and Fletcher, J.C., "An International Survey of Attitudes of Medical Geneticists Toward Mass Screening and Access to Results," *Public Health Reports* 104:35-44, 1989.
92. Wertz, D.C., Rosenfield, J.M., Janes, S.R., et al., "Attitudes Toward Abortion Among Parents of Children With Cystic Fibrosis," *American Journal of Public Health* 81:992-996, 1991.
93. Wilfond, B.S., and Fost, N., "The Cystic Fibrosis Gene: Medical and Social Implications for Heterozygote Detection," *Journal of the American Medical Association* 263:2777-2783, 1990.

Chapter 3

# Medical Aspects

# Contents



## *Box*



## *Figures*



## *Tables*

# Chapter 3
# Medical Aspects

For a 17th century mother, tasting salt on her baby's brow portended death; according to folk wisdom of the period, the infant was hexed. Modern medicine now recognizes this as an indicator of cystic fibrosis (CF), the most common, life-shortening, recessive genetic disorder among Caucasians. CF also occurs in other races, but at a 4- to 36-fold lower incidence. A multifaceted condition, CF compromises many functions throughout the body, but varies from patient to patient in severity of symptoms and the extent to which different organs are affected. No cure exists for CF, but treatment of the digestive and respiratory symptoms lengthens lifespan considerably.

While the devastating consequences of CF have been apparent for centuries, an understanding of its underlying mechanisms is relatively new. Since the 1940s, doctors and scientists have known that CF is a heritable condition—i.e., one transmitted from parents to children through their genes. New scientific developments have revealed the nature of the genetic defect and offer insight into the relationship between genetics and disordered function, clearing the way to increased comprehension of the manifestations of CF and better management of the condition. Today, new therapeutic possibilities for affected patients exist, as well as technologies to detect people who are asymptomatic carriers of CF mutations.

This chapter provides an overview of the medical principles important to understanding the context of carrier screening for CF. It describes CF's manifestations, outlines how the condition is diagnosed, and summarizes methods of treatment. Additionally, this chapter looks forward, considering new medical techniques that could improve therapy and prognosis for people with CF.

## PATHOLOGY

CF affects the respiratory, gastrointestinal (GI), and reproductive systems, as well as the sweat glands. Although the disorder is present at birth in affected persons, the symptoms vary among individuals. Approximately 10 percent of people with CF are born with a detectable intestinal blockage called meconium ileus. In general, diagnosis occurs by age 3, although some individuals do not develop symptoms until later in childhood, adolescence, or even adulthood (10,15,54,62).

CF generally involves dysfunction of exocrine glands, the glands that secrete into ducts or onto specific organ surfaces. Exocrine glands include lacrimal (tear) glands, sweat glands, and part of the pancreas. Mucus-producing cells lining the respiratory and GI tracts are also part of the exocrine system. Although specific glands are impaired to differing degrees, CF affects both major classes of exocrine glands—the serous and mucous types. In CF, secretions from serous glands have an increased salt content. In contrast, secretions from mucous glands have a normal or diminished salt content, but the disorder causes them to be thicker than normal secretions, leading to obstruction of the gland's ducts (49).

### *Respiratory System*

CF affects both the lower respiratory tract (the lungs) and the upper respiratory tract (the nose and sinuses), although the upper tract is less involved (figure 3-1). CF produces thick, sticky mucus that obstructs breathing passages and interferes with normal gas exchange and removal of bacteria, viruses, and other particles from the airways. Thus, individuals with CF often suffer chronic lung infections, followed by inflammation, then subsequent lung damage (8,24,30,54). What often begins as coughing and wheezing can progress over time to shortness of breath, limited lung function, chronic lung infections, and numerous pulmonary complications that often include respiratory and heart failure.

Three types of bacteria, *Staphylococcus aureus, Haemophilus influenzae,* and *Pseudomonas aeruginosa*, generally colonize the lungs of CF patients. The former two are usually the first bacteria found in CF-affected bronchi, while *Pseudomonas* more frequently occurs as the disease progresses. *Pseudomonas* infection poses additional problems because it is often resistant to antibiotics.

The severity of respiratory problems often determines the quality of life and survival of CF patients (15,37,54). At one extreme, some infants develop chronic lung obstruction and infection soon after

Figure 3-1—Organ Systems Affected in Cystic Fibrosis

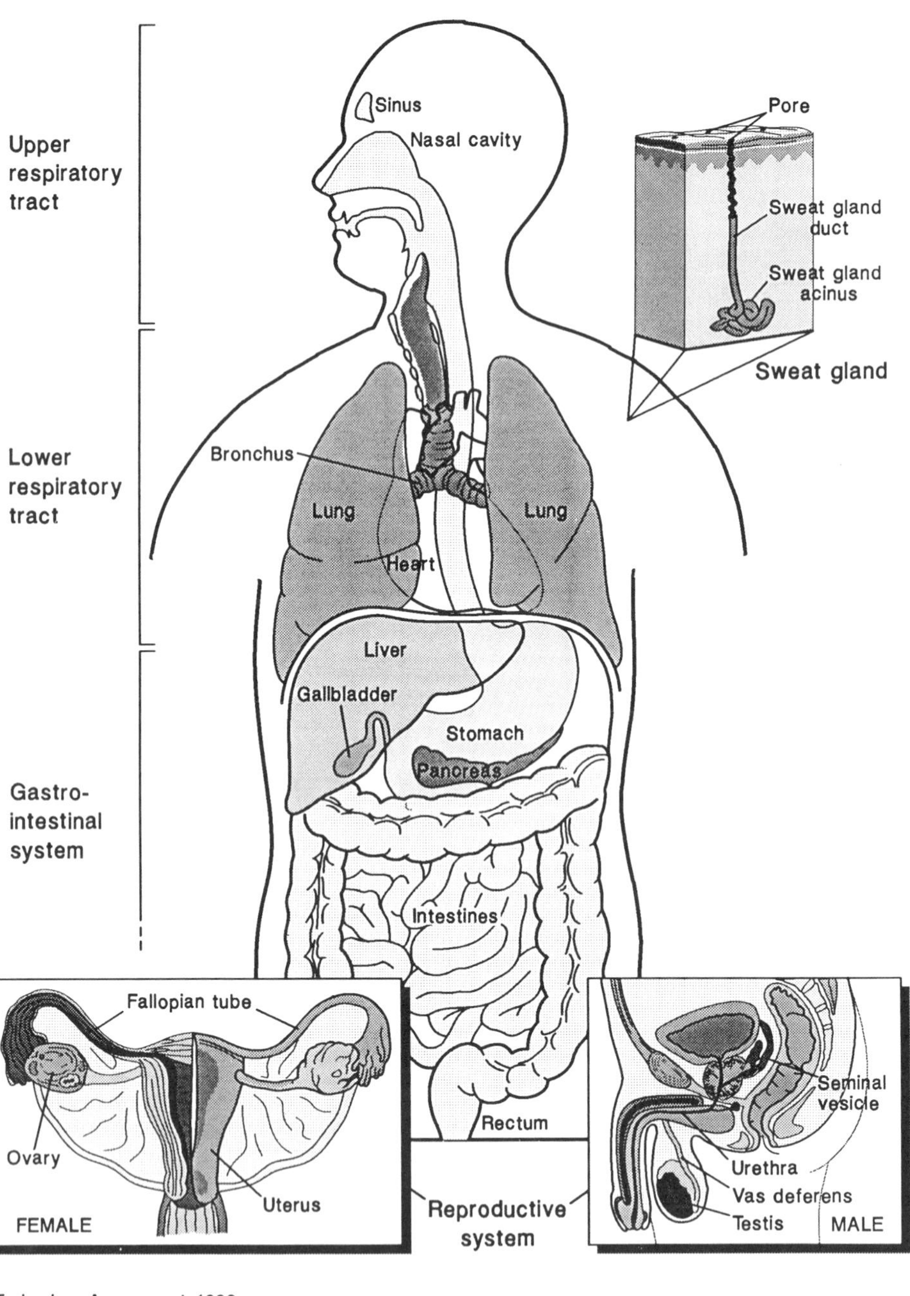


SOURCE: Office of Technology Assessment, 1992.

birth, resulting in impaired pulmonary function and early death. Other patients experience only mild symptoms, living several decades before they succumb to progressive lung disease and heart failure.

## *Gastrointestinal System*

Digestive difficulties are common in CF and often predominate over respiratory symptoms early in life. The pancreas, liver, gallbladder, stomach, and intestines can be affected, but if treated properly the problems generally are not life-threatening (figure 3-1).

About 85 to 90 percent of individuals with CF experience some pancreatic problems, primarily because inadequate quantities of pancreatic enzymes are released to digest food (22,54). As in the lungs, overly thick secretions are produced by the exocrine pancreatic ducts. Digestive enzymes are trapped, leading to destruction of pancreatic tissue and preventing the enzymes from reaching the intestines, where they are needed for digestion. Poor nutrition and impaired growth result because food—particularly fat and protein—is not broken down appropriately and cannot be absorbed by the body.

Nutritional status and pulmonary disease appear related; adequate nutrition helps alleviate the symptoms of pulmonary disease, whereas poor nutrition reinforces pulmonary disease and worsens the prognosis (37,54). The inability to digest fat interferes with intestinal absorption of fat-soluble vitamins—A, D, E, and K. Vitamin deficiencies lead to complications such as impaired blood clotting and scaly skin. Restricted protein digestion also causes serious problems, including edema in infants due to lack of blood proteins. Insufficient release of pancreatic enzymes can cause large, greasy, malodorous stools, abdominal pain or discomfort, and excessive gas.

Pancreatic manifestations also include diabetes and pancreatitis, although both are less frequent complications than exocrine pancreatic insufficiency. Diabetes, rare in CF patients under age 10, occurs in 10 percent of patients between ages 10 and 20, and is found in an additional 10 percent with each decade thereafter. CF-associated diabetes is mild compared to juvenile-onset diabetes; it tends not to cause the severe manifestations seen with the latter disease, such as nerve lesions and skin ulcers. Pancreatitis, inflammation of the pancreas, occurs in 1 percent of CF patients (54).

In the GI tract and associated organs, thick mucus accounts for many of CF's clinical signs. As mentioned earlier, meconium ileus occurs in 10 percent of newborns with CF. In these babies, meconium (fetal stool) obstructs part of the small intestine, the distal ileum. Older CF patients can develop a condition akin to meconium ileus—distal intestinal obstruction syndrome—where intestinal contents either partially or completely block the intestine (10,54,62).

CF patients suffer, less frequently, two additional GI complications: intussusception, or the folding of a piece of the intestine within itself, and rectal prolapse, or projection of the rectum through the anus (10,54). Finally, the liver is generally unaffected, although biliary cirrhosis—caused by blockage of the ducts that transport bile into the intestine—and fatty liver can occur, generally late in the course of the disease (10,54). In 1990, liver complications caused death in approximately 4 percent of people with CF (27).

### *Reproductive System*

CF manifests itself in the reproductive systems of males and females (figure 3-1); for both, sexual development can be delayed. CF damages the Wolffian duct, the embryological precursor of the male reproductive organs, in 95 percent of CF-affected males (10). The vas deferens is often absent (57), incompletely formed, or blocked by mucus, resulting in an effect similar to vasectomy. Additionally, sperm might be improperly formed in men with CF (39,65). As a result of these factors, only 2 to 3 percent of CF males are fertile (10). DNA analysis of men with congenital absence of the vas deferens reveals that many with this disorder might have CF, although they have no apparent symptoms of CF other than infertility (5).

In women with CF, thick, dehydrated mucus containing abnormal electrolytes often plugs the opening to the uterus, impeding sperm migration and reducing the pregnancy rate. Additionally, women patients can develop amenorrhea secondary to pulmonary disease and poor nutritional status, further reducing the chances of conception. Fertility in women with CF is estimated at 2 to 20 percent; this figure, however, might be an underestimate as many women with CF use contraceptives (10,15,27,42,54). In patients with advanced lung disease, the physical strain of pregnancy poses a health risk. Nevertheless, with proper care, increasing numbers of women with CF are successfully having children. Women with milder symptoms tolerate pregnancy better than those with advanced lung disease.

### *Sweat Glands*

CF also manifests in the sweat glands of affected individuals (figure 3-1). Patients lose excessive amounts of salt in their sweat, predisposing them to episodes of salt depletion. Although not a major concern for children and young adults—who can take salt tablets to compensate—infants can suffer from potentially fatal salt loss, particularly during periods of warm weather (10,54,62).

### *Skeletal System*

Skeletal system problems occur in some CF patients, probably secondary to the pulmonary and digestive malfunctions. Many—but not all—children with CF are short in stature and also have a delayed growth spurt. Other skeletal affects include joint pain, spine curvature (kyphosis), and clubbing (swelling) of fingers and toes (10,54,62).

## DIAGNOSIS

Physicians combine clinical criteria and laboratory tests to diagnose CF. Early signs and symptoms can include recurrent wheezing, persistent cough and excessive mucus, recurrent pneumonia, intestinal obstruction, low weight gain despite normal eating (failure to thrive), abnormal bowel movements, rectal prolapse, salty taste to the skin, nasal polyps, and enlargement of the fingertips. Since many childhood ailments share symptoms with CF and since its symptoms vary in severity from individual to individual, CF is often undiagnosed or misdiagnosed (10,15). Physicians identify some newborns before they develop symptoms through assays performed because of a family history of CF. Family history of CF also contributes to diagnosis for older individuals.

### *Sweat Test*

The sweat test is the most common method for confirming CF. CF affects exocrine glands, including sweat glands. In these glands, excess sodium ($Na^+$) and chloride ($Cl^-$) ions are lost to the sweat. This indicator was first discovered in 1953; by 1959, it was being used to diagnose CF (29).

To measure the salt content in the sweat of an individual, sweating is induced by placing a pilocarpine-soaked gauze pad or filter paper on the person's arm or back. (Pilocarpine activates sweat glands.) A low electric current is passed through the area (iontophoresis) for 4 to 6 minutes to drive the pilocarpine into the sweat glands. Next, the skin is cleaned and a sterile, preweighed, dry gauze pad is taped to it. Sweat is collected for up to an hour to obtain an adequate amount (100 mg), then the pad is weighed and sweat volume determined. Finally, the sweat is rinsed out of the pad and the pad weighed to ascertain salt content. Elevated $Cl^-$ levels confirm a diagnosis of CF (17) (table 3-1). The test is generally repeated in positive cases, as well as in borderline or negative cases where symptoms still strongly suggest CF. Some laboratories measure both $Na^+$ and $Cl^-$ content in sweat.

**Table 3-1—Sweat Chloride Levels in Normal and Cystic Fibrosis-Affected Individuals**

| Sweat chloride (mmol/L) | Status |
|---|---|
| < 40 | Normal |
| 40 to 60 | Borderline |
| > 60 | Presumptive case of CF |

SOURCE: Office of Technology Assessment, 1992.

Although painless and of moderate expense, the sweat test has several drawbacks. As with all diagnostic tests, accuracy depends on how the test is performed and interpreted. Pilocarpine iontophoresis is difficult to perform on newborns under 8 weeks of age, making sweat testing in newborns difficult. Moreover, complications of CF—such as edema (swelling)—or recent use of corticosteroids can confound the results. Other conditions also yield elevated $Cl^-$ levels in perspiration, although these can generally be distinguished from CF by their symptoms. Additionally, normal adults, in rare instances, have increased levels of $Cl^-$ in their sweat. One study found approximately 40 percent of persons referred to CF centers were sent there as a result of false sweat test results by the initial tester (10). Though the sweat test is imperfect, when performed properly and considered in conjunction with typical clinical findings, it can be used to diagnose CF with better than 98 percent accuracy (9). Increasingly, DNA analysis is used to establish a positive CF diagnosis in people with borderline sweat test results (68).

### *Immunoreactive Trypsin Test*

A protocol for newborn CF screening, the immunoreactive trypsin (IRT) test, measures levels of pancreatic trypsin, a digestive enzyme (13,54). In newborns with CF, obstruction of the exocrine pancreatic ducts causes this enzyme to back up into the circulatory system. In the IRT test, a drop of blood is isolated on a card, dried, and chemically analyzed to detect elevated levels of the enzyme (71). This use of dried spots—known as "Guthrie spots"—parallels the method of newborn screening for a range of genetic disorders (e.g., phenylketonuria and hypothyroidism) performed by a number of States (4).

While sweat testing is intended to be diagnostic, the IRT test is not. The IRT test yields a higher number of false positives (unaffected individuals incorrectly suggested to have CF) and false negatives (affected individuals incorrectly identified as not having CF) than the sweat test (7,55). For this reason, it requires followup with other procedures for conclusive diagnosis.

Current pilot studies combine the use of the IRT test with other tests to enhance its diagnostic value. In a Colorado study, the false positive rate of IRT alone was reduced by sequential use of the IRT test with a stricter threshold and the sweat test (34,36). Programs in Wisconsin, Pennsylvania, and Australia combine the IRT test with followup DNA analysis for a single mutation, known as ΔF508 (53,56,70). The efficacy of presymptomatic identification of the disease in newborns remains uncertain (1,36).

### *DNA Analysis*

Both sweat testing and IRT analysis measure phenomena secondary to the genetic cause of CF. In contrast, DNA analysis directly examines the genetic material, DNA, to reveal whether an individual has CF mutations. Currently, DNA analysis is used for prenatal detection, carrier screening, and, increasingly, determining the status of borderline cases (3,68). DNA analysis has also been used in combination with the IRT test for newborn screening (56,70).

As detailed in chapter 4, more than 170 different mutations cause CF, complicating prospects for routine diagnosis using DNA assays. While direct analysis of DNA is theoretically the most precise diagnostic test, at present it can only be used to diagnose positive cases. Due to the number of different mutations—many of which are not included in test panels—DNA analysis cannot confirm that a person does not have CF. However, DNA-based CF mutation tests, while not yet ready for sole use as a diagnostic test in negative cases, are accepted as conclusive in positive cases. In its 1991 patient registry, the Cystic Fibrosis Foundation (CFF) accepted positive results from DNA tests without requiring sweat test confirmation (68). (Chapter 4 discusses the correlation of molecular diagnosis with clinical outcome.)

## TREATMENT AND PROGNOSIS

The goal of CF treatment is to maintain a stable condition for long periods of time and allow affected individuals to lead relatively normal lives. General therapy involves home treatment, with occasional hospital stays as needed. The regularity of hospitalization varies with the individual, and ranges from infrequent to once every 2 to 3 months, often increasing in frequency in the last few years of life. Treatment seeks to control infection, promote mucus clearance, and improve nutrition.

The daily therapeutic regimen for a person with CF depends on the severity of the disease, but the principal components of treatment focus on managing the pulmonary and digestive manifestations of CF. Even in asymptomatic cases, regular visits to a physician are generally advised to prevent problems and to detect early problems when complications do arise. Today, treating CF has become increasingly specialized. Currently, more than 110 major clinical centers and a number of satellite centers and outreach clinics are devoted to delivering CF care (16,17) (figure 3-2). The CFF, founded in 1955, has played a large role in advancing CF treatment and care.

### *Medical Management*

Medical management of CF focuses on:

- alleviating blockage in the airways,
- fighting lung infection,
- managing airways inflammation, and
- facilitating proper nutrition.

#### Lung Therapy

Lung therapy for CF requires both medical and physical approaches. Medical management of lung-related problems involves the use of aerosols, bronchodilators, corticosteroids, and antibiotics (table 3-2). Aerosols deliver medications and water to the lower respiratory tract; most use a saline solution as the vehicle. Mucolytics, agents (generally N-acetyl cysteine) designed to break up mucus, can be used in aerosol form. Efficacy and safety of currently available mucolytics are controversial, but new mucolytics under development might prove safer and more appropriate for routine use. Bronchodilators are also used to reverse airway obstruction. These drugs are delivered by aerosol, injection, or orally. Bronchodilators used to treat CF include metaproterenol, isoetharine, and albuterol. In some instances, physicians prescribe corticosteroids to treat allergic reaction to a fungus that grows in CF airways. Long-term corticosteroid therapy is not generally beneficial (9). Finally, nonsteroidal anti-inflammatory medications for CF therapy are cur-

**Figure 3-2—Distribution of Cystic Fibrosis Centers in the United States**

SOURCE: Office of Technology Assessment, 1992.

rently being evaluated. Most of these treatments involve some degree of controversy as to efficacy and long-term safety (11,25,64).

As with all therapies, each of these treatments is useful in some patients and not in others. In some patients, for example, currently available mucolytic aerosols induce cough, cause bronchospasm, and promote inflammation; bronchodilators, while often providing immediate relief of symptoms, might not be effective in long-term use and could even prove harmful. Short-term use of corticosteroids in adults can cause detrimental effects such as immune system suppression, hyperglycemia, and glycosuria—requiring their use be evaluated on a patient-by-patient basis (25).

In contrast to the use of aerosols, bronchodilators, and corticosteroids (which ameliorate the physiological effects of CF on the lung), antibiotic treatment reduces or eliminates infection, thus slowing inflammatory responses that can lead to progressive lung disease. Unlike other therapies, antibiotic treatment is not controversial. The major antibiotics for countering CF-related infections include the penicillins and the aminoglycosides (e.g., gentamicin). Quinolones (e.g., ciprofloxacin and norfloxacin), cephalosporins, and other antibiotic families are also used. Antibiotics can also be aerosolized.

Treatment with antibiotics varies from intermittent use to repeated, sustained use. CF patients often need much larger doses than persons without CF. As with all long-term antibiotic treatment, bacterial resistance to treatment is a major problem. While antibiotic therapy is generally provided on an outpatient basis, hospitalization is sometimes necessary. Lengths of stay vary, but can range as long as 2 to 3 weeks. Hospitalization allows more careful management of patients, as well as the use of antibiotics that must be given intravenously. The

**Table 3-2—Medications Used in Management of Cystic Fibrosis**

| Type of medication | Purpose |
|---|---|
| **Antibiotics** | **Fight bacteria in lungs** |
| Penicillins | |
| penicillin | *Streptococcus* |
| ampicillin | *Haemophilus* |
| ticarcillin | *Pseudomonas* |
| piperacillin | *Pseudomonas* |
| methicillin | *Staphylococcus* |
| Cephalosporins | |
| cephalexin, cephaclor | *Streptococcus, Staphylococcus, Haemophilus* |
| ceftazidime | *Pseudomonas* |
| Tetracyclines | *Streptococcus, Staphylococcus, Haemophilus, Pseudomonas* |
| Erythromycin | *Streptococcus, Staphylococcus, Haemophilus* |
| Chloramphenicol | *Staphylococcus, Haemophilus* |
| Sulfa drugs (e.g., sulfisoxazole, trimethoprim-sulfamethoxazole) | *Haemophilus* |
| Aminoglycosides (e.g., gentamicin, tobramycin, amikacin) | *Pseudomonas* |
| Quinolones (e.g., ciprofloxacin, norfloxacin) | *Pseudomonas* |
| Aztreonam | *Pseudomonas* |
| **Bronchodilators** | **Open air passages, facilitate breathing** |
| Theophyllines | |
| β-agonists (e.g., albuterol, metaproterenol, isoetharine) | |
| Cromolyn | |
| Corticosteroids | |
| **Mucolytics** | **Dissolve mucus** |
| N-acetyl cysteine | |
| **Digestive enzymes** | **Digest food** |
| **Dietary supplements** | **Maintain nutrition** |
| Vitamins A,D,E,K, | |
| High-calorie supplements | |
| **Diuretics** | **Stimulate kidneys to remove water; relieve fluid accumulation throughout body; ease pumping burden on heart** |

SOURCE: Office of Technology Assessment, 1992, based on D.M. Orenstein, *Cystic Fibrosis: A Guide for Patient and Family* (New York, NY: Raven Press, Ltd., 1989).

necessity of hospitalization, however, remains controversial. Recent evidence indicates that less expensive home intravenous treatment could be as effective as identical hospital-based treatment if the patient complies. Although antibiotics are standard treatment, the method of delivery, selection of particular antibiotics, and the usefulness of prophylactic treatment are debated. Finally, in addition to aerosols, bronchodilators, corticosteroids, and antibiotics, several new methods to manage CF are foreseeable (box 3-A).

## Digestive Therapy

Digestive therapy for CF has several goals—achieving ideal weight, insuring normal growth and maturation, sustaining respiratory muscle strength, and maintaining adequate immunity. The major components of digestive therapy are pancreatic enzyme replacement, administration of fat-soluble vitamins, and dietary supplementation.

For pancreatic enzyme replacement therapy, extracts of animal pancreas are ingested orally. Replacing pancreatic enzymes enables CF patients to properly digest food and absorb nutrients. Enzymes must be taken each time a person eats. Dietary supplementation is important to compensate for increased energy needs associated with CF. Many people with CF need approximately 150 percent of normal caloric intake. Recommended supplements are high-calorie, consisting of high-protein foods, medium-chain triglycerides, and simple carbohydrates. In severe cases of malnutrition or during acute periods of lung infection, nasogastric feeding or total parenteral nutrition can be necessary.

### *Box 3-A—Cystic Fibrosis Therapies on the Horizon*

Standard CF pulmonary treatments of the past few decades focused on fighting infection and clearing mucus in the airways, but new therapies target preventing the process of infection and subsequent inflammatory response (12,19). These therapies attempt to intervene at specific junctures in the disease process by:

- decreasing the viscosity of lung secretions;
- protecting the airway from destruction and preventing infection;
- correcting the ionic imbalance; and
- compensating for the genetic defect through gene therapy.

Two promising drugs—DNase and amiloride—thin CF lung secretions through different mechanisms. Both are in clinical trials for approval by the U.S. Food and Drug Administration (FDA) (42,58). DNase is an enzyme that breaks down DNA that accumulates from the debris of inflammation-fighting cells in the lungs of CF patients. DNase loosens CF mucus in vitro and in vivo, and is safe for short-term use (2,38,60,61); long-term studies are underway (58). In short-term studies, patients using DNase show noticeable improvement in breathing ability, suggesting that it might be an effective CF therapy in the near future (40). Company officials anticipate that DNase might be on the market in early 1993 (6). Amiloride is a diuretic that loosens lung secretions by blocking sodium ion reabsorption (43). Clinical trials demonstrating safety and efficacy of its use in aerosol form are in progress (42). The FDA has announced plans to streamline its drug approval process, a decision that could make treatments such as DNase and amiloride commercially available soon (69).

Ironically, the body's natural infection fighting mechanism contributes to the massive destruction of airways in CF patients. Substances known as antiproteases can protect the airway epithelium from injury caused by innate bacterial defense substances. Included in this category are alpha-1-antitrypsin, secretory leukocyte protease inhibitor, and a compound known as ICI 200,880, all undergoing clinical trials for FDA approval (8,42,51).

Other potential therapies circumvent the environment that leads to infection and subsequent buildup of cells. These new therapies attempt to correct the ion imbalance characteristic of CF (ch. 4). Administration of adenosine triphosphate and uridine triphosphate in conjunction with the diuretic amiloride stimulates chloride ion ($Cl^-$) secretion, and could mitigate the effects of decreased $Cl^-$ conductance in individuals with CF; FDA clinical studies are in progress (42,44). Recent in vitro evidence that proper chloride conductance can be induced in some CF cells by applying large doses of cyclic-AMP-stimulating drugs suggests this as a future avenue for pharmaceutical intervention (18,20,73).

Unlike treatments that attack symptoms of CF, gene therapy is designed to rectify the deficits of the disease by directly altering DNA. In theory, new DNA can be inserted into faulty cells to compensate for the genetic defect. Gene therapy for CF (discussed in greater detail in ch. 4) is in the animal experiment stage at present. Using a crippled virus specific to airway epithelial cells, the normal human CF gene was administered directly to the lungs of rats by aerosol spray. After 6 weeks, the transferred DNA continued to function (14,59). Aerosolized liposomes, fatty capsules that can transport drugs directly into cells, have been used to deliver alpha-1-antitrypsin genes into rabbit lungs, and a similar mechanism might be used to deliver the CF gene to the lungs (35). Despite significant experimental progress, however, many hurdles remain for CF gene therapy to be feasible in humans. Long-term safety of the procedure needs to be demonstrated, and questions need to be answered regarding the most appropriate means of transferring the gene, the number of cells that need to be corrected, and the duration of treatment.

SOURCE: Office of Technology Assessment, 1992.

## *Chest Physical Therapy*

The cornerstone of CF treatment is chest physical therapy (PT) to move the mucus that blocks major air passages out of the lungs (10,15,54). Bronchial drainage, also called postural drainage, is a specific form of chest PT commonly used for CF, and involves turning the body in various positions to align air passages in the lung to optimize the effect of gravity. It is usually performed leaning on a table or lying over a couch. Either the individual or an assistant (e.g., a professional therapist or a family member) claps on the chest or back to loosen mucus while the patient coughs it up. Mechanical percussors and vibrators can facilitate PT. People with CF generally require PT one to four times daily, depending on their clinical status. When flareups in CF symptoms or acute infection occur, the frequency of chest PT generally increases. In light of the time commitment required to accomplish PT—at least once a day for 20 minutes or longer—compliance is often poor.

## *Lung Transplants*

Since the 1980s, two new options have become available to treat the terminal stages of CF: heart-lung and double-lung replacement surgery (23,26,47,66). Through 1991, 140 heart-lung transplants have been performed in CF patients internationally, with 89 patients still alive (26,47). Since 1986, 58 double-lung transplants have been performed in CF patients; 40 patients were surviving in 1991 (26). As with all organ transplantation, rejection is the major obstacle to survival. Availability of donor organs and cost are also limiting factors. When lungs were available, the operation and aftercare cost approximately $200,000 in 1989 (47).

Heart-lung transplant was pioneered first, and appears better suited for CF patients with right heart failure. Double-lung transplants are increasing in frequency (66), and they appear to reduce the incidence of acute and chronic infection. While the sputum of CF patients after transplant can become colonized with *Pseudomonas* bacteria, it generally clears in the succeeding months (47,66). Since most CF patients do not require new hearts, leaving the original heart intact avoids coronary artery disease associated with rejection of the transplant. It also removes the CF patient from the competition for donor hearts (31,47).

To be considered for transplant therapy, a patient must be a nonsmoker who has no signs of other systemic or pleural disease, does not take high doses of steroids, is within 15 percent of ideal weight, and has an emotionally supportive family (23,26,47,66). Neither approach can cure CF: Both types of transplants leave the pancreatic, GI, reproductive, and sweat gland manifestations of CF unresolved.

## *Prognosis*

Survival data for individuals with CF change rapidly (figure 3-3). Fifty years ago, most infants born with CF died within the first 2 years of life. In 1959, the median age at death was 4 years. Median age at death has increased one year per year for the past decade (33). In 1990, the median age at death had progressed to 28 years—meaning that one-half of babies born with CF in 1963 were alive in 1990 (27,28). The CFF reports that estimating the life expectancy of an infant born with CF today would require data from considerably more years than have been collected and is thus not feasible (52). Some research indicates past improvements in survival

**Figure 3-3—Median Survival of U.S. Cystic Fibrosis Patients Over Time**

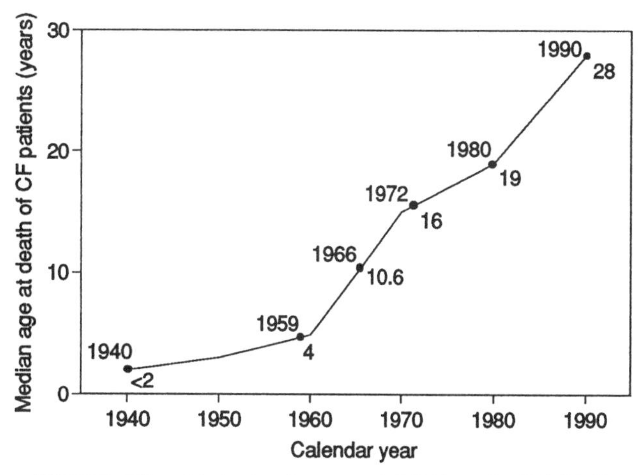


SOURCE: Office of Technology Assessment, 1992, based on S.C. FitzSimmons, "Cystic Fibrosis Patient Registry, 1990: Annual Data Report," Cystic Fibrosis Foundation, Bethesda, MD, January 1992; and S.C. FitzSimmons, Cystic Fibrosis Foundation, Bethesda, MD, personal communication, February 1992.

might not continue without further advances in therapy: Data from Canada show a plateau in median survival age at 28 years over the last decade (21). Nevertheless, a few postulate median survival of an infant born in 1992 with CF might be 40 years (12).

Three medical factors account for the increased life expectancy of persons with CF: the advent of antibiotics in the 1940s, the introduction of chest PT in the 1950s, and advances in nutrition in the 1970s and 1980s. Past failure to diagnose mild cases of CF also probably negatively skewed median survival data. And, the development of CF centers providing specialized and comprehensive care likely has contributed to longevity. Currently, people with CF can often lead full lives for long periods and can pursue college educations, maintain careers, marry, and have families.

As noted earlier, CF follows a varied clinical course. A few affected babies die from meconium ileus within the first days of life, while some people can be largely asymptomatic for 10 years (figure 3-4). Certain prognostic factors help predict survival. Lung function measured by a clinical test helps predict short-term survival, particularly when considered with respect to age and gender (41). Mortality rates in children and adolescents are higher than in adults with similar measured lung function (41). Likewise, mortality rates are higher in females than males with comparable lung function (41). In general, males live a few years longer than

**Figure 3-4—Age Distribution of U.S. Cystic Fibrosis Patients in 1990**

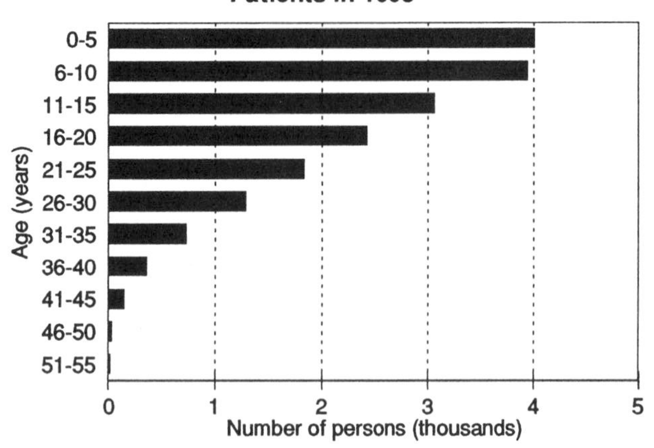


SOURCE: Office of Technology Assessment, 1992, based on S.C. FitzSimmons, "Cystic Fibrosis Patient Registry, 1990: Annual Data Report," Cystic Fibrosis Foundation, Bethesda, MD, January 1992.

females (27). African Americans have more difficulty during the early years of life, but fare better than Caucasians later (10). Psychosocial aspects of the disease also affect prognosis. Denial, dependence, and depression—on the part of either parent or patient—result in poor compliance with therapeutic regimens and poor prognosis. Other signs of poor prognosis include respiratory complications before age 1, abnormal chest x-ray at diagnosis, or inadequate nutritional status. Clinically mild symptoms at the time of diagnosis and pancreatic sufficiency both predict longer survival (10). Certain CF mutations associate with severe lung disease, while others cause milder forms (9,67,72).

The effect of early diagnosis on clinical outcome is unclear, and whether neonatal screening for CF is useful remains controversial. Some clinical experience supports the belief that earlier diagnosis, and hence earlier treatment, improve prognosis. However, when and where patients are treated has a significant effect on outcome (10). Long-term data from neonatal screening programs in Colorado and Wisconsin could settle this issue (1,34,36,70).

### *Psychological Aspects*

Beyond physical ailments, people with CF and their families experience emotional and social effects. As in any chronic illness, being sick adds additional strain to life. In particular, because CF requires daily treatments—even mildly affected individuals require digestive enzymes and usually chest PT—the disease affects the daily lives of the affected person and his or her family. Psychological response to the demands of the disease and outlook on life vary considerably (45,50,63). Not all individuals with CF and their families experience all—or even most—negative emotional aspects of the disease; in particular, those with milder cases grow up with little disruption to their lives. Specific feelings and reactions, however, typify particular periods in both the life of the individual and the course of CF (15,54,62).

Children, adolescents, and adults with CF react differently to the condition. Children sometimes resist home care, such as chest PT, because it intrudes on play time and interferes with their desire for independence. Children can be sensitive to being different from their peers, and can be embarrassed by physical manifestations of the disease such as flatulence, coughing, and dietary restrictions. Missing school due to frequent lung infections or hospitalizations interferes with school work; hospitalization can be a stressful and frightening experience. Children above age 10 start worrying about the seriousness of CF and the possibility of death, feelings that are exacerbated when they see friends succumb to the disease (48,62).

CF accentuates the fears, conflicts, and insecurities associated with adolescence. At a time when they most want to be independent of their parents, teenagers with CF frequently depend on them for daily chest PT and other medical necessities. Some teenagers refuse to take enzyme supplements or medication to rebel against parental attempts to keep them healthy. The short stature, delayed onset of secondary sex characteristics, and sometimes impaired athletic ability characteristic of CF can be traumatic for adolescents who feel acutely aware of their bodies and seek peer approval. Nonetheless, many teenagers with CF are as psychologically healthy as their peers (15,50,54,62).

As the prognosis for CF improves, it increasingly becomes a disease of young adulthood rather than a disease of childhood (figure 3-4). Due to the progressive nature of CF lung complications, many adults are physically more restricted than they were as children. If they go to college, young adults with CF must factor the demands of daily chest PT, often performed by family, into the decision of where to attend. CF also influences career choice, since individuals face frequent bouts of infection and can be limited in physical activity. The need for health insurance also can shape job choice (ch. 8). The threat of early death influences family planning and marriage decisions as well. Within the medical community, a need exists for increased attention to CF care of adults, whose medical and psychological concerns differ from those of children (17,46,48,54,63).

For the family of a child with CF, the disease often shapes family time and interactions (54,62). With the initial diagnosis, family members experience a range of emotions: anger, shock, and grief that their child has a life-threatening illness; denial of the disease; relief to finally know definitely what is wrong and what can be done (15,48,54,62). Having a child with CF can cause conflicts in marriage and social life. CF is time consuming, can limit family vacations, and demands that parents have jobs that provide adequate health insurance. For siblings, CF can inspire resentment of the attention focused on the sick child. When a sibling with CF dies, siblings can feel guilt that they were responsible (15,48,54,62).

While the burden of CF can be emotionally difficult, many individuals and their families persevere, leading full lives. The medical condition is only part of the social context and environment that creates the whole life. Since studies focus on the negative impact of CF, it is much easier to enumerate the difficult parts of the experience than it is to document that many CF patients and families are happy (15,62). For them, CF is just a part of life. As one mother of a son with CF notes, no one—with or without CF—is guaranteed a "perfect life or a perfect child" (32).

## SUMMARY AND CONCLUSIONS

CF is the most common, life-shortening, recessive genetic disorder among Caucasians. Age of diagnosis and severity of symptoms vary greatly from individual to individual. CF generally involves exocrine gland dysfunction. It affects multiple organ systems in the body, including the lungs, digestive tract, reproductive organs, and sweat glands, as well as the skeletal framework. The condition wreaks its most severe toll, however, on the respiratory and digestive systems, where it is characterized by chronic obstruction and infection of the airways and insufficiency of the pancreas in providing digestive enzymes. Ultimately, the majority of people with CF die from heart failure stemming from the respiratory consequences of CF.

Diagnosis of CF usually follows from its clinical symptoms, with confirmation by laboratory tests. Currently, analysis of sweat is the simplest and most reliable method for confirming CF. Measurement of a pancreatic enzyme, trypsin, is used—imperfectly—to identify CF. DNA-based CF mutation tests are not yet ready as the sole diagnostic test in negative cases, but can be used to accurately diagnose positive incidences.

Treating CF focuses on managing the respiratory and digestive symptoms to maintain a stable condition. At present, lung obstruction is ameliorated by daily chest PT and is sometimes augmented by aerosols, bronchodilators, and corticosteroids. Antibiotics are used to fight infection. The drugs DNase and amiloride, currently in clinical trials, offer hope of attenuating obstruction as well, although they will not be a substitute for chest PT. To attain adequate nutrition, pancreatic enzyme replacement and dietary supplementation are required. In a few end-stage cases, double-lung or heart-lung replacement alleviates CF-related pulmonary complications, although digestive difficulties remain and rejection of the transplanted organs can limit longevity.

Research into the underlying basis responsible for CF progresses at a rapid pace. In the coming decade, new approaches to therapy, including amiloride and nucleotide triphosphates, could decrease lung damage by combating primary manifestations at the level of ion transport. Treatments that correct the molecular deficit rather than ameliorating symptoms seem feasible. Eventually, gene therapy to supplement the gene itself could be possible.

Over the past half-century, CF has evolved from an illness nearly always fatal in childhood to one in which numerous individuals now survive into adulthood. Currently, the median survival age is 28 years. Advances in new therapies and comprehensive approaches to patient care have all contributed to longer lives for people with CF. The search for new treatments, and ultimately a cure, might improve future prognosis.

## CHAPTER 3 REFERENCES

1. Abman, S.H., Ogle, J.W., Harbeck, R.J., et al., "Early Bacteriologic, Immunologic and Clinical Courses of Young Infants With Cystic Fibrosis Identified by Neonatal Screening," *Journal of Pediatrics* 119:211-217, 1991.

2. Aitken, M.L., Burke, W., McDonald, G., et al., "Recombinant Human DNase Inhalation in Normal Subjects and Patients With Cystic Fibrosis: A Phase I Study," *Journal of the American Medical Association* 267:1947-1951, 1992.
3. Amos, J., James, S., and Erbe, R., "DNA Analysis of CF Genotypes in Relatives With Equivocal Sweat Test Results," *Clinical and Investigative Medicine* 13:1-5, 1990.
4. Andrews, L.B. (ed.), *State Laws and Regulations Governing Newborn Screening* (Chicago, IL: American Bar Foundation, 1985).
5. Anguiano, A., Oates, R.D., Amos, J.A., et al., "Congenital Bilateral Absence of the Vas Deferens: A Primarily Genital Form of Cystic Fibrosis," *Journal of the American Medical Association* 267: 1794-1797, 1992.
6. Associated Press, "Therapy Found Effective for Cystic Fibrosis," *Washington Post*, p. A3.
7. Beaudet, A.L., Howard Hughes Medical Institute, Baylor College of Medicine, Houston, TX, personal communication, September 1991.
8. Berger, M., "Inflammation in the Lung in Cystic Fibrosis: A Vicious Cycle That Does More Harm Than Good?," *Clinical Reviews in Allergy* 9:119-142, 1991.
9. Boat, T.F., University of North Carolina at Chapel Hill, Chapel Hill, NC, personal communication, January 1992.
10. Boat, T.F., Welsh, M.J., and Beaudet, A.L., "Cystic Fibrosis," *The Metabolic Basis of Inherited Disease*, C.R. Scriver, A.L. Beaudet, W.S. Sly, et al. (eds.) (New York, NY: McGraw Hill, 1989).
11. Burrows, B., and Lebowitz, M.D., "The β-Agonist Dilemma," *New England Journal of Medicine* 326: 560-561, 1992.
12. Collins, F.S., "Cystic Fibrosis: Molecular Biology and Therapeutic Implications," *Science* 256:774-779, 1992.
13. Crossley, J.R., Elliott, R.B., and Smith, P.A., "Dried-Blood Spot Screening for Cystic Fibrosis in the Newborn," *Lancet* 1:472, 1979.
14. Crystal, R.G., National Heart, Lung, and Blood Institute, National Institutes of Health, Bethesda, MD, presentation before the U.S. Congress, Office of Technology Assessment advisory panel on *Cystic Fibrosis and DNA Tests: Implications of Carrier Screening*, September 1991.
15. Cunningham, J.C., and Taussig, L.M., *A Guide to Cystic Fibrosis for Parents and Children* (Bethesda, MD: Cystic Fibrosis Foundation, 1989).
16. Cystic Fibrosis Foundation, *Accredited Cystic Fibrosis Care Centers* (Bethesda, MD: Cystic Fibrosis Foundation, 1992).
17. Cystic Fibrosis Foundation, Center Committee and Guidelines Subcommittee, "Cystic Fibrosis Foundation Guidelines for Patient Services, Evaluation, and Monitoring in Cystic Fibrosis Centers," *Arizona Journal of Diseases of Childhood* 144:1311-1312, 1990.
18. Dalemans, W., Barbry, P., Champigny, G., et al., "Altered Chloride Ion Channel Conductance Associated With ΔF508 Cystic Fibrosis Mutation," *Nature* 354:526-528, 1991.
19. Davis, P.B., "Cystic Fibrosis From Bench to Bedside," *New England Journal of Medicine* 325:575-577, 1991.
20. Drumm, M.L., Wilkinson, D.J., Smit, L.S., et al., "Chloride Conductance Expressed by ΔF508 and Other Mutant CFTRs in *Xenopus* Oocytes," *Science* 254:1797-1799, 1991.
21. Durie, P.R., Department of Pediatrics, University of Toronto, Toronto, Canada, personal communication, June 1992.
22. Durie, P.R., and Forstner, G.G., "Pathophysiology of the Exocrine Pancreas in Cystic Fibrosis," *Journal of the Royal Society of Medicine* 82(Supp.):2-10, 1989.
23. Egan, T.M., Westerman, J., Mill, M.R., et al., "Lung Transplantation at UNC," *North Carolina Medical Journal* 52:325-329, 1991.
24. Elborn, J.S., and Shale, D.J., "Cystic Fibrosis. 2. Lung Injury in Cystic Fibrosis," *Thorax* 45:970-973, 1990.
25. Fick, R.B., and Stillwell, P.C., "Controversies in the Management of Pulmonary Disease Due to Cystic Fibrosis," *Chest* 95:1319-1327, 1989.
26. Fiel, S.B., "Mechanical Ventilation in the Severely Ill Adult With Cystic Fibrosis," *Pediatric Pulmonology 1991 North American Cystic Fibrosis Conference* 6(Supp.):217-218, 1991.
27. FitzSimmons, S.C., "Cystic Fibrosis Patient Registry, 1990: Annual Data Report," Cystic Fibrosis Foundation, Bethesda, MD, January 1992.
28. FitzSimmons, S.C., Cystic Fibrosis Foundation, Bethesda, MD, personal communication, February 1992.
29. Gibson, L.E., and Cooke, R.E., "A Test for Concentration of Electrolytes in Sweat in Cystic Fibrosis of the Pancreas Utilizing Pilocarpine by Iontophoresis," *Pediatrics* 23:545-549, 1959.
30. Gilligan, P.H., "Microbiology of Airway Disease in Patients With Cystic Fibrosis," *Clinical Microbiology Reviews* 4:35-51, 1991.
31. Goldsmith, M.F., "Which Transplant Technique Will Let Cystic Fibrosis Patients Breathe Easier?," *Journal of the American Medical Association* 264:9-10, 1990.
32. Greenberg, G., Cystic Fibrosis Foundation, Metropolitan Washington, DC chapter, personal communication, December 1991.
33. Habbersett, C., Cystic Fibrosis Foundation, Bethesda, MD, personal communication, November 1991.

34. Hammond, K.B., Abman, S.H., Sokol, R.J., et al., "Efficacy of Statewide Neonatal Screening for Cystic Fibrosis by Assay of Trypsinogen Concentrations," *New England Journal of Medicine* 325:769-774, 1991.
35. Holden, C. (ed.), "Aerosol Gene Therapy," *Science* 253:964-965, 1991.
36. Holtzman, N.A., "What Drives Neonatal Screening Programs?," *New England Journal of Medicine* 325:802-804, 1991.
37. Huang, N.N., Schidlow, D.V., Szatrowski, T.H., et al., "Clinical Features, Survival Rate, and Prognostic Factors in Young Adults With Cystic Fibrosis," *American Journal of Medicine* 82:871-879, 1987.
38. Hubbard, R.C., McElvaney, N.G., Birrer, P., et al., "A Preliminary Study of Aerosolized Recombinant Human Deoxyribonuclease in the Treatment of Cystic Fibrosis," *New England Journal of Medicine* 326:812-815, 1992.
39. "Infertility in Cystic Fibrosis," *Lancet* 338:1008, 1991.
40. Jacoby, J.Z., III, Cystic Fibrosis Center, St. Vincent's Hospital and Medical Center, New York, NY, personal communication, March 1992.
41. Kerem, E., Reisman, J., Corey, M., et al., "Prediction of Mortality in Patients With Cystic Fibrosis," *New England Journal of Medicine* 326:1187-1191, 1992.
42. Knowles, M.R., University of North Carolina at Chapel Hill, Chapel Hill, NC, personal communication, February 1992.
43. Knowles, M.R., Church, N.L., Waltner, W.E., et al., "A Pilot Study of Aerosolized Amiloride for the Treatment of Lung Disease in Cystic Fibrosis," *New England Journal of Medicine* 322:1189-1194, 1990.
44. Knowles, M.R., Clarke, L.L., and Boucher, R.C., "Activation by Extracellular Nucleotides of Chloride Secretion in the Airway Epithelia of Patients With Cystic Fibrosis," *New England Journal of Medicine* 325:533-538, 1991.
45. Lavigne, J.V., "Psychological Functioning of Cystic Fibrosis Patients," *Textbook of Cystic Fibrosis*, J.D. Lloyd-Still (ed.) (Boston, MA: John Wright-PSG Inc., 1983).
46. Lewiston, N.J., "The CF 'Young Adult' Comes of Age," *Chest* 97:1282-1283, 1990.
47. Lewiston, N., Starnes, V., and Theodore, J., "Heart-Lung and Lung Transplantation for Cystic Fibrosis," *Clinical Reviews in Allergy* 9(1-2):231-247, 1991.
48. Lloyd-Still, D.M., and Lloyd-Still, J.D., "The Patient, the Family, and the Community," *Textbook of Cystic Fibrosis*, J.D. Lloyd-Still (ed.) (Boston, MA: John Wright-PSG Inc., 1983).
49. MacLusky, I., McLaughlin, F.J., and Levison, H.R., "Cystic Fibrosis: Part I," *Current Problems in Pediatrics*, J.D. Lockhart (ed.) (Chicago, IL: Year Book Medical Publishers, 1985).
50. Mador, J.A., and Smith, D.H., "The Psychosocial Adaptation of Adolescents With Cystic Fibrosis," *Journal of Adolescent Health Care* 10:136-142, 1988.
51. McElvaney, N.G., Hubbard, R.C., Birrer, P., et al., "Aerosol Alpha-1-Antitrypsin Treatment for Cystic Fibrosis," *Lancet* 337:793-798, 1991.
52. McPherson, L., Cystic Fibrosis Foundation, Bethesda, MD, personal communication, November 1991.
53. Naylor, E.W., Jr., "New Technologies in Newborn Screening," *Yale Journal of Biology and Medicine* 64:21-24, 1991.
54. Orenstein, D.M., *Cystic Fibrosis: A Guide for Patient and Family* (New York, NY: Raven Press, Ltd., 1989).
55. Orenstein, D.M., Cystic Fibrosis Center, University of Pittsburgh School of Medicine, Pittsburgh, PA, personal communication, December 1991.
56. Ranieri, E., Ryall, R.G., Morris, C.P., et al., "Neonatal Screening Strategy for Cystic Fibrosis Using Immunoreactive Trypsinogen and Direct Gene Analysis," *British Medical Journal* 302:1237-1240, 1991.
57. Rigot, J.-M., Lafitte, J.-J., Dumur, V., et al., "Cystic Fibrosis and Congenital Absence of the Vas Deferens," *New England Journal of Medicine* 325:64-65, 1991.
58. Rose, M., Genentech, Inc., Washington, DC, personal communication, January 1992.
59. Rosenfeld, M.A., Yoshimura, K., Trapnell, B.C., et al., "In Vivo Transfer of the Human Cystic Fibrosis Transmembrane Conductance Regulator Gene to the Airway Epithelium," *Cell* 68:143-155, 1992.
60. Shak, S., Genentech, Inc., San Francisco, CA, personal communication, September 1991.
61. Shak, S., Capon, D.J., Hellmiss, R., et al., "Recombinant Human DNase I Reduces the Viscosity of Cystic Fibrosis Sputum," *Proceedings of the National Academy of Sciences, USA* 87:9188-9192, 1990.
62. Shapiro, B.L., and Heussner, R.C., Jr., *A Parent's Guide to Cystic Fibrosis* (Minneapolis, MN: University of Minnesota Press, 1991).
63. Sheperd, S.L., Melbourne, F.H., Harwood, I.R., et al., "A Comparative Study of the Psychosocial Assets of Adults With Cystic Fibrosis and Their Healthy Peers," *Chest* 97:1310-1316, 1990.
64. Spitzer, W.O., Suissa, S., Ernst, P., et al., "The Use of β-Agonists and the Risk of Death and Near Death From Asthma," *New England Journal of Medicine* 326:501-506, 1992.

65. Trezise, A.E., and Buchwald M., "In Vivo Cell-Specific Expression of the Cystic Fibrosis Transmembrane Conductance Regulator," *Nature* 353:434-437, 1991.
66. Trulock, E.P., Cooper, J.D., Kaiser, L.R., et al., "The Washington University-Barnes Hospital Experience With Lung Transplantation," *Journal of the American Medical Association* 266:1943-1946, 1991.
67. Tsui, L.-C., and Buchwald, M., "Biochemical and Molecular Genetics of Cystic Fibrosis," *Advances in Human Genetics*, vol. 20, H. Harris and K. Hirschorn (eds.) (New York, NY: Plenum Press, 1991).
68. Valverde, K.D., Cystic Fibrosis Center, St. Vincent's Hospital and Medical Center, New York, NY, personal communication, February 1992.
69. Verncai, R.L., "FDA Announces Faster Approval for Breakthrough Drugs," *Associated Press*, Apr. 9, 1992.
70. Wilfond, B., Gregg, R., Laxova, A., et al., "Mutation Analysis for CF Newborn Screening: A Two-Tiered Approach," *Pediatric Pulmonology 1991 North American Cystic Fibrosis Conference* 6(Supp.):238, 1991.
71. Williams, C., Weber, L., and Williamson, R., "Guthrie Spots for DNA-Based Carrier Testing in Cystic Fibrosis," *Lancet* 2:693, 1988.
72. Wine, J.J., "No CFTR: Are CF Symptoms Milder?," *Nature Genetics* 1:10, 1992.
73. Wine, J.J., "The Mutant Protein Responds," *Nature* 354:503-504, 1991.

Chapter 4

# The State-of-the-Art in Genetics

# Contents



## *Boxes*



## *Figures*



## *Tables*

# Chapter 4
# The State-of-the-Art in Genetics

Molecular biology is being integrated into genetics and medicine at a rapid pace. In the laboratory, more than 300 human conditions can be analyzed with today's molecular genetic technology (31). With clinical application of these discoveries comes public expectations for new means of diagnosis, screening, and even cure. This chapter provides an overview of the genetic principles important to understanding the issues involved in carrier screening for cystic fibrosis (CF). It outlines basic tenets of human genetics and molecular biology and summarizes the technical aspects of CF carrier screening. Finally, this chapter looks forward, exploring how advances in automation could affect both carrier screening for CF and testing and screening for additional genetic disorders. Earlier OTA reports provide further background in human genetics and biotechnology (128-134).

## BASIC GENETIC PRINCIPLES

Genetics explores how specific traits are passed from one generation to the next. Except for certain specialized cells, each of the trillions of cells in a human being contains a complete set of genetic instructions—the genome—for the individual. A person's genome governs everything from the structure of a single molecule, such as a protein, to the expression of identifiable traits, such as eye color. An intricate hierarchy of instructions determines which cells act on this genetic information, as well as when they do so.

Scientists study human genetics on many levels. They assess, for example, the molecular basis for inheritance by examining the specific structure and function of the genetic material, DNA. Geneticists also observe how the environment influences the expression of genetic traits, and trace the clustering of biological characteristics in populations.

### *Function and Organization of DNA*

As in all higher organisms, genetic information is stored in humans in DNA, a double stranded structure resembling a twisted ladder. This double helix consists of a genetic alphabet of four different nucleic acid molecules, or bases—adenine (A), thymine (T), guanine (G), and cytosine (C)—each of which is attached to a deoxyribose sugar group and a phosphate molecule (figure 4-1). The bases pair predictably—A with T, and G with C—to form the DNA double helix structure.

Lengths of DNA ranging from 1,000 to 2 million base pairs comprise a gene, the fundamental physical and functional unit of heredity. About 50,000 to 100,000 structural genes, spread over 3.3 billion base pairs, make up the human genome. DNA associates with proteins to form chromosomes, tightly coiled structures located in the cell nucleus.

In humans, genes are arrayed on 46 chromosomes—22 pairs of autosomes and 1 pair of sex chromosomes (figure 4-2). Females have two X chromosomes, and males have one X and one Y chromosome. Egg and sperm cells contain just one copy of each chromosome; fusion of the sperm and egg at fertilization creates a full genetic complement.

The physical location of a gene on a chromosome is called its locus. Some genes have been mapped—plotted at a specific locus on a chromosome—and cloned—generated in multiple copies in the laboratory. Alternative forms of a gene at a particular locus are called alleles. At each locus along pairs of autosomes, an individual can have two identical or two different alleles, one copy inherited from the mother and one from the father. If the alleles are the same, the person is said to be homozygous for that particular locus. If the versions differ, the person is said to be heterozygous. Even though a normal individual has at most two alleles at a given locus—again, one copy inherited from the mother and one from the father—additional alleles can exist in other individuals.

### *DNA Replication*

Through DNA replication, a full genome of DNA is regenerated each time a cell undergoes division to yield two daughter cells. In DNA replication, each chain of the double helix is used as a template to synthesize copies of the original DNA (figure 4-3). During cell division, the DNA double helix unwinds, the weak bonds between base pairs break, and the DNA strands separate. A series of enzymes insert a complementary base opposite each base in the original strand—A opposite T, and G opposite

**Figure 4-1—The Structure of DNA**

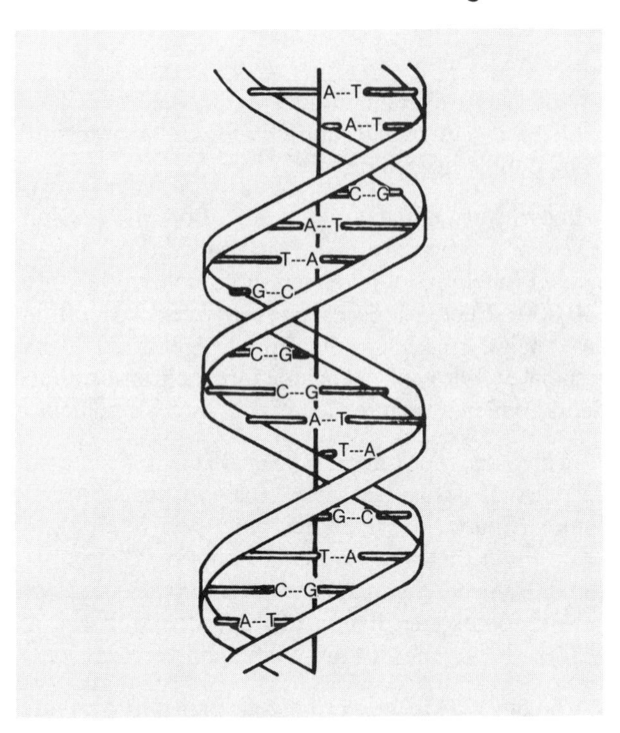


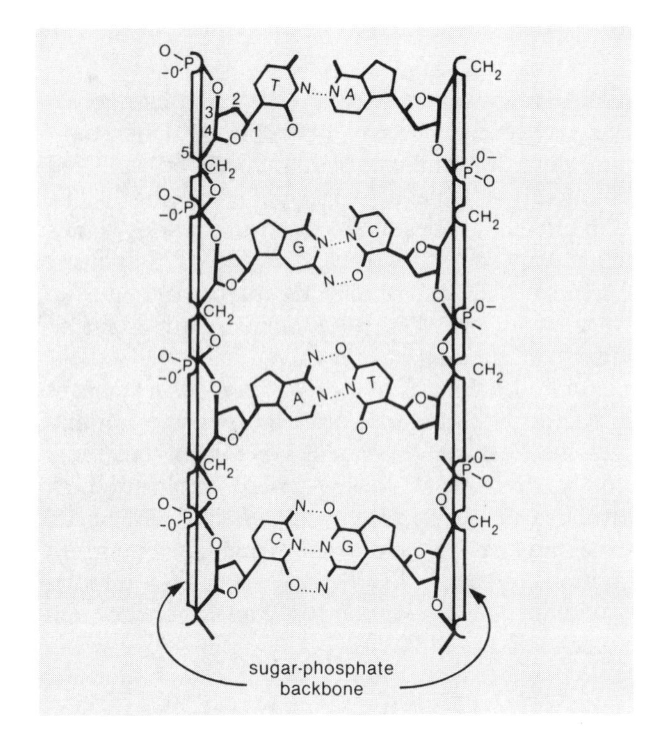


**Left:** A schematic diagram of the DNA double helix. **Right:** The four bases form the letters in the alphabet of the genetic code. The sequence of the bases along the sugar-phosphate backbone encodes the genetic information.

SOURCE: Office of Technology Assessment, 1992.

C—creating two identical copies of the original DNA.

## *Proteins*

The bridge between DNA's chemical information and physical realization of its instructions consists of steps that convert the DNA code into biologically active products. Through a process known as gene expression, a DNA sequence for a structural gene ultimately results in formation of a molecule called a protein (figure 4-4). Proteins are required for the structure, function, and regulation of all cells, tissues, and organs in the body.

First, the bases in the DNA sequence are copied, or transcribed, into messenger ribonucleic acid (mRNA), a single-stranded molecule that carries the genetic information out of the nucleus. The bases in the mRNA are read as triplets, or codons, that specify 1 of 20 different amino acids, the building blocks of a protein molecule. Then, in accordance with the instructions in the mRNA, amino acids are assembled into a specific protein molecule. Thus, the information encoded by DNA is transcribed to mRNA and translated from the DNA code into a protein that has a particular function in a cell.

## *Genetics and Disease*

Hereditary variation is the result of changes—or mutations—in DNA. Mutations present in germ cells (egg or sperm) are inherited by offspring, whereas those that occur in somatic cells (other body cells) are not passed on to future generations. Most mutations exist in both cell types. Mutations arise from the substitution of one DNA base for another, from rearrangements (e.g., small insertions or deletions) within the gene, or from duplication or deletion of the entire gene. Approximately 4,000 known human disorders result from genetic causes (88). Disorders arising from a mutation in only one gene are known as monogenic. CF is a monogenic condition.

Since proteins are produced from the instructions in genes, a mutation in a gene that codes for a specific protein can affect the structure, regulation, function, or synthesis of the protein. A particular mutation in a gene can produce a benign or mild effect, while a different mutation in the same gene

**Figure 4-2—Chromosome Complement of a Normal Human Female**

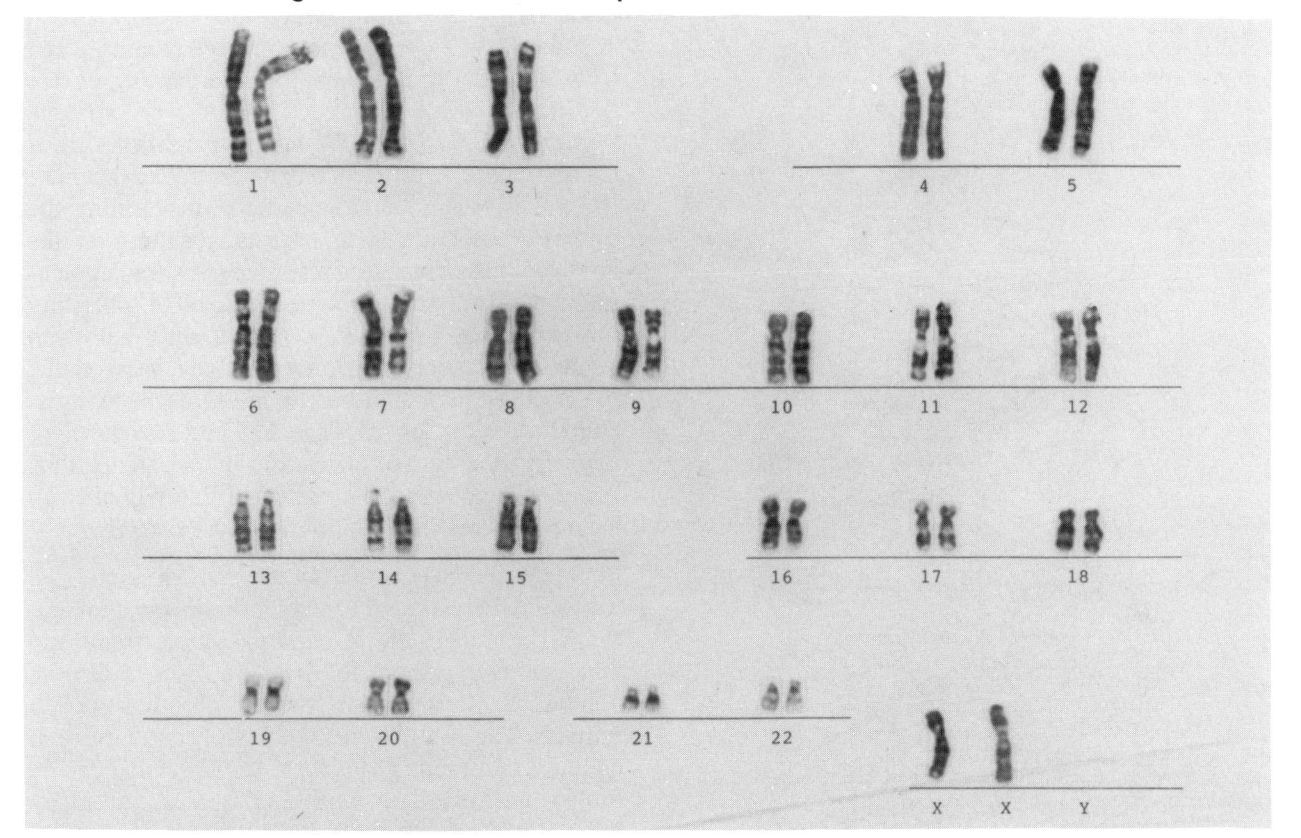


SOURCE: Genetics & IVF Institute, Fairfax, VA, 1990.

can result in gross reduction or complete loss of activity of the resulting protein. The genetic constitution of an individual is its genotype, while the observable expression of the genes is its phenotype.

### *Modes of Inheritance*

Although DNA's structure was first determined in 1953, the inheritance of genetic traits was studied long before then. In 1865, Austrian monk Gregor Mendel postulated that discrete biological units—later named genes—were responsible for both the maintenance and variation of certain characteristics from one generation to the next. Since then, understanding of the processes underlying inheritance has been refined.

In human monogenic—i.e., single gene—disorders, the altered gene can be on any one of the 22 autosomal chromosomes or on a sex chromosome. Traditionally, modes of inheritance are classed in three basic categories:

- autosomal dominant,
- autosomal recessive, and
- X-linked.

In the past decade, several less common heredity patterns—e.g., genomic imprinting and mitochondrial inheritance—have been identified in humans and tied to specific diseases such as Fragile X syndrome and some respiratory enzyme deficiencies (55,85,104).

In an autosomal dominant disease, a single mutant allele causes the trait to be expressed, even though the corresponding allele is normal. A heterozygous individual with a mutation usually is symptomatic at some level, although many disorders, such as neurofibromatosis, vary in the severity and age of onset of the condition. Marfan syndrome, Huntington disease, and adult polycystic kidney disease are inherited in an autosomal dominant manner. Every affected individual generally has an affected parent (except for cases arising from a spontaneous, or de novo, mutation). If the affected individual has an unaffected spouse, each potential child will have a

**Figure 4-3—DNA Replication**

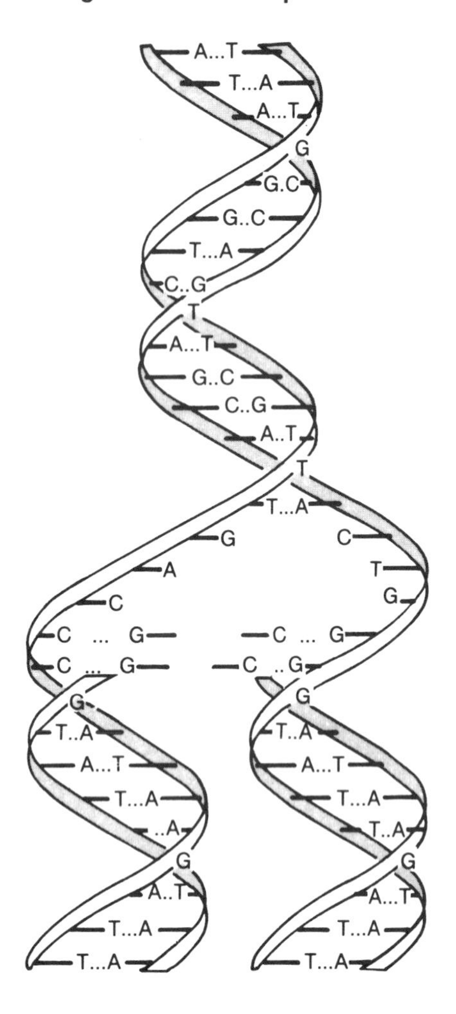


When DNA replicates, the original strands unwind and serve as templates for the building of new, complementary strands. The daughter molecules are exact copies of the parent, each daughter having one of the parent strands.

SOURCE: Office of Technology Assessment, 1992.

50 percent chance of inheriting the mutant allele and having the disease, and a 50 percent chance of not inheriting the mutant allele (figure 4-5). Male and female offspring are equally likely to inherit the mutation. Barring de novo mutations, children who do not receive the abnormal allele will not have the disorder and cannot pass it on to future generations.

Recessive traits result in illness only if a person has two mutant copies of the gene. That person is said to be homozygous for the gene. Heterozygous carriers of a single copy of the defective gene are usually clinically asymptomatic. CF is inherited in an autosomal recessive manner. Other common disorders inherited this way include sickle cell anemia, Tay-Sachs disease, phenylketonuria, and β-thalassemia. Individuals with the disease receive one mutant allele from each parent, who are each asymptomatic carriers. When both parents are carriers of a recessive trait, each potential son or daughter has a 1 in 4 chance (25 percent) of inheriting the mutant gene from both parents, resulting in the homozygous affected state (figure 4-5). Each potential child also has a 50 percent chance of inheriting one mutation from either parent, thus being an unaffected carrier. They, in turn, can pass on the mutation to their children. Finally, each potential child of two carriers also has a 25 percent chance of inheriting the normal allele from both parents, thus being a homozygous unaffected individual who cannot pass on the mutation to future offspring.

In sex-linked disorders, the mutant gene can theoretically be on either sex chromosome, the X or the Y. In reality, the Y chromosome is small and contains few genes. To date, no known disease conditions transmit via the Y chromosome in humans. The X chromosome, however, is large and contains numerous genes for many traits that can be mutant and result in disease. Diseases caused by aberrations in genes on the X chromosome are called X-linked disorders. Duchenne muscular dystrophy and hemophilia A and B are X-linked conditions.

Genes on the X chromosome can also be dominant or recessive, but because females have two X chromosomes, and males have one X and one Y, male and female offspring show different patterns of inheritance. Sons of a carrier female—who is often asymptomatic—have a 50 percent probability of inheriting the mutant gene from their mother. These sons will be affected (figure 4-5). Sons who do not inherit the abnormal gene are unaffected and cannot transmit the gene. Daughters of a carrier mother each have a 50 percent chance of inheriting the defective gene—thus being unaffected carriers—and a 50 percent chance of not inheriting the gene and being unaffected noncarriers. A male with an X-linked recessive condition will transmit the gene to all of his daughters, who will be carriers, but to none of his sons, who will be unaffected. An X-linked dominant disease affects the mother, who can also pass it on to both her sons and daughters with a 50 percent chance. Affected males pass X-linked dominant conditions to all of their daughters, but not their sons. X-linked dominant diseases are relatively rare.

**Figure 4-4—Gene Expression**

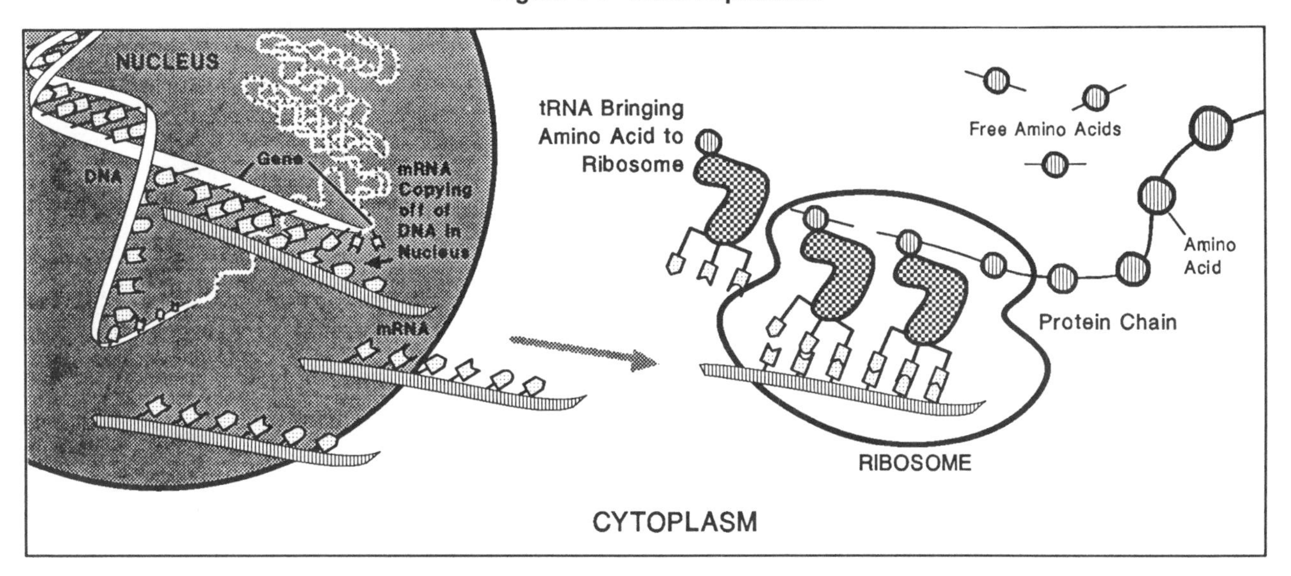


In the first step of gene expression, messenger RNA (mRNA) is synthesized, or transcribed, from genes by a process somewhat similar to DNA replication. In higher organisms, this process takes place in the nucleus of a cell. In response to certain signals (e.g., association with a particular protein), sequences of DNA adjacent to, or sometimes within, genes control the synthesis of mRNA. Protein synthesis, or translation, is the second major step in gene expression. Messenger RNA molecules are known as such because they carry messages specific to each of the 20 different amino acids that make up proteins. Once synthesized, mRNAs leave the nucleus of the cell and go to another cellular compartment, the cytoplasm, where their messages are translated into the chains of amino acids that make up proteins. A single amino acid is coded by a sequence of three nucleotides in the mRNA, called a codon. The main component of the translation machinery is the ribosome—a structure composed of proteins and another class of RNAs, ribosomal RNAs. The ribosome reads the genetic code of the mRNA, while a third kind of RNA molecule, transfer RNA (tRNA), mediates protein synthesis by bringing amino acids to the ribosome for attachment to the growing amino acid chain. Transfer RNAs have three nucleotides that are complementary to the codons in the mRNA.

SOURCE: Office of Technology Assessment, 1992.

**Figure 4-5—Modes of Inheritance of Single Gene Disorders**

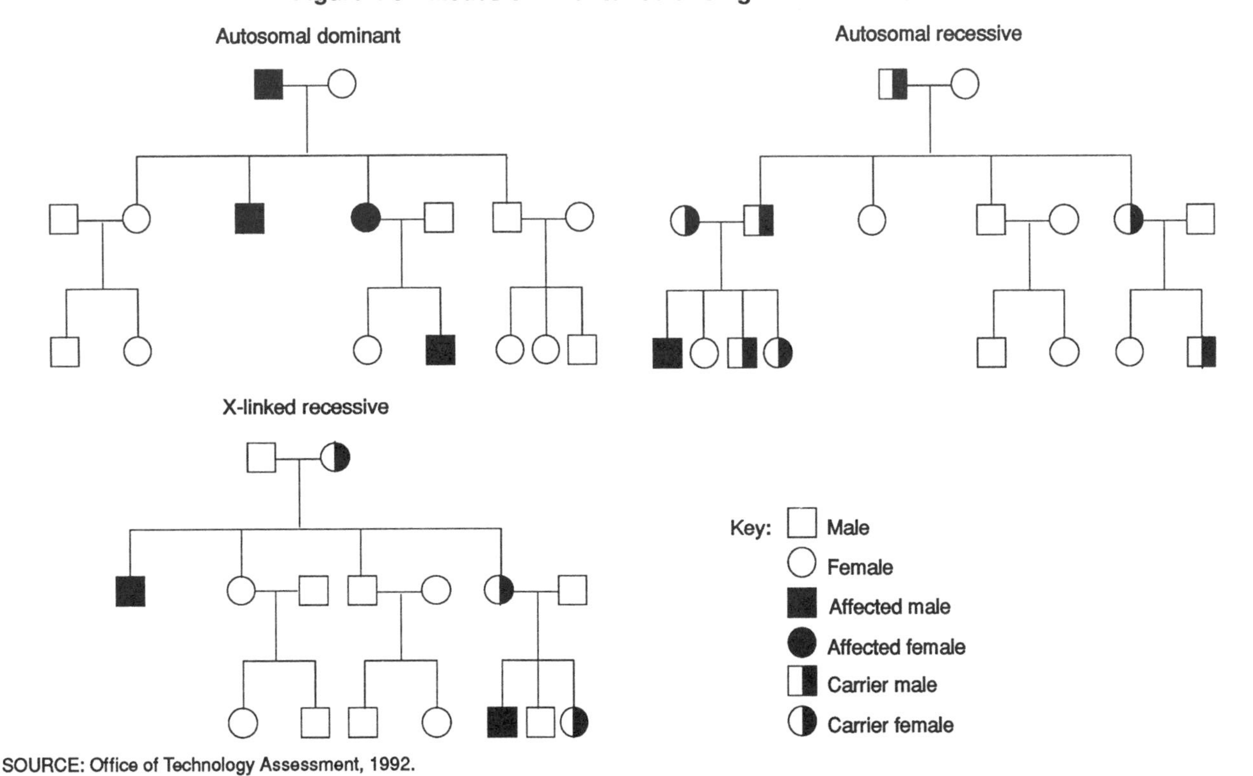


SOURCE: Office of Technology Assessment, 1992.

## THE GENETICS OF CYSTIC FIBROSIS

As mentioned earlier, CF is an autosomal recessive disorder. To have CF, an individual must inherit a mutant CF gene from each parent. If a couple produces a child with CF, each parent (excluding the possibility of a new mutation, nonpaternity, or a rare genetic event called uniparental disomy) must possess one of the 170+ CF mutations, but not necessarily the same ones. Each parent, then, is an asymptomatic carrier. For these couples, the chance of having a child with CF is 1 in 4 for each pregnancy (figure 4-6). Furthermore, if two carriers have an unaffected child, there is a 2 in 3 possibility that the unaffected child is a carrier. Again, because CF is an autosomal recessive disorder, it equally affects males and females.

CF occurs in all racial and ethnic groups, although more frequently in some than in others (table 4-1). It is the most common, life-shortening, recessive genetic disorder in Caucasians of Northern and Central European descent. In the United States, the incidence of CF in Caucasians is about 1 in 2,500 live births (17,56,81). An incidence of 1 in 2,500 implies a carrier frequency of approximately 4 percent. In other words, 1 in 25 Caucasians in the United States—about 8 million Americans—possess 1 chromosome with a CF mutation and 1 chromosome with a normal CF gene, and hence are carriers.

**Figure 4-6—Inheritance of Cystic Fibrosis**

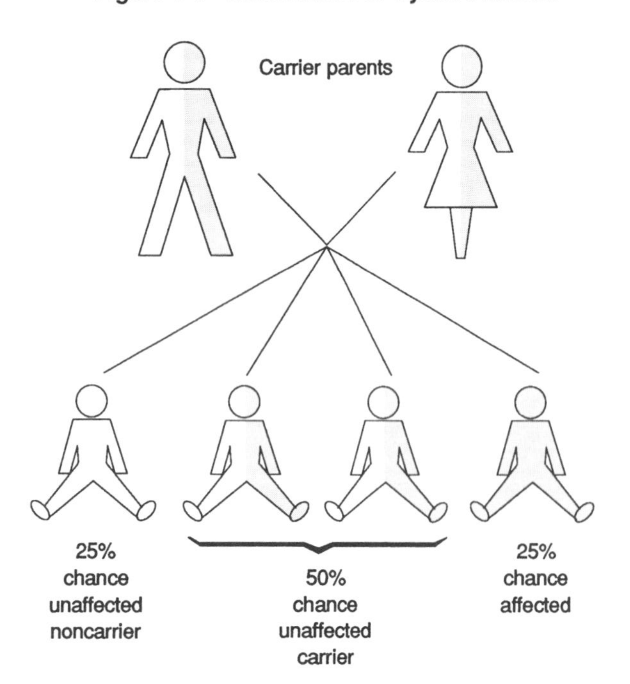


SOURCE: Office of Technology Assessment, 1992.

**Table 4-1—Incidence of Cystic Fibrosis Among Live Births in the United States**

| Population | Incidence (births) |
|---|---|
| Caucasian | 1 in 2,500[a,b,c] |
| Hispanic | 1 in 9,600[d] |
| African American | 1 in 17,000[a,e] to 1 in 19,000[f] |
| Asian American | 1 in 90,000[f] |

[a]T.F. Boat, M.J. Welsh, and A.L. Beaudet, "Cystic Fibrosis," *The Metabolic Basis of Inherited Disease*, C.R. Scriver, A.L. Beaudet, W.S. Sly, et al. (eds.) (New York, NY: McGraw Hill, 1989).
[b]K.B. Hammond, S.H. Abman, R.J. Sokol, et al., "Efficacy of Statewide Neonatal Screening for Cystic Fibrosis by Assay of Trypsinogen Concentrations," *New England Journal of Medicine* 325:769-774, 1991.
[c]W.K. Lemna, G.L. Feldman, B.-S. Kerem, et al., "Mutation Analysis for Heterozygote Detection and the Prenatal Diagnosis of Cystic Fibrosis," *New England Journal of Medicine* 322:291-296, 1990.
[d]S.C. FitzSimmons, remarks at Fifth Annual North American Cystic Fibrosis Conference, Dallas, TX, October 1991.
[e]J.C. Cunningham and L.M. Taussig, *A Guide to Cystic Fibrosis for Parents and Children*, (Bethesda, MD: Cystic Fibrosis Foundation, 1989).
[f]I. MacLusky, F.J. McLaughlin, and H.R. Levinson, "Cystic Fibrosis: Part 1," *Current Problems in Pediatrics*, J.D. Lockhart (ed.) (Chicago, IL: Year Book Medical Publishers, 1985).

SOURCE: Office of Technology Assessment, 1992.

In 1989, the gene responsible for CF was isolated and precisely located among the 22 autosomes, its most common mutation identified, and its DNA sequence determined (74,101,107). Prior to this discovery, the CF gene was localized to a specific chromosome—chromosome 7—by examining markers in the area of DNA believed to surround the gene. Using a technique known as restriction fragment length polymorphism (RFLP) analysis, scientists could follow a DNA pattern of inheritance through a family. The DNA pattern itself was not the CF gene, but was close in location—or linked—to it. This method of linkage analysis enabled researchers to hone in on the exact location of the gene on chromosome 7 (77,125,126,137,143). While some of the underlying problems in CF could be studied with linkage analysis, the exact identification of the gene opened up new avenues of pursuit in understanding the nature of the biochemical defect, in elucidating possibilities for treatment and cure, and in developing assays to detect carriers of CF mutations.

## *The Cystic Fibrosis Gene*

The CF gene is located on the long arm of chromosome 7, where it is distributed over 250,000 base pairs (250 kb) of genomic DNA (figure 4-7). At this locus, regions of the gene that code for the CF protein product are separated into 27 fragments, or exons, interspersed with portions of DNA that do not get translated—i.e., stretches, called introns, that are not decoded into proteins. During a process called transcription, introns are spliced out and the exons are pieced together into a precisely ordered string of 6,100 base pairs that codes for a protein comprised of 1,480 amino acids (74,101,107).

Ultimately, the CF gene product—known as the cystic fibrosis transmembrane conductance regulator (CFTR)—links DNA instructions with a critical biochemical function. When the CFTR comes from a mutant CF gene, that function is impaired and produces the medical manifestations of the disorder (box 4-A). The exact biochemical malfunction responsible for CF remains unknown, but CFTR regulates chloride ion ($Cl^-$) conductance in affected cell types and appears to be a $Cl^-$ channel (the structure that governs $Cl^-$ entry and exit in the cell). Efforts to understand the underlying pathogenesis of CF, and to develop treatment for it, focus on the structure and function of the protein product (30). In particular, studies of CFTR examine its role in ion transport, the key disturbance in the disorder. One major avenue of intervention under development is gene therapy, which involves inserting DNA that codes for CFTR into cells with mutant CF alleles in order to restore physiological function (box 4-B).

## *Mutations in the Gene*

Concomitant with the elucidation of the CF gene was the identification of a three base-pair deletion that resulted in a faulty CF gene product (i.e., a flawed CFTR). This mutation—abbreviated as ΔF508—results in the deletion of one amino acid, phenylalanine, at position number 508 in CFTR. Approximately 70 percent of all mutant CF genes in Caucasians in the United States and Canada exhibit this mutation (74,81). However, international studies reveal ethnic and regional variation in the distribution of this mutation (36,105). Overall, the ΔF508 mutation is most frequent in Northern European populations and less prevalent in persons from Southern Europe. Not surprisingly, the multicultural nature of the United States and Canada reflects this variation. The international epidemiology of the ΔF508 mutation is summarized in appendix A, as is the ethnic distribution of the mutation in the United States.

No other mutation accounts for a similarly large fraction of mutant CF genes among the remaining 30 percent of CF gene defects. To date, more than 170 additional mutations have been identified, and the number continues to grow (30). Most of these appear only in single individuals or families, although a few mutations account for 1 to 3 percent of CF mutations in the United States. The more common CF mutations and their epidemiology are also described in appendix A; some of these lesions also appear to vary among populations. Among Jewish persons of Central and Eastern European descent (Ashkenazic Jews), ΔF508 is relatively rare, but one mutation, W1282X, accounts for approximately 60 percent of CF mutations in this group (114).

## *Correlation Between Genotype and Phenotype*

The severity of CF symptoms differs greatly among individuals. To some extent, this is due to DNA differences and resultant alterations in the CFTR protein. Some mutations correlate with particular symptoms—primarily status of pancreatic enzyme sufficiency—and can be considered either mild (conferring pancreatic enzyme sufficiency) or severe (conferring pancreatic enzyme insufficiency) for this criterion (7,11,19,20,25,38,74,76,78,110-114,116,124). ΔF508 is considered a severe mutation, generally resulting in pancreatic insufficiency in homozygotes or in conjunction with a different severe defect in the other CF gene, although some exceptions exist (20,74,76,78,110-112,124). Other mutations appear to correlate with milder clinical outcome, but more data need to be collected (11,25,78,113,116). Pancreatic sufficient patients tend to have better respiratory function.

Pulmonary function generally does not correspond to specific mutations (74,78,110-112,124). Correlations between other mutations and particular phenotypes are also being studied (11,33,44,91, 99,119,124,145). Overall the course of the disease depends on both genetic and environmental features, and complete clinical outcome cannot be predicted on the basis of DNA analysis alone.

327-046 - 92 - 4 : QL 3

Figure 4-7—The Cystic Fibrosis Gene

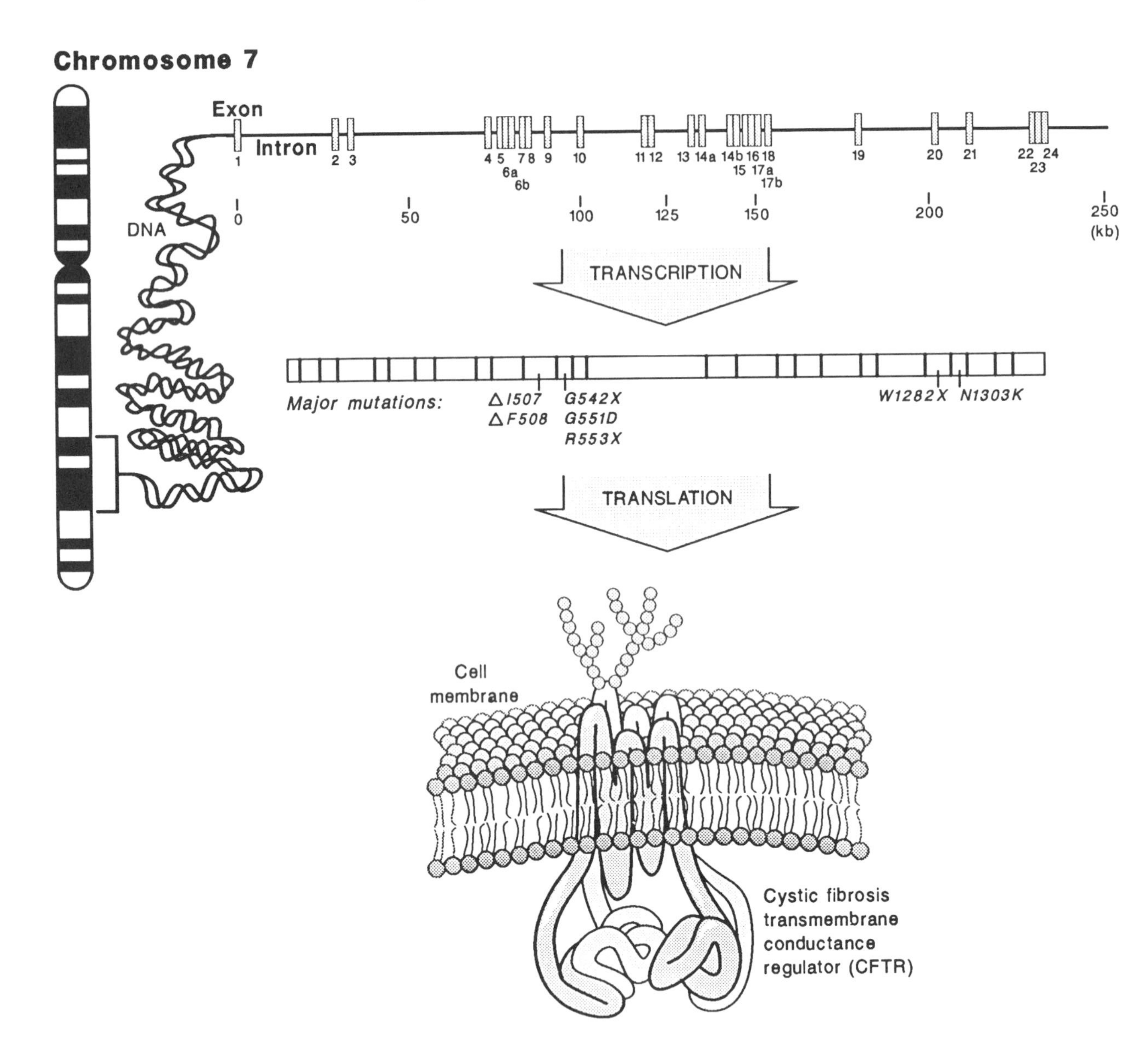


The CF gene is located on the long arm of chromosome 7, where it is spread over 250,000 base pairs (250 kb) of DNA. Coding regions of the DNA, or exons, are separated by noncoding regions, or introns. After the DNA is transcribed into messenger RNA (mRNA) comprised of all 27 exons of the gene, the mRNA is exported from the cell nucleus. Finally, instructions in the mRNA are translated, using special structures in the cell to assemble 1,480 amino acids into the final protein product.

SOURCE: Office of Technology Assessment, 1992, based on M.C. Iannuzzi and F.S. Collins, "Reverse Genetics and Cystic Fibrosis," *American Journal of Respiratory Cellular and Molecular Biology* 2:309-316, 1990.

### Box 4-A—The Gene Product: Cystic Fibrosis Transmembrane Conductance Regulator

Since the discovery of the principal mutation responsible for cystic fibrosis (CF), scientists have begun to relate this DNA error to the defect in ion transport long known to characterize the disorder. Cells cannot pump water, but must move fluids across their membranes through osmosis, a process that depends largely on ion movement either through pores in the membrane (channels), or transport systems designed to convey ions from one side of the membrane to the other. In CF affected individuals, regulation of chloride ion ($Cl^-$) transport is defective (46,95,140,142). Like most cellular activities, $Cl^-$ movement requires communication among different parts of the cell. Intricate networks of messenger systems accomplish this. Two common mediators of these systems, calcium ions ($Ca^{2+}$) and cyclic adenosine monophosphate (cAMP), regulate $Cl^-$ channels in epithelial cells. In individuals with CF, channels regulated by $Ca^{2+}$ function properly, but those channels controlled by cAMP and its intermediaries do not (95).

The product of the CF gene, a protein called the cystic fibrosis transmembrane conductance regulator (CFTR), regulates this cAMP-mediated $Cl^-$ conductance across the epithelial membrane. With current techniques, direct observation does not reveal the structure of CFTR. Rather, increasingly detailed examination of the deficits accompanying the defect, elucidation of the effect of conferring normal activity to CF cells, and speculations about both structure and function of the molecule based on predictive modeling techniques and manipulation of the gene are slowly providing insight into CFTR's nature.

Current research suggests that CFTR functions as a $Cl^-$ channel (4,13,47,49,96,98,118), although it may have other functions as well (10,22,59). On the basis of predicted structure, CFTR might belong to a large family of energy-dependent membrane transport systems that consists of several membrane-spanning regions and segments that bind the cellular energy source, adenosine triphosphate, or ATP (2,49,58,59,65,93,100,122). Activation of CFTR is mediated by a regulatory domain (3,27,98). The ΔF508 mutation and numerous other mutations are located in regions that code for portions of the protein likely to be of functional importance (35,38,50,71,75,92,98).

Numerous experiments have established $Cl^-$ channel regulation by cAMP through introduction of the normal CF gene into CFTR-defective epithelial cell lines (39,52,69,96,97,102). Moreover, $Cl^-$ conductance was induced in nonepithelial cell lines—theoretically without previously existing $Cl^-$ channels—when the CF gene was introduced, suggesting that CFTR might be a channel (4,5,72,106). Conversely, $Cl^-$ conductance was blocked in normal cells by preventing production of the CFTR product (115). However, recent discoveries of new types of chloride channels in CF-affected tissues complicate a straightforward interpretation that CFTR is a chloride channel (121,135).

Scientists are attempting to correlate symptoms of CF with a single molecular cause. CFTR has been demonstrated in all organs affected by the condition (123). One hypothesis to explain the multitude of CF traits attributes multiple symptoms to improper modification of a host of non-CFTR proteins due to decreased $Cl^-$ permeability in cells of CF patients (8,141).

As the workings of CFTR are better understood, new avenues for treatment open. If the underlying causes of the disease are understood, therapies can be directed at correcting the molecular deficits. For example, administering large quantities of cAMP analogs elicits $Cl^-$ conductance in cells with mutant CFTR (37,40,144). This suggests cAMP analogs might be an effective pharmacological intervention.

Elucidation of the structure and function of the CFTR protein could facilitate a means of assaying for CF mutations that would not require DNA analysis and the need to examine multiple mutations. Such a functional test theoretically would measure either the presence, absence, or altered state of the protein product. A functional test might, for example, evaluate $Cl^-$ conductance. Indications that $Cl^-$ conductance can be induced in cells with CF mutations, but with a recognizably different pattern than normal cells, might enable such a test (37,40). This pattern, however, appears to differ among cell types and methods of measurement (4,144). Likewise, understanding the extent to which protein processing is affected remains elusive (26,37,40,147). Although knowledge about the nature of CFTR and its structure and function continues to advance rapidly, answers that would render feasible a functional test for CF carriers are lacking. It might be, for example, that the more than 170 different CF mutations lead to a range of activities at the cellular level.

SOURCE: Office of Technology Assessment, 1992.

### *Box 4-B—Gene Therapy and Cystic Fibrosis*

Recent advances have moved gene therapy from theory to clinical and therapeutic experimental application (6,57). Protocols for gene therapy in humans must be approved by the Recombinant DNA Advisory Committee (RAC) of the National Institutes of Health (NIH), the NIH director, and the Food and Drug Administration in what can be a lengthy process, although the procedure has been streamlined (48). The first human gene therapy clinical trial was approved by the RAC and the NIH director in July 1990. By June 1992, nine other protocols for human gene therapy were in various stages of approval (146). For gene transfer—experiments that mark cells to trace the course of a treatment or the disease but are not therapeutic—an additional 15 protocols were in various stages of approval (146). Current human gene therapy trials include alteration of:

- white blood cells to treat a rare genetic disorder, severe combined immune deficiency due to adenosine deaminase deficiency, begun in September 1990 (84);
- immune system cells to produce an anticancer agent, begun in January 1991 (21,62); and
- liver cells to correct hypercholesterolemia, a genetic disorder of fatal cholesterol buildup, approved in October 1991, begun in June 1992 (56 FR 58800; 146).

The theory of gene therapy is straightforward: The normal gene is inserted into the cellular DNA either to code for a functioning protein product, or, in the case of cancer therapies, to confer disease-fighting properties. Experimentally, delivering normal genes into desired cells can be accomplished through physical or chemical means that disturb the cell membrane and allow DNA to enter, including specially modified viruses, liposomes (fatty materials able to transport drugs directly into cells), and direct injection (43,136). Somatic cell therapy—the only approach approved for human trials—changes only the DNA of the person receiving the therapy and cannot be inherited by offspring. In contrast, germ line gene therapy would alter the genetic material that is passed on to future generations. To date, no germ line therapy in humans has been proposed. For most conditions, cells are removed from the patient, genetically altered, and replaced.

Correction of abnormal $Cl^-$ transport through insertion of the normal CF gene into defective cells suggested that gene therapy was a viable consideration for treating CF (39,52,97,102); currently, the respiratory deficits of the disease are being targeted for correction by gene therapy. Unlike the RAC-approved protocols in progress, however, lung cells are generally inaccessible for removal and redelivery after gene transfer, making other means of administering the DNA necessary. Several systems are under investigation for efficacy of delivery in vivo without side effects. In one system, DNA is removed from an adenovirus, the type of virus responsible for some forms of the common cold and other respiratory ailments, and the inactivated virus shell is used as a vector to deliver the CF gene directly into the lungs of rats (108,109). The CF gene has also been delivered into cells isolated from the lungs of CF patients by bronchial brushing (32). In vivo and in vitro, significant amounts of messenger RNA for the CF gene are still present 6 weeks later, suggesting that long-term expression of the gene will be feasible. It is not yet known, however, how frequently new doses of the CF gene would have to be administered (32,109). An alternative delivery mechanism, aerosolized liposomes, has been used to deliver alpha-1-antitrypsin genes into rabbit lungs (61). A similar system might be applicable to delivery of the CF gene to human lungs.

Many questions about the safety and efficacy of gene therapy for CF must be answered before it will be suitable for human trials (30,32,138,146). Scientists do not yet know how much corrected protein product is needed to restore normal function to a patient with CF. Neither do they know whether adverse health effects will result from placing too much CFTR in a patient. Further, even though the virus has been fully debilitated in theory, using a viral vector raises concerns about expression of contaminating normal virus. Crippled viruses could also join with genetic material already in the cell and allow expression of a new virus or activate cells to a cancerous state.

Ethical considerations are also raised by some (6). Because gene therapy involves altering the genetic makeup of an individual, some express concerns about eugenic overtones, although only somatic cell therapy is under consideration. The general public, however, is enthusiastic. A 1986 OTA survey found that 83 percent of the American public approved of human cell manipulation to cure usually fatal diseases, 78 percent would be willing to undergo gene therapy personally to correct a genetic proclivity to a serious or fatal disease, and 86 percent of respondents would be willing to have his or her child undergo gene therapy for a usually fatal disease (131).

Gene therapy clearly offers the promise of treatment for some disorders. On the other hand, heightened attention to genetics in general—and gene therapy in particular—in the popular press can raise false hopes for cures for diseases long before they will be feasible or readily available (53,70). For CF, critical steps have been made towards the first attempt at gene therapy in humans; clinical applications, however, are still on the horizon.

SOURCE: Office of Technology Assessment, 1992.

# TESTS FOR CYSTIC FIBROSIS MUTATIONS

Localization of the CF gene and determination of its sequence enabled direct analysis of DNA for the presence of CF mutations. However, as mentioned in chapter 2, carrier screening is hindered by the multitude of mutations, particularly those too rare to be used practically in a CF carrier screening panel. Moreover, not all CF mutations have been discovered. As additional mutations are elucidated and incorporated into carrier screening protocols, the detection limitation decreases incrementally, although the inability to detect all mutations remains. This section explains the process and limits of direct DNA screening as applied to CF.

## *Techniques Used in Direct DNA Analysis*

Multiple techniques are used to analyze DNA. Four principal processes are employed, though each technique is not performed on all samples. These methods, depicted in figure 4-8, are:

- DNA amplification;
- restriction enzyme digestion;
- gel electrophoresis; and
- Southern transfer, dot-blotting, and probe hybridization.

### DNA Amplification

DNA amplification increases the amount of DNA to be analyzed by making copies of the original sequence from the sample. A process called the polymerase chain reaction (PCR) is typically used (box 4-C). Using PCR, selected areas of a gene can be amplified through repeated cycles to yield large quantities of the sample for rapid diagnosis.

While PCR amplifies a stretch of DNA between two primers, a new technique known as ligase chain reaction (LCR) amplifies only the region of DNA directly beneath the known sequence. Like PCR, millions of copies of the original sequence are made, but LCR's advantage is that it lends itself easily to detecting mutations differing by even a single base (82,139)—although it is not yet practical for general use.

### Restriction Enzyme Digestion

Restriction enzymes act as molecular scissors, cutting the DNA into fragments at specific sequences. Different enzymes recognize and cut different sequences. Mutations in a gene alter the DNA sequence and sometimes create or destroy specific sequences known as restriction sites. Thus, when changes occur in the DNA, restriction enzyme digestion can yield different sizes of DNA fragments in samples taken from an affected versus an unaffected individual. To distinguish between alleles, a restriction enzyme site can be intentionally created if a mutation has not created one (45,54).

### Gel Electrophoresis

In gel electrophoresis, PCR-derived or restriction enzyme digested DNA is separated into its different sized fragments. The sample is placed in a semisolid matrix, called a gel, and exposed to an electric field. In response to the electrical field, the DNA migrates toward one edge of the gel. In doing so, the gel acts as a sieve, with large DNA fragments passing through the gradient more slowly than small ones, allowing the DNA mixture to be separated according to size.

After gel electrophoresis is performed, the fragments can either be viewed directly using ultraviolet light plus a dye called ethidium bromide, or transferred onto a membrane for analysis with specific DNA probes (described in the following section).

### Southern Transfer, Dot Blotting, and Probe Hybridization

DNA fragments from gel electrophoresis can be transferred to a nylon or other type of membrane, forming a "Southern blot." A specific probe—or short sequence of single-stranded DNA complementary to the DNA sequence being sought—can be washed onto a membrane, where it will bind to complementary sequences on the membrane. Although most probes are generic and identify all alleles of a gene, allele specific oligonucleotide (ASO) probes refine diagnostic accuracy by perfectly matching the nucleotide sequence of a portion of the gene in question (24). Sequences differing by only one base can be detected. Before use, the probe is labeled with a fluorescent or radioactive marker so the region of DNA binding to the probe can be detected. The hybridized molecules can be viewed either by exposure to x-ray film for a radioactive probe or by other methods, such as colorimetric dyes.

A variation of Southern transfer, called dot blotting, involves directly spotting DNA into discrete spots on the nylon membrane. A probe, or

### Figure 4-8—Techniques for DNA Analysis of Cystic Fibrosis Mutations

Living Cell

Intact DNA is chemically extracted from the sample

DNA

**RESTRICTION ENZYMES** (V)

act like molecular scissors and cut the DNA into fragments

Each individual restriction enzyme cuts at its own specific sequence whenever found along the DNA chain

...GC CTAG AGTCT...

*DNA fragments*

**AMPLIFICATION**

(Molecular photocopying of DNA)

Each sample is amplified manually or in a machine

Original DNA sample

Denature and synthesize

New DNA

PCR primer

Denature and synthesize

Multiple copies of DNA sample

(20-25 cycles of the PCR yields about one million-fold reproduction)

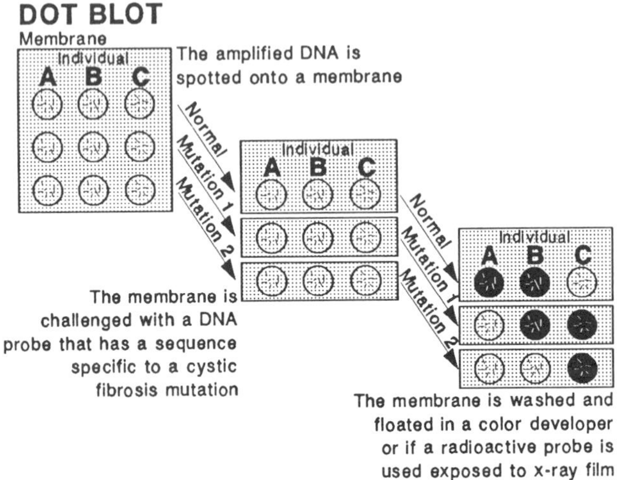


There are over 170 mutations at the cystic fibrosis locus (the most common mutation is Δ F508)

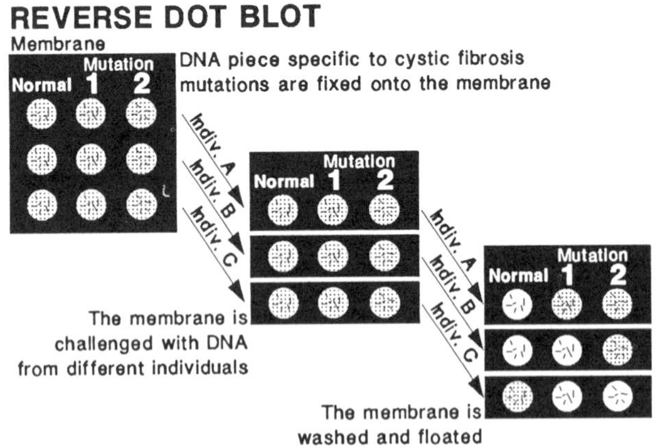


There are over 170 mutations at the cystic fibrosis locus (the most common mutation is Δ F508)

larger -

smaller +

Individual

A B C

**ELECTROPHORESIS**

The DNA fragments are separated by size into bands in a gel and visualized directly or through a process called Southern blotting

SOURCE: Office of Technology Assessment, 1992.

### *Box 4-C—The Polymerase Chain Reaction: Step-by-Step*

For the most part, all body cells within an individual contain the same DNA. Thus, DNA molecules in cells must regenerate copies of themselves each time a cell divides; DNA reproduces through a process called replication. During replication, the original strands in the DNA double helix unwind and serve as templates for the building of new, complementary strands, resulting in two identical copies of the original DNA molecule.

The polymerase chain reaction (PCR) is an in vitro technology based on the principles of replication. First described in 1985, PCR is now widely performed in research and clinical laboratories, and it is a critical technology for DNA diagnostics.

PCR involves the repeated duplication of a specific area of DNA to increase the amount of that DNA available to be used for research or test purposes. For example, consider the following sequence of double-stranded DNA to be amplified using PCR:

*[-G-G-C-T]*
-**T-T-C-G**-A-T-G-G-A-T-A-A-A-C-C-G-A
-A-A-G-C-T-A-C-C-T-A-T-T-T-**G-G-C-T**
*[-T-T-C-G-]*

In order to perform PCR, the sequences of the DNA at both ends of the region of interest must be known, and their complementary sequences available as short pieces of purified DNA called primers. One primer must be complementary to the end of one strand, the second to the opposite end of the other strand as indicated in the diagram above. These two specific sequences flank the area the scientist wants to copy, and serve as the foundation to which bases can be added and the DNA strand copied.

In PCR, the temperature of the solution containing the DNA to be amplified is raised to about 95 °C, which results in the separation, or melting, of the double helix to yield single-stranded pieces:

-**T-T-C-G**-A-T-G-G-A-T-A-A-A-C-C-G-A-

-A-A-G-C-T-A-C-C-T-A-T-T-T-**G-G-C-T**-

Copies of the primers are then added to the solution and allowed to hybridize to the DNA of interest, by lowering the temperature:

-**T-T-C-G**-A-T-G-G-A-T-A-A-A-C-C-G-A-
←*[-G-G-C-T]*

*[-T-T-C-G-]*→
-A-A-G-C-T-A-C-C-T-A-T-T-T-**G-G-C-T**-

The scientist then sets conditions in the reaction that allow new copies of the DNA of interest to be synthesized from the ends of the primers (referred to as primer elongation). That is, DNA polymerase (a heat-stable version of an enzyme from *Thermus aquaticus*, a thermophilic microorganism isolated from a hot spring in Yellowstone National Park) starts at the end of the primers and, using bases (G,A,T,C) that are part of the reaction mixture, synthesizes complementary strands of each of the two single strands to yield two strands from the original one:

-T-T-C-G-A-T-G-G-A-T-A-A-A-C-C-G-A
-A-A-G-C-T-A-C-C-T-A-T-T-T-G-G-C-T---

---T-T-C-G-A-T-G-G-A-T-A-A-A-C-C-G-A-
-A-A-G-C-T-A-C-C-T-A-T-T-T-G-G-C-T-

Thus one cycle of PCR occurs, doubling the number of DNA copies from the original area of interest. After this first round of synthesis, and for each subsequent cycle, the temperature of the reaction is raised to approximately 95 °C to separate the DNA strands. Primers are again allowed to hybridize to the strands, and DNA synthesis allowed to occur. After a second cycle of PCR, the two strands become four, and after 20 to 25 cycles of PCR, the original DNA area of interest has been amplified about a millionfold.

SOURCE: Office of Technology Assessment, 1992.

probes, can be hybridized to the membrane and diagnosis made rapidly (42). Reverse dot blotting is similar in principle to Southern transfer and dot blotting. However, in reverse dot blotting, the unique ASOs (which would be used as probes in conventional Southern analysis or dot blotting) are immobilized on the membrane. Key segments of the individual's uncharacterized DNA are then amplified, labeled, and hybridized to the probes on the membrane.

## *Carrier Assays for Cystic Fibrosis*

CF carrier tests involve direct DNA analysis, and all laboratories currently performing them directly analyze the gene. In individuals with a family history of CF or an affected child, an indirect assay—called linkage analysis—sometimes can be used to obtain additional information in the event of a negative test result. The following sections describe the techniques used to detect CF mutations.

### Direct Analysis of the Cystic Fibrosis Gene

DNA is generally obtained from white blood cells in blood samples, although it can be obtained from almost all nucleated cells in the body. Some groups in the United Kingdom and one project in the United States use buccal cells from mouthwash samples (ch. 10). In 1992, most laboratories in the United States assay DNA samples for the ΔF508 mutation and 6 to 12 other common mutations (ΔF508+6-12). Some laboratories screen for additional mutations.

After DNA is extracted from a blood sample, key segments containing mutations are amplified with PCR. When multiple segments are amplified at once, it is referred to as multiplexing. For some of the mutations, the amplified DNA can be electrophoresed and the migration pattern—specific to each mutation—visualized directly on the gel. Digestion with restriction enzymes followed by gel electrophoresis and visualization, or blotting followed by ASO hybridization, can also be used to analyze the mutations.

Reverse dot blotting using ASOs can be used to simultaneously analyze multiple CF mutations, and test kits that do so are under development in the United States (42) (figure 4-9).

### The Limits of DNA-Based Tests

DNA analysis for CF is limited by the diversity of mutations and the variation in frequency among different racial and ethnic groups (34). At present, ΔF508+6-12 other mutations account for approximately 85 to 95 percent of all mutant CF genes in Caucasians, depending on ethnicity (14,89). The range in detection rates presented by different commercial vendors likely results from their using tests that detect different mutations and assaying different populations (29,66).

The presence of a specific mutation establishes CF carrier status; a negative test, however, does not preclude carrier status, since not all mutations are known or assayed. About 5 to 15 percent of carriers remain undetected because their mutation is not included in the assay.

### Linkage Analysis

In some cases where one partner is a known carrier and the other does not test positive for the most common mutations, linkage analysis can be used to gather more information about an individual's risk. Linkage analysis can only be performed in families with living members with CF or on DNA samples stored from a deceased individual; frequently, such samples are unavailable. In this procedure, DNA markers are studied to trace the transmission pattern in a particular family of the CF gene and a specific mutation. For families with the necessary DNA samples, linkage analysis is generally informative. In rare instances, recombination events, which can alter marker patterns, can lead to erroneous conclusions from linkage results.

## *Automation of DNA Diagnostic Procedures*

The ability to test quickly and accurately will be crucial to widespread, inexpensive CF carrier screening, particularly if batteries of genetic assays are to be developed: Automation will be key. At present, the Human Genome Project serves as the primary impetus for automating DNA analysis, and the National Laboratories engaged in this aspect of the project are at the forefront in developing these technologies (1,51,64,103). As such, most advances in automated technology are specific to DNA sequencing, not diagnosis. Facets of the two processes overlap considerably, however, and some sequencing technologies can be made directly applicable to, or adapted for, DNA diagnosis.

Over the past few years, a number of instruments have been developed to increase the speed and volume of routine DNA diagnostic procedures (79).

**Figure 4-9—Reverse Dot Blot Analysis for Six Common Mutations**

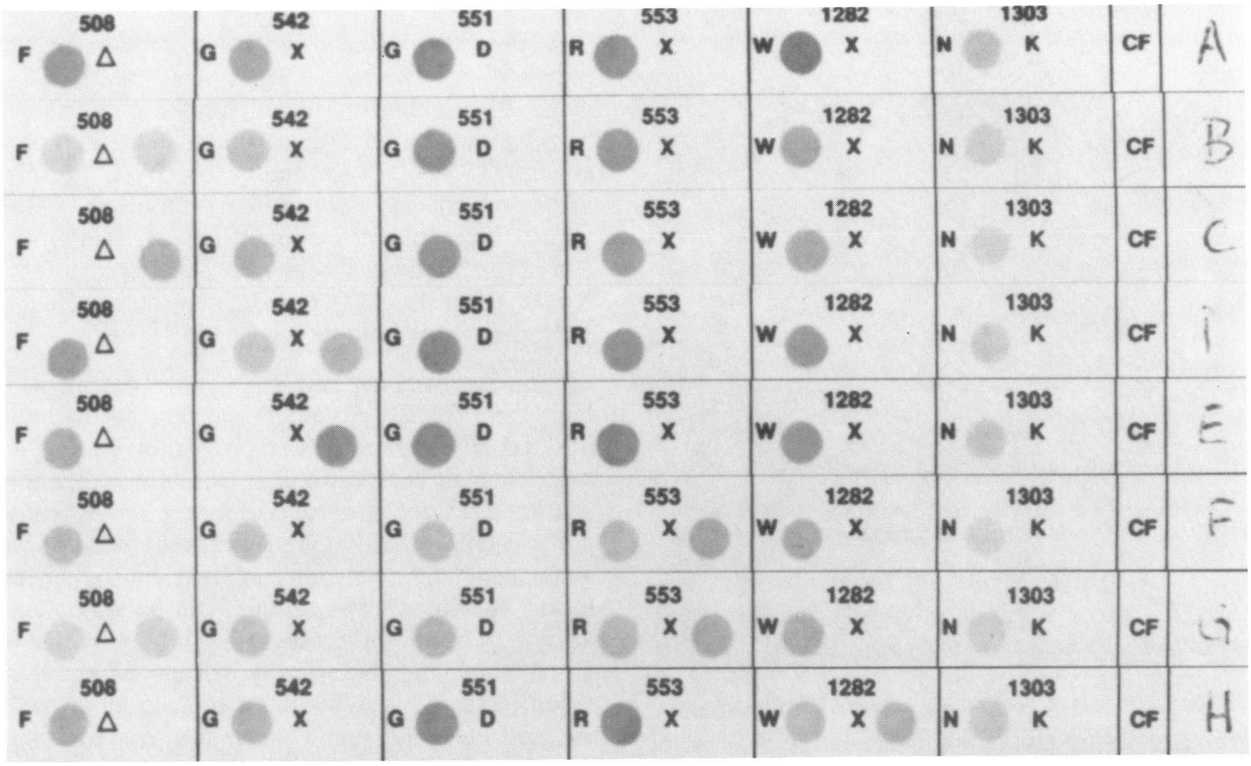

Allele specific oligonucleotide probes are bound to the test strip to detect six common CF mutations; in this photograph, each individual strip runs horizontally. DNA samples from individuals of unknown CF status are PCR-amplified and hybridized to separate test strips. Here, test strips for eight different individuals are shown (rows A through H). Following hybridization and colorimetric analysis, the patterns of dots on the strips are revealed—and hence the CF status of the individuals.

For each mutation on the strip (ΔF508, G542X, G551D, R553X, W1282X, and N1303K) the left dot, if present, indicates the person has a normal DNA sequence at that part of the CF gene. The right dot, if present, indicates the person has a CF mutation at that site. Individual A, then, has no CF mutations at the loci tested, as demonstrated by single dots on the left side for all mutations. In contrast, individuals B,D,F, and H are carriers, as demonstrated by the presence of two dots for one of the CF mutations. Individual C has CF, as demonstrated by a single dot on the right side of the ΔF508 panel; individual E has CF, as demonstrated by the single dot on the right side of the G542X panel. Individual G also has CF, but this person's CF arises from two different mutations—ΔF508 and R553X—as indicated by the pairs of dots in each of these panels.

SOURCE: Roche Molecular Systems, Inc., Emeryville, CA, 1992.

To the extent that the Human Genome Project is federally funded, integration of existing technologies specifically for clinical diagnostics will depend on Federal research priorities. Given the state of already automated procedures used in diagnosis and the rapid development of new DNA technologies under the auspices of the Human Genome Project, DNA automation is advancing at a pace that could realize entirely automated DNA diagnosis in the next few years.

Already, a nonelectrophoretic method of mutation detection has been automated and applied to detection of ΔF508 (90). Most DNA diagnostic techniques, however, still depend on electrophoresis, especially in the early stages of gene identification. Currently, instrumentation for such DNA diagnosis, except probe hybridization to a membrane, are automated or being automated. At a minimum, these steps include:

- DNA isolation from source material (e.g., blood, mucosal, or buccal cells),
- DNA amplification,
- gel preparation and loading for electrophoresis, and
- visualization and interpretation of results.

Automated instruments for DNA extraction, gene amplification by PCR, sequence gel loading, and visualization of sequence gel results already exist

Photo credit: Tony J. Beugelsdijk, Los Alamos National Laboratory

Automated robotic system used in DNA analysis at Los Alamos National Laboratory.

(42,51,64,79,103). While not faster than humans, these machines are designed to standardize procedures and decrease human error. They are not, however, always reliable (18,73,103). At Lawrence Berkeley Laboratory, scientists are creating separate robotic systems to allow simultaneous amplification and analysis of 96 different samples (94,127). At Los Alamos National Laboratory, a robotic system can produce high-density filters that increase the amount of information on one 96-sample filter by a factor of 16—i.e., 1,536 results can be obtained from one filter (15). To visualize and interpret results, scientists have designed an image plate reader that eliminates x-ray film (51,94,127).

Many new techniques for automated DNA analyses have been proposed, although at present they are not viable for large-scale clinical application (51,60, 64). One new procedure, DNA sequencing by hybridization, involves coating a DNA chip with short pieces of known DNA to which DNA of unknown sequence can be hybridized. Based on the hybridization pattern, mathematical algorithms can be used to decode the unknown sequence (9,23,51,64). A technique known as flow cytometry sequences DNA one molecule at a time by identifying individual, consecutive bases through base-specific fluorescence (67,68,83). Several other approaches to increasing sequencing speed are under development, including mass spectroscopy, atomic probe microscopy, and x-ray diffraction (28,51,64,79,103). As DNA sequencing becomes more rapid and less expensive, the prospect of determining CF (and other) mutations by direct DNA sequence analysis might someday be realized (86,87).

At present, most components of DNA analysis necessary for DNA-based genetic tests are automated as individual units, though some scientists are working on coordinating sequential steps into one system (15,94,127). In Japan, HUGA-1, a fully integrated robotic system that automates DNA sequencing from purification of DNA to interpretation of the final sequence, was to begin operating in April 1992 (41,64,117). Research into automation is also being carried out in Europe (63,120).

Clearly, crucial steps in DNA-based carrier screening assays for CF are automated now in research settings: DNA extraction, amplification, gel preparation, loading for electrophoresis, and visualization and interpretation of results. Only the step where a probe is hybridized to the membrane remains to be automated. The availability of high density filters means numerous DNA samples can be stored and analyzed concurrently, suggesting improved methods for handling the quantity of samples in high-volume screening. While currently the procedures are automated as single units, integrated robotic systems are being developed. Taken together, these advances in instrumentation indicate that automating the components of rapid carrier screening for CF is already technologically feasible, although an integrated system incorporating all of the processes has yet to be created.

# RESEARCH FUNDING

Federal funding for CF research is principally through the National Institutes of Health (NIH). Unlike most biomedical research, private-sector funds, which are made available through grants by the Cystic Fibrosis Foundation (CFF) and direct conduct of research by Howard Hughes Medical Institute (HHMI), account for a large portion of CF-related research. Some investigators receive funding through one or more of these sources. All three sources fund basic biomedical research. NIH and CFF support specific projects, while HHMI awards salaries and research support to individual investigators.

## *Federal Efforts*

In fiscal year 1991, NIH allocated $46,937,000 for all research related to CF. Of that, $20,662,724 was provided for intramural research. Extramurally,

$19,753,956 was granted to investigators for whom CF is the primary focus of laboratory or clinical research and an additional $6,520,320 to those for whom CF research is a secondary component. Nine of 12 institutes of the NIH and 3 of 4 research centers apportion these monies. In fiscal year 1991, two institutes—the National Institute of Diabetes and Digestive and Kidney Diseases (NIDDK) and the National Heart, Lung, and Blood Institute (NHLBI)—accounted for 73 percent of Federal funding for CF-related research: NIDDK disbursed $10,226,070 (39 percent) and NHLBI disbursed $8,847,051 (34 percent). One institute of the Alcohol, Drug Abuse, and Mental Health Administration provided $66,236 in fiscal year 1991 for CF-related research as a secondary research focus. Table 4-2 lists funding by institute or center. The U.S. Department of Energy (DOE) appropriated $48 million in fiscal year 1991 and $59 million in fiscal year 1992 for the human genome initiative. Indirectly, technology development—e.g., in the form of automation of DNA diagnostic technologies—from DOE's funding of the Human Genome Project could affect DNA diagnostics if it is determined to be a program priority.

Beginning in fiscal year 1991, eight coordinated CF carrier screening pilot studies were funded by NIH at a total of $1,657,086 for the first fiscal year and $4,442,568 over a 3-year period (80). Six of these NIH pilots were funded by the National Center for Human Genome Research's Ethical, Legal, and Social Issues (ELSI) Program (ch. 2), one by the National Center for Nursing Research, and one by the National Institute of Child Health and Human Development, although all eight are considered a single consortium (ch. 6). ELSI alone targeted $1,340,963 in fiscal year 1991 and $3,200,178 through the end of fiscal year 1993 for these pilots (80).

### Private Efforts

The Cystic Fibrosis Foundation is a nonprofit organization whose mission is to further research, medical care, public policy, and education for CF. It dedicates 33 percent of its annual budget to research towards a cure. CFF maintains research centers at medical schools and universities throughout the United States and provides grants to individual researchers. For calendar years 1991 and 1992, CFF earmarked $20 million per year for biomedical research, nearly equivalent to NIH's extramural

*REQUEST FOR APPLICATIONS*

**New Therapies for the Treatment and Cure of Cystic Fibrosis**

**The National Institute of Diabetes and Digestive and Kidney Diseases (NIDDK), the National Heart, Lung and Blood Institute (NHLBI) and the Cystic Fibrosis Foundation (CFF)** invite investigator-initiated research applications to develop and characterize new therapeutic approaches for the treatment and ultimate cure of cystic fibrosis (CF).

Research areas include but are not limited to:

- Development of viral and nonviral delivery systems to achieve gene therapy for CF;
- Identification of cell types that are appropriate targets for gene therapy;
- Characterization of the regulatory mechanism which confers stable and efficient gene transfer into CF cells;
- Evaluation of the safety issues associated with the development of gene therapy;
- Development of pharmacological interventions based upon knowledge of the basic defect and function of the CFTR protein;
- Development of therapeutic strategies involving CFTR protein.

Research approaches must be directly relevant to cystic fibrosis. Applications must be submitted to the NIH. **Proposals which are judged meritorious but are not funded by the NIH will be considered for support by the CFF.**

NIH support will be provided as grants-in-aid. CFF support is available for up to two years.

The deadline for letters of intent is Jan. 17, and for applications is Feb. 21, 1992. For the full text of the RFA, contact **Judith Fradkin, M.D.,** Chief, Endocrinology & Metabolic Diseases Branch, NIDDK, Westwood Bldg, Rm. 603, Bethesda, Md. 20892 or **Susan Banks-Schlegel, Ph.D.,** Director, Asthma and CF Program, NHLBI, Westwood Bldg, Rm. 6A15. For information about CFF award, contact **Robert J. Beall, Ph.D.,** CFF Executive Vice President for Medical Affairs, 6931 Arlington Road, Bethesda, Md. 20814.

*Photo credit: Cystic Fibrosis Foundation*

Advertisement seeking grant proposals for CF-related research. The Cystic Fibrosis Foundation funds proposals that are deemed meritorious by the National Institutes of Health peer review panels, but that do not receive support due to lack of funds.

**Table 4-2—Federal Extramural Funding for Cystic Fibrosis Research, Fiscal Year 1991**

| Institute or center[a] | CF-related research funding (dollars) | Number of projects | Total budget (dollars) | Percent of budget for CF-related research |
|---|---|---|---|---|
| NIDDK | 8,413,847[b] | 66 | | |
| | 1,812,223[c] | 12 | | |
| | 10,226,070[d] | 78 | 615,300,000 | 1.7 |
| NHLBI | 7,620,247 | 45 | | |
| | 1,226,804 | 9 | | |
| | 8,847,051 | 54 | 1,126,900,000 | 0.8 |
| NIAID | 1,106,789 | 6 | | |
| | 753,823 | 3 | | |
| | 1,860,612 | 9 | 907,300,000 | 0.2 |
| NCHGR | 1,340,963 | 6 | | |
| | 449,074 | 3 | | |
| | 1,790,037 | 9 | 87,400,000 | 2.0 |
| NCRR | 773,673 | 37 | | |
| | 208,347 | 3 | | |
| | 982,020 | 40 | 335,800,000 | 0.3 |
| NICHD | 361,381 | 2 | | |
| | 254,018 | 1 | | |
| | 615,399 | 3 | 479,000,000 | 0.1 |
| NIDR | 0 | 0 | | |
| | 558,351 | 4 | | |
| | 558,351 | 4 | 148,900,000 | 0.4 |
| NEI | 0 | 0 | | |
| | 478,044 | 2 | | |
| | 478,044 | 2 | 253,200,000 | 0.2 |
| NIGMS | 0 | 0 | | |
| | 330,200 | 3 | | |
| | 330,200 | 3 | 760,000,000 | 0.04 |
| NCNR | 274,110 | 1 | | |
| | 0 | 0 | | |
| | 274,110 | 1 | 39,900,000 | 0.7 |
| NIAMS | 0 | 0 | | |
| | 220,862 | 1 | | |
| | 220,862 | 1 | 193,200,000 | 0.1 |
| NINDS | 0 | 0 | | |
| | 162,338 | 1 | | |
| | 162,338 | 1 | 541,700,000 | 0.03 |
| NIH | 19,753,956 | 163 | | |
| | 6,520,320 | 43 | | |
| | 26,274,276 | 206 | 8,276,700,000 | 0.3 |
| ADAMHA | | | | |
| NIMH | 0 | 0 | | |
| | 66,236 | 1 | | |
| | 66,236 | 1 | 622,714,000 | 0.01 |

[a] Institute and center abbreviations refer, in order, to the following: NIDDK, National Institute of Diabetes and Digestive and Kidney Diseases; NHLBI, National Heart, Lung, and Blood Institute; NIAID, National Institute of Allergy and Infectious Diseases; NCHGR, National Center for Human Genome Research; NCRR, National Center for Research Resources; NICHD, National Institute of Child Health and Human Development; NIDR, National Institute of Dental Research; NEI, National Eye Institute; NIGMS, National Institute of General Medical Sciences; NCNR, National Center for Nursing Research; NIAMS, National Institute of Arthritis and Musculoskeletal and Skin Diseases; NINDS, National Institute of Neurological Disorders and Stroke; NIH, National Institutes of Health; ADAMHA, Alcohol, Drug Abuse, and Mental Health Administration; NIMH, National Institute of Mental Health.
[b] CF-related research is primary focus of grant.
[c] CF-related research is secondary focus of grant.
[d] Total CF-related research (primary plus secondary).

SOURCE: Office of Technology Assessment, 1992.

funding level (12). Approximately 30 to 40 percent of the research budget supports studies in gene therapy, and the balance is spent on research into the pathophysiological basis of the disease and other therapies (12). In collaboration with NIH, CFF funds projects deemed meritorious by the NIH peer review process but not supported because of lack of funds. A total of $2 million in calendar year 1991 and $3.6 million in calendar year 1992 has been designated under this program (12). CFF does not support research into carrier screening.

HHMI is a philanthropic organization that sponsors biomedical research through support of individual investigators. For fiscal year 1992, $68.9 million, or 34 percent of its research budget, has been dedicated towards this purpose (16). No breakdown was available for the amount of spending specifically for CF, but six HHMI investigators who carry out such research as all or part of their activities receive a total of $5,743,093 to cover salary, research materials, and all operative costs (16).

## SUMMARY AND CONCLUSIONS

The 1989 isolation of the CF gene—and a single mutation responsible for about 70 percent of mutations in CF patients and families—opened new possibilities for understanding the basic defect, finding a cure, and testing and screening for carriers. Since then, more than 170 mutations of the CF gene have been identified. Approximately 13 mutations account for 85 to 95 percent of CF mutations in the United States in Caucasians.

One hallmark of CF—its varied symptoms and severity—sometimes correlates with differences in genetic mutations. Pancreatic insufficiency, for example, appears to be associated with the most common mutation (ΔF508), although other aspects of the disease have not yet been shown to correspond to specific mutations.

One outgrowth of identifying the CF gene has been the ability to directly analyze DNA to detect carriers of the condition. Although the presence of multitudinous mutations that vary in frequency among ethnic and racial groups confounds screening and DNA diagnosis on clinical, ethical, and legal levels, new technologies promise to surmount at least some of the technical difficulties. Coordinating existing automation and developing new automated techniques would facilitate rapid, large-scale detection of CF mutations. DNA automation is advancing at a pace that would enable entirely automated DNA diagnosis to be realized in the next few years, if this should be deemed a priority.

Another significant research result has been elucidation and understanding of the CF gene product, cystic fibrosis transmembrane conductance regulator, which could lead to improved diagnosis and treatment. As research in the underlying mechanism of CF progresses, new molecular-based treatments could further improve the health and quality of life of affected individuals. Future therapies, for example, might be targeted at correcting the deficits in $Cl^-$ flow or overcoming the defects in the gene through gene therapy.

## CHAPTER 4 REFERENCES

1. American Society of Human Genetics, Human Genome Committee Project Report, "The Human Genome Project: Implications for Human Genetics," *American Journal of Human Genetics* 49:687-691, 1991.
2. Ames, G.F.-L., Mimura, C.S., and Shyamala, V., "Bacterial Periplasmic Permeases Belong to a Family of Transport Proteins Operating From *Escherichia coli* to Human: Traffic ATPases," *FEMS Microbiology Reviews* 75:429-226, 1990.
3. Anderson, M.P., Berger, H.A., Rich, D.P., et al., "Nucleoside Triphosphates Are Required to Open the CFTR Chloride Channel," *Cell* 67:775-784, 1991.
4. Anderson, M.P., Gregory, R.J., Thompson, S., et al., "Demonstration that CFTR Is a Chloride Channel by Alteration of Its Anion Selectivity," *Science* 253:202-205, 1991.
5. Anderson, M.P., Rich, D.P., Gregory, R.J., et al., "Generation of cAMP-Activated Chloride Currents by Expression of CFTR," *Science* 251:679-682, 1991.
6. Anderson, W.F., "Human Gene Therapy," *Science* 256:808-813, 1992.
7. Bal, J., Sturhman, M., Schloesser, M., et al., "A Cystic Fibrosis Patient Homozygous for the Nonsense Mutation R553X," *Journal of Medical Genetics* 28:715-717, 1991.
8. Barasch, J., Kiss, B., Prince, A., et al., "Defective Acidification of Intracellular Organelles in Cystic Fibrosis," *Nature* 352:70-73, 1991.
9. Barinaga, M., "Will 'DNA Chip' Speed Genome Initiative?," *Science* 253:1489, 1991.
10. Barinaga, M., "Novel Function Discovered for the Cystic Fibrosis Gene," *Science* 256:444-445, 1992.

11. Barreto, C., Pinto, L.M., Duarte, A., et al., "A Fertile Male With Cystic Fibrosis: Molecular Genetic Analysis," *Journal of Medical Genetics* 28:420-421, 1991.
12. Beall, R., Cystic Fibrosis Foundation, Bethesda, MD, personal communication, November 1991.
13. Bear, C.E., Li., C., Kartner, N., et al., "Purification and Functional Reconstitution of the Cystic Fibrosis Transmembrane Conductance Regulator (CFTR)," *Cell* 68:809-818, 1992.
14. Beaudet, A.L., Howard Hughes Medical Institute, Baylor College of Medicine, Houston, TX, personal communication, February 1992.
15. Beugelsdijk, T.J., Los Alamos National Laboratory, Los Alamos, NM, personal communication, 1991.
16. Blanchard, F., Howard Hughes Medical Institute, Bethesda, MD, personal communications, November 1991, February 1992.
17. Boat, T.F., Welsh, M.J., and Beaudet, A.L., "Cystic Fibrosis," *The Metabolic Basis of Inherited Disease*, C.R. Scriver, A.L. Beaudet, W.S. Sly, et al. (eds.) (New York, NY: McGraw Hill, 1989).
18. Bobrow, M., Paediatric Research Unit, Division of Medical and Molecular Genetics, United Medical and Dental Schools of Guy's and St. Thomas's Hospital, London, England, personal communication, December 1991.
19. Bonduelle, M., Lissens, W., Liebaers, I., et al., "Mild Cystic Fibrosis in Child Homozygous for G542 Non-Sense Mutation in CF Gene," *Lancet* 338:189, 1991.
20. Borgo, G., Mastella, G., Gasparini, P., et al., "Pancreatic Function and Gene Deletion F508 in Cystic Fibrosis," *Journal of Medical Genetics* 27:665-669, 1990.
21. Borman, S., "Gene Therapy Eyed to Elicit Cancer Immunity," *Chemical and Engineering News* 69:6, 1991.
22. Bradbury, N.A., Jilling, T., Berta, G., et al., "Regulation of Plasma Membrane Recycling by CFTR," *Science* 256:530-532, 1992.
23. Cantor, C., Southern, E., and Mirzabekov, A., "Moscow Workshop on Sequencing by Hybridization," *Human Genome News* 3:12-13,16, 1992.
24. Caskey, C.T., "Disease Diagnostics by Recombinant DNA Methods," *Science* 236:1223-1228, 1987.
25. Chalkley, G., and Harris, A., "A Cystic Fibrosis Patient Who Is Homozygous for the G85E Mutation Has Very Mild Disease," *Journal of Medical Genetics* 28:875-877, 1991.
26. Cheng, S.H., Gregory, R.J., Marshall, J., et al., "Defective Intracellular Transport and Processing of CFTR is the Molecular Basis of Most Cystic Fibrosis," *Cell* 63:827-834, 1990.
27. Cheng, S.H., Rich, D.P., Marshall, J., et al., "Phosphorylation of the R-Domain by cAMP-Dependant Protein Kinase Regulates the CFTR Chloride Channel," *Cell* 66:1027-1036, 1991.
28. Coghlan, A., "Dyes Could Speed Up the Genome Project. . .," *New Scientist* 133:25, 1992.
29. Collaborative Research Inc., Waltham, MA, "Now . . . A Test to Identify Greater than 90% of Cystic Fibrosis Carriers," advertisement, 1991.
30. Collins, F.S., "Cystic Fibrosis: Molecular Biology and Therapeutic Implications," *Science* 256:774-779, 1992.
31. Cooper, D.N., and Schmidtke, J., "Diagnosis of Genetic Disease Using Recombinant DNA. Third Edition," *Human Genetics* 87:519-560, 1991.
32. Crystal, R.G., National Institutes of Health, Bethesda, MD, presentation before the U.S. Congress, Office of Technology Assessment, Advisory Panel on Cystic Fibrosis and DNA Tests: Implications of Carrier Screening, September 1991.
33. Cuppens, H., Maryen, P., De Boeck, C., et al., "A Child, Homozygous for a Stop Codon in Exon 11, Shows Milder Cystic Fibrosis Symptoms Than Her Heterozygous Nephew," *Journal of Medical Genetics* 27:717-719, 1990.
34. Cutting, G.R., Curristin, S.M., Nash, E., et al., "Analysis of Four Diverse Population Groups Indicates That a Subset of Cystic Fibrosis Mutations Occur in Common Among Caucasians," *American Journal of Human Genetics* 50:1185-1194, 1992.
35. Cutting, G.R., Kasch, L.M., Rosenstein, B.J., et al., "A Cluster of Cystic Fibrosis Mutations in the First Nucleotide-Binding Fold of the Cystic Fibrosis Conductance Regulator Protein," *Nature* 346:366-369, 1990.
36. Cystic Fibrosis Genetic Analysis Consortium, "Worldwide Survey of the ΔF508 Mutation—Report from the Cystic Fibrosis Genetic Analysis Consortium," *American Journal of Human Genetics* 47:354-359, 1990.
37. Dalemans, W., Barbry, P., Champigny, G., et al., "Altered Chloride Ion Channel Kinetics Associated With the ΔF508 Cystic Fibrosis Mutation," *Nature* 354:526-528, 1991.
38. Dean, M., White, M.B., Amos, J., et al., "Multiple Mutations in Highly Conserved Residues Are Found in Mildly Affected Cystic Fibrosis Patients," *Cell* 61:863-870, 1990.
39. Drumm, M.L., Pope, H.A., Cliff, W.H., et al., "Correction of the Cystic Fibrosis Defect In Vitro by Retrovirus-Mediated Gene Transfer," *Cell* 62:1227-1233, 1990.
40. Drumm, M.L., Wilkinson, D.J., Smit, L.S., et al., "Chloride Conductance Expressed by ΔF508 and Other Mutant CFTRs in *Xenopus* Oocytes," *Science* 254:1797-1799, 1991.

41. Endo, I., Soeda, E., Murakami, Y., et al., "Human Genome Analysis System," *Nature* 352:89-90, 1991.
42. Erlich, H.A., Gelfand, D., and Sninsky, J.J., "Recent Advances in the Polymerase Chain Reaction," *Science* 252:1643-1651, 1991.
43. Felgner, P.L., and Rhodes, G., "Gene Therapeutics," *Nature* 349:351-352, 1991.
44. Ferrari, M., Colombo, C., Sebastio, G., et al., "Cystic Fibrosis Patients With Liver Disease Are Not Genetically Distinct," *American Journal of Human Genetics* 48:815-816, 1991.
45. Friedman, K.J., Highsmith, W.E., and Silverman, L.M., "Detecting Multiple Cystic Fibrosis Mutations by Polymerase Chain Reaction-Mediated Site-Directed Mutagenesis," *Clinical Chemistry* 37:753-755, 1991.
46. Frizzell, R.A., "Cystic Fibrosis: A Disease of Ion Channels?," *Trends in Neursoscience* 10:190-193, 1987.
47. Frizzell, R.A., and Cliff, W.H., "Back To the Chloride Channel," *Nature* 350:277-278, 1991.
48. Gershon, D., "NIH Merger to Shorten Review," *Nature* 355:664, 1992.
49. Gibson, A.L., Wagner, L.M., Collins, F.C., et al., "A Bacterial System for Investigating Transport Effects of Cystic Fibrosis-Associated Mutations," *Science* 254:109-111, 1991.
50. Grannell, R., Solera, J., Carrasco, S., et al., "Identification of a Nonframeshift 84-bp Deletion in Exon 13 of the Cystic Fibrosis Gene," *American Journal of Human Genetics* 50:1022-1026, 1992.
51. Gray, J., "Instrumentation Is the Key to Mapping, Sequencing," *Human Genome News* 3:1-4, September 1991.
52. Gregory, R.J., Cheng, S.H., Rich, D.P., et al., "Expression and Characterization of the Cystic Fibrosis Transmembrane Conductance Regulator," *Nature* 347:382-386, 1990.
53. Griffin, R.D., "Gene Therapy," *Congressional Quarterly Researcher* 1:779-797, 1991.
54. Haliassos, A., Chomel, F.C., Tesson, L., et al., "Modification of Enzymatically Amplified DNA for the Detection of Point Mutations," *Nucleic Acids Research* 17:3606, 1989.
55. Hall, J.G., "Genomic Imprinting and Its Clinical Implications," *New England Journal of Medicine* 326:827-829, 1992.
56. Hammond, K.B., Abman, S.H., Sokol, R.J., et al., "Efficacy of Statewide Neonatal Screening for Cystic Fibrosis by Assay of Trypsinogen Concentrations," *New England Journal of Medicine* 325:769-774, 1991.
57. Herman, R., "Gene Therapy Is No Longer a Rarity," *Washington Post*, Jan. 21, 1992, Health 7.
58. Higgins, C.F., "Protein Joins Transport Family," *Nature* 341:103, 1989.
59. Higgins, C.F., and Hyde, S. C., "Channelling Our Thoughts," *Nature* 352:194-195, 1991.
60. Higuchi, R., Dollinger, G., Walsh, P.S., et al., "Simultaneous Amplification and Detection of Specific DNA Sequences," *Bio/Technology* 10:413-417, 1992.
61. Holden, C. (ed.), "Aerosol Gene Therapy," *Science* 253:964-965, 1991.
62. Holden, C. (ed.), "Gene Therapy Trials on the Move," *Science* 254:372, 1991.
63. Holden, C. (ed.), "Genetics 'Force de Frappe'," *Science* 251:623, 1991.
64. Hunkapiller, T., Kaiser, R.J., Koop, B.F., et al., "Large-Scale and Automated DNA Sequence Determination," *Science* 254:59-67, 1991.
65. Hyde, S.C., Emsley, P., Hartshorn, M.J., et al., "Structural Model of ATP-Binding Proteins Associated With Cystic Fibrosis, Multidrug Resistance and Bacterial Transport," *Nature* 346:362-365, 1990.
66. Integrated Genetics Laboratories, Inc., Framingham, MA, "CF/12 Mutation Test for Cystic Fibrosis," brochure, fall 1991.
67. Jett, J.H., Los Alamos National Laboratory, Los Alamos, NM, personal communication, September 1991.
68. Jett, J.H., Keller, R.A., Martin, J.C., et al., "High-Speed DNA Sequencing: An Approach Based Upon Fluorescence Detection of Single Molecules," *Journal of Biomolecular Structure and Dynamics* 7:301-309, 1989.
69. Jilling, T., Cunningham, S., Barker, P.B., et al., "Genetic Complementation in Cystic Fibrosis Pancreatic Cell Lines by Somatic Cell Fusion," *American Journal of Physiology* 259:C1010-C1015, 1990.
70. Johnson, J.A., "Human Gene Therapy," *Congressional Research Service Issue Brief*, IB90108, Apr. 1, 1991.
71. Jones, C.T., McIntosh, I., Keston, M., et al., "Three Novel Mutations in the Cystic Fibrosis Gene Detected by Chemical Mismatch Cleavage: Analysis of Variant Splicing and a Nonsense Mutation," *Human Molecular Genetics* 1:11-17, 1992.
72. Kartner, N., Hanrahan, J.W., Jensen, T.J., et al., "Expression of the Cystic Fibrosis Gene in Non-Epithelial Cells Produces a Regulated Anion Conductance," *Cell* 64:681-691, 1991.
73. Kazazian, H.H., Center for Medical Genetics, Johns Hopkins Hospital, Baltimore, MD, personal communication, December 1991.
74. Kerem, B.-S., Rommens, J.M., Buchanan, J.A., et al., "Identification of the Cystic Fibrosis Gene: Genetic Analysis," *Science* 245:1073-1080, 1989.

75. Kerem, B.-S., Zielenski, J., Markiewicz, D., et al., "Identification of Mutations in Regions Corresponding to the Two Putative Nucleotide (ATP)-Binding Folds of the Cystic Fibrosis Gene," *Proceedings of the National Academy of Sciences, USA* 87:8447-8451, 1990.
76. Kerem, E., Corey, M., Kerem, B.-S., et al., "The Relation Between Genotype and Phenotype in Cystic Fibrosis—Analysis of the Most Common Mutation," *New England Journal of Medicine* 323:1517-1522, 1990.
77. Knowlton, R.G., Cohen-Haguenauer, O., Van Cong, N., et al., "A Polymorphic DNA Marker Linked to Cystic Fibrosis Is Located on Chromosome 7," *Nature* 318:380-382, 1985.
78. Kristidis, P., Bozon, D., Corey, M., et al., "Genetic Determination of Exocrine Pancreatic Function in Cystic Fibrosis," *American Journal of Human Genetics* 50:1178-1184, 1992.
79. Landegren, U., Kaiser, R., Caskey, C.T., et al., "DNA Diagnostics: Molecular Techniques and Automation," *Science* 242:229-237, 1988.
80. Langfelder, E.J., National Center for Human Genome Research, National Institutes of Health, Bethesda, MD, personal communication, February 1992.
81. Lemna, W.K., Feldman, G.L., Kerem, B.-S., et al., "Mutation Analysis for Heterozygote Detection and the Prenatal Diagnosis of Cystic Fibrosis," *New England Journal of Medicine* 322:291-296, 1990.
82. Lewis, R., "Innovative Alternatives to PCR Technology Are Proliferating," *The Scientist* Jan. 21, 1991, p. 23.
83. Marrone, B.L., Los Alamos National Laboratory, Los Alamos, NM, personal communication, September 1991.
84. Marwick, C., "Gene Replacement Therapy Enters Second Year," *Journal of the American Medical Association* 266:2193, 1991.
85. McBride, G., "Nontraditional Inheritance—II: The Clinical Implications," *Mosaic* 22:12-25, 1991.
86. McCabe, E.R.B., "Genetic Screening for the Next Decade: Application of Present and New Technologies," *Yale Journal of Biology and Medicine* 64:9-14, 1991.
87. McCabe, E.R.B., "Implementation of DNA Technology" *Yale Journal of Biology and Medicine* 64:19-20, 1991.
88. McKusick, V.A., *Mendelian Inheritance in Man: Catalogs of Autosomal Dominant, Autosomal Recessive, and X-Linked Phenotypes, 9th ed.* (Baltimore, MD: The Johns Hopkins University Press, 1990).
89. Ng, I.S.L., Pace, R., Richard, M.V., et al., "Methods for Analysis of Multiple Cystic Fibrosis Mutations," *Human Genetics* 87:613-617, 1991.
90. Nickerson, D.A., Kaiser, R., Lappin, S., et al., "Automated DNA Diagnostics Using an ELISA-Based Oligonucleotide Ligation Assay," *Proceedings of the National Academy of Sciences, USA* 87:8923-8927, 1990.
91. Nunes, V., Bonizzato, A., Gaona, A., et al., "A Frameshift Mutation (2869insG) in the Second Transmembrane Domain of the CFTR Gene: Identification, Regional Distribution, and Clinical Presentation," *American Journal of Human Genetics* 50:1140-1142, 1992.
92. Osborne, L., Knight, R., Santis, G., et al., "A Mutation in the Second Nucleotide Binding Fold of the Cystic Fibrosis Gene," *American Journal of Human Genetics* 48:608-612, 1991.
93. Parham, P., "Transporters of Delight," *Nature* 348:674-675, 1990.
94. Pollard, M., Lawrence Berkeley Laboratory, Berkeley, CA, personal communication, September 1991.
95. Quinton, P.M., "Cystic Fibrosis: A Disease in Electrolyte Transport," *FASEB Journal* 4:2709-2717, 1990.
96. Quinton, P.M., "Righting the Wrong Protein," *Nature* 347:226, 1990.
97. Rich, D.P., Anderson, M.P., Gregory, R.J., et al., "Expression of Cystic Fibrosis Transmembrane Conductance Regulator Corrects Defective Chloride Channel Regulation in Cystic Fibrosis Airway Epithelial Cells," *Nature* 347:358-363, 1990.
98. Rich, D.P., Gregory, R.J., Anderson, M.P., et al., "Effect of Deleting the R Domain on CFTR-Generated Chloride Channels," *Science* 253:205-207, 1991.
99. Rigot, J.-M., Lafitte, J.-J., Dumur, V., et al., "Cystic Fibrosis and Congenital Absence of the Vas Deferens," *New England Journal of Medicine* 325:64-65, 1991.
100. Ringe, D., and Petsko, G.A., "A Transport Problem?," *Nature* 346:312-313, 1990.
101. Riordan, J.R., Rommens, J.M., Kerem, B.-S., et al., "Identification of the Cystic Fibrosis Gene: Cloning and Characterization of Complementary DNA," *Science* 245:1066-1072, 1989.
102. Roberts, L., "Cystic Fibrosis Corrected in Lab," *Science* 249:1503, 1990.
103. Roberts, S.S., "Push Buttons and the Genome: The Role of Automation in DNA Sequencing," *Journal of NIH Research* 3:89-92, 1991.
104. Rogers, J., "Nontraditional Inheritance—I: Mechanisms Mendel Never Knew," *Mosaic* 22:3-11, 1991.
105. Romeo, G., and Devoto, M. (eds.), "Population Analysis of the Major Mutation in Cystic Fibrosis," *Human Genetics* 85:391-445, 1990.

106. Rommens, J.M., Dho, S., Bear, C.E., et al., "Cyclic-AMP-Inducible Chloride Conductance in Mouse Fibroblast Lines Stably Expressing Human Cystic Fibrosis Transmembrane Conductance Regulator (CFTR)," *Proceedings of the National Academy of Sciences, USA* 88:7500-7504, 1991.
107. Rommens, J.M., Iannuzzi, M.C., Kerem, B.-S., et al., "Identification of the Cystic Fibrosis Gene: Chromosome Walking and Jumping," *Science* 245:1059-1065, 1989.
108. Rosenfeld, M.A., Siegfried, W., Yoshimura, K., et al., "Adenovirus-Mediated Transfer of a Recombinant Alpha-1-Antitrypsin Gene to the Lung Epithelium In Vivo," *Science* 252:431-434, 1991.
109. Rosenfeld, M.A., Yoshimura, K., Trapnell, B.C., et al., "In Vivo Transfer of the Human Cystic Fibrosis Transmembrane Conductance Regulator Gene to the Airway Epithelium," *Cell* 68:143-155, 1992.
110. Santis, G., Osborne, L., Knight, R.A., et al., "Genetic Influences on Pulmonary Severity in Cystic Fibrosis," *Lancet* 334:294, 1990.
111. Santis, G., Osborne, L., Knight, R.A., et al., "Independent Determinants of Pancreatic and Pulmonary Status in Cystic Fibrosis," *Lancet* 336:1081-1084, 1990.
112. Santis, G., Osborne, L., Knight, R.A., et al., "Linked Marker Haplotypes and the ΔF508 Mutation in Adults With Mild Pulmonary Disease and Cystic Fibrosis," *Lancet* 335:1426-1429, 1990.
113. Schloesser, M., Arleth, S., Lenz, U., et al., "A Cystic Fibrosis Patient With the Nonsense Mutation G542X and the Splice Site Mutation 171-1," *Journal of Medical Genetics* 28:878-880, 1991.
114. Shoshani, T., Augarten, A., Gazit, E., et al., "Association of a Nonsense Mutation (W1282X), the Most Common Mutation in the Ashkenazi Jewish Cystic Fibrosis Patients in Israel, With Presentation of Severe Disease," *American Journal of Human Genetics* 50:222-228, 1992.
115. Sorscher, E.J., Kirk, K., Weaver, M.L., et al., "Antisense Oligodeoxynucleotide to the Cystic Fibrosis Gene Inhibits Anion Transport in Normal Cultured Sweat Duct Cells," *Proceedings of the National Academy of Sciences, USA* 88:7759-7762, 1991.
116. Strong, T.V., Smit, L.S., Turpin, S.V., et al., "Cystic Fibrosis Gene Mutation in Two Sisters With Mild Disease and Normal Sweat Electrolyte Levels," *New England Journal of Medicine* 325:1630-1634, 1991.
117. Swinbanks, D., "The New Automated DNA Sequencer," *Nature* 351:593, 1991.
118. Tabcharani, J.A., Chang, X.-B., and Riordan, J.R., "Phosphorylation-Regulated $Cl^-$ Channel in CHO Cells Stably Expressing the Cystic Fibrosis Gene," *Nature* 352:628-631, 1991.
119. Taylor, C.J., Hughes, H., Hardcastle, P.T., et al., "Genotype and Secretory Response in Cystic Fibrosis," *Lancet* 339:67-58, 1992.
120. Thiel, E., Lawrence Berkeley Laboratory, Berkeley, CA, personal communication, September 1991.
121. Thiemann, A., Grunder, S., Pusch, M., et al., "A Chloride Channel Widely Expressed in Epithelial and Non-Epithelial Cells," *Nature* 356:57-60, 1992.
122. Thomas, P.J., Shenbagamurthi, P., Ysern, X., et al., "Cystic Fibrosis Transmembrane Conductance Regulator: Nucleotide Binding to a Synthetic Peptide," *Science* 251:555-557, 1991.
123. Trezise, A.E., and Buchwald, M., "In Vivo Cell-Specific Expression of the Cystic Fibrosis Transmembrane Conductance Regulator," *Nature* 353:434-427, 1991.
124. Tsui, L.-C., and Buchwald, M., "Biochemical and Molecular Genetics of Cystic Fibrosis," *Advances in Human Genetics*, vol. 20, H. Harris and K. Hirschorn (eds.) (New York, NY: Plenum Press, 1991).
125. Tsui, L.-C., Buchwald, M., Barker, D., et al., "Cystic Fibrosis Locus Defined by a Genetically Linked Polymorphic DNA Marker," *Science* 230: 1054-1057, 1985.
126. Tsui, L.-C., Zengerling, S., Willard, H.F., et al., "Mapping of the Cystic Fibrosis Locus on Chromosome 7," *Cold Spring Harbor Symposia on Quantitative Biology* 51:325-335, 1986.
127. Uber, D., Lawrence Berkeley Laboratory, Berkeley, CA, personal communication, September 1991.
128. U.S. Congress, Office of Technology Assessment, *Human Gene Therapy*, OTA-BP-BA-32 (Washington, DC: U.S. Government Printing Office, December 1984).
129. U.S. Congress, Office of Technology Assessment, *Techniques for Detecting Heritable Mutations in Human Beings*, OTA-H-298 (Washington, DC: U.S. Government Printing Office, September 1986).
130. U.S. Congress, Office of Technology Assessment, *New Developments in Biotechnology: Ownership of Human Tissues and Cells*, OTA-BA-337 (Washington, DC: U.S. Government Printing Office, March 1987).
131. U.S. Congress, Office of Technology Assessment, *New Developments in Biotechnology: Public Perceptions of Biotechnology*, OTA-BP-BA-45 (Washington, DC: U.S. Government Printing Office, May 1987).
132. U.S. Congress, Office of Technology Assessment, *Mapping Our Genes—The Genome Projects: How Big, How Fast?*, OTA-BA-373 (Washington, DC: U.S. Government Printing Office, April 1988).

133. U.S. Congress, Office of Technology Assessment, *Genetic Witness: Forensic Uses of DNA Tests*, OTA-BA-438 (Washington, DC: U.S. Government Printing Office, July 1990).
134. U.S. Congress, Office of Technology Assessment, *Genetic Monitoring and Screening in the Workplace*, OTA-BA-455 (Washington, DC: U.S. Government Printing Office, September 1990).
135. Valverde, M.A., Diaz, M., Sepulveda, F.V., et al., ''Volume-Regulated Chloride Channels Associated With the Human Multidrug-Resistance P-glycoprotein,'' *Nature* 355:830-833, 1992.
136. Verma, I.M., ''Gene Therapy,'' *Scientific American* 264:68-84, 1990.
137. Wainwright, B.J., Scambler, P.J., Schmidtke, J., et al., ''Localization of Cystic Fibrosis Locus to Human Chromosome 7cen-q22,'' *Nature* 318:384-385, 1985.
138. Weatherall, D.J., ''Gene Therapy in Perspective,'' *Nature* 349:275-276, 1991.
139. Weiss, R., ''Hot Prospect for New Gene Amplifier,'' *Science* 254:1292-1293, 1991.
140. Welsh, M.J., ''Abnormal Regulation of Ion Channels in Cystic Fibrosis Epithelia,'' *FASEB Journal* 4:2718-2725, 1990.
141. Welsh, M.J., ''Acidification Identification,'' *Nature* 352:23-24, 1991.
142. Welsh, M.J., and Fick, R.B., ''Cystic Fibrosis,'' *Journal of Clinical Investigation* 80:1523-1526, 1987.
143. White, R., Woodward, S., Leppert, M., et al., ''A Closely Linked Genetic Marker for Cystic Fibrosis,'' *Nature* 318:382-384, 1985.
144. Wine, J.J., ''The Mutant Protein Responds,'' *Nature* 354:503-504, 1991.
145. Wine, J.J., ''No CFTR: Are CF Symptoms Milder?,'' *Nature Genetics* 1:10, 1992.
146. Wivel, N., Office of Recombinant DNA Activities, National Institutes of Health, Bethesda, MD, personal communications, March 1992, June 1992.
147. Zeitlin, P.I., Crawford, I., Lu, L., et al., ''CFTR Protein Expression in Primary and Cultured Epithelia,'' *Proceedings of the National Academy of Sciences, USA* 89:344-347, 1992.

# Chapter 5

# Quality Assurance

# Contents



## *Boxes*



## *Figure*

# Chapter 5
# Quality Assurance

Quality assurance for cystic fibrosis (CF) carrier screening is multifaceted. In particular, three aspects of quality assurance are important to ensuring the safety, efficacy, and accurate interpretation of DNA-based CF assays:

- the quality of clinical laboratory services;
- the quality of genetic diagnostic kits, reagents, assays, and instrumentation; and
- the quality of professional services, including diagnostic and counseling services.

Oversight of quality assurance extends to Federal, State, and local governments. It includes the judiciary, professional societies, and clinical laboratories as well. All have a stake in ensuring high-quality diagnostic services, although the extent of involvement varies. For example, all play a part in oversight of laboratory performance, but the Federal Government has primary responsibility for ensuring the safety and efficacy of clinical laboratory devices (e.g., DNA test kits). Professional societies and courts, on the other hand, have a large impact on the quality of professional services.

This chapter concentrates on the roles of all interested parties in ensuring that both private and public facilities provide high-quality DNA-based genetic analysis, especially CF mutation screening. It discusses voluntary versus mandatory standards, and how both regulatory and nonregulatory mechanisms can facilitate efforts to guarantee high quality.

## QUALITY ASSURANCE FOR CLINICAL LABORATORIES

Laboratories use quality control to ensure that a laboratory's results meet predetermined criteria. It includes the steps taken by a laboratory to produce valid, reproducible, and reliable results each time the test is performed. Quality assurance programs document the satisfactory performance of quality control, and can include proficiency testing and external inspections (83,84,85). Quality control and quality assurance are essential components of good laboratory practice.

In 1991, Congress reviewed progress toward overcoming longstanding difficulties with ensuring the accuracy of diagnostic laboratory tests performed by facilities across the Nation. Congressional concern persists that quality problems could remain unresolved, despite recent changes in Federal law (167). Questions about laboratory quality are important to CF mutation analysis.

First, the quality of a laboratory's performance affects the quality of counseling services. Accurate reporting and interpretation of the mutations used by a laboratory are necessary if used by genetics and other health professionals are to convey accurate results to their clients. Failure to assay a less common mutation (or to properly interpret the results of the battery of mutations used) could result in clients mistakenly believing themselves to be at negligible risk of conceiving a child with CF. Conversely, misinterpreting test results could also mislead individuals to think they are at increased risk and to decide against conception. Second, the technical skills of both the technician and laboratory are essential for maintaining an acceptable standard of practice to allow a laboratory to conduct DNA analysis of CF mutations. Today's assays, for example, use the polymerase chain reaction (PCR), and some observers have concerns about the proper controls to ensure against potential mishaps—chiefly contamination—using PCR-based techniques (54,74,169).

Because the intensity of Federal interest in clinical laboratory performance is new and evolving, and because Congress has expressly involved itself by taking action in this area, this section focuses on recent congressional action, chiefly the Clinical Laboratory Improvement Amendments of 1988 (CLIA) (Public Law 100-578). States and professional organizations, however, also play key roles in certain aspects of laboratory quality assurance. Thus, this section also examines how each has been involved in specific debates surrounding quality assurance for clinical facilities performing DNA-based diagnostic procedures, which include carrier screening and testing for CF.

### *The Clinical Laboratory Improvement Amendments of 1988*

To remedy problems of inadequate and inconsistent clinical diagnostic testing, the 100th Con-

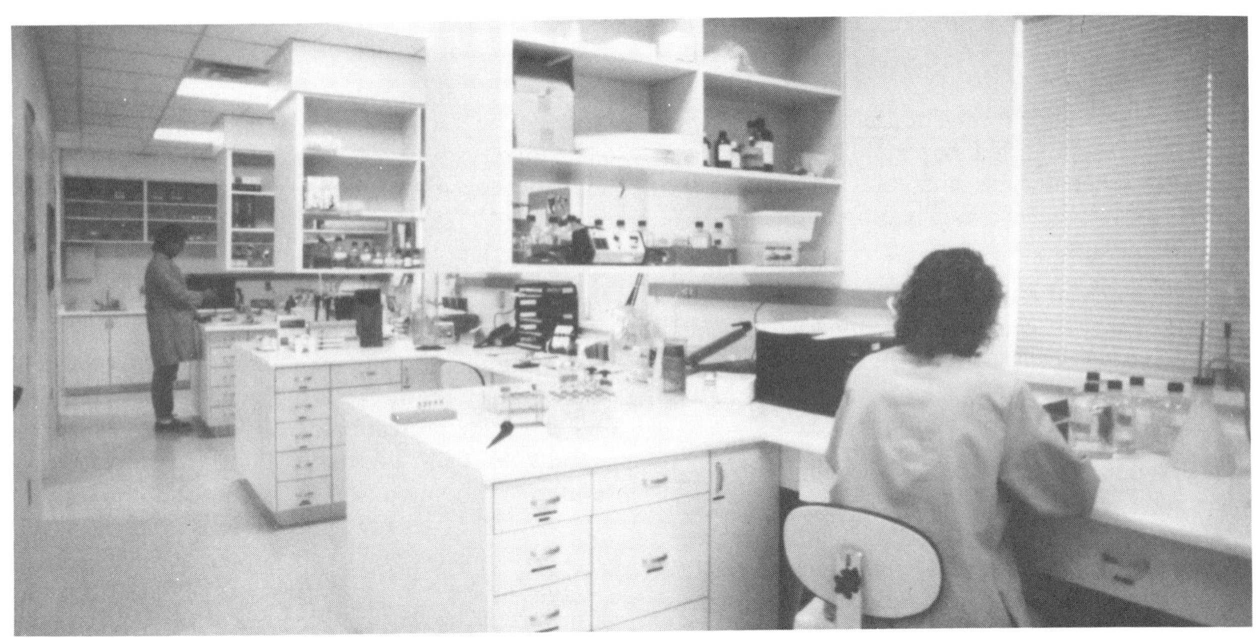

Photo credit: Vivigen, Inc.

Facilities that perform DNA-based diagnostic tests are subject to the Clinical Laboratory Improvements Amendments of 1988.

gress passed legislation that subjects most clinical laboratories to a number of requirements, including qualifications for the laboratory director, standards for the supervision of laboratory testing, qualifications for technical personnel, management requirements, and an acceptable quality control program. Many of these same standards were already in place prior to 1988 with regard to laboratories doing testing for Medicare or accepting samples across State lines, but CLIA represents the congressional response to national concern over shortcomings in the stringency and coverage of the original 1967 law. Designed to strengthen Federal oversight of laboratories to ensure that test results are accurate and reliable, CLIA creates a national, unified mechanism that regulates virtually every laboratory in the country—not just those involved in interstate commerce or participating in Medicare. Another impact of CLIA, beyond its extension to all laboratories, is the integration of the previously separate inspection and enforcement systems.

Prior to enactment of CLIA, Federal regulations covered the approximately 12,000 laboratories that either transported samples between States or performed tests billed to Medicare. In 1990, however, the Health Care Financing Administration (HCFA) of the U.S. Department of Health and Human Services (DHHS) began exercising sweeping regulatory authority over clinical laboratories. HCFA's mandate is to set standards for staffing and maintaining all medical laboratories, including physician office testing. HCFA is also directed to manage a comprehensive program to police the facilities, and it can impose sanctions.

Under CLIA, the Secretary of DHHS (hereinafter the Secretary) shall establish national standards for quality assurance in clinical laboratory services. The Secretary must implement recordkeeping, inspection, and proficiency testing programs, and report to Congress on a range of issues gauging the impact of various quality assurance mechanisms. Regulatory requirements will vary according to whether the facility performs tests considered "simple," "moderately complex," or "highly complex" (42 CFR 493). For example, cytogenetic testing—examining chromosome profiles—is likely to be considered "highly complex" (108). DNA-based genetic tests are not yet covered by the cytogenetics category, but unless specifically categorized, a test is considered "highly complex" (57 FR 7245). Tests similar to DNA-based genetic assays—i.e., DNA analysis to detect viruses—have been classified "highly complex" (57 FR 7288).

CLIA and the Omnibus Budget Reconciliation Act of 1989 (Public Law 101-239) grant HCFA the power to suspend or revoke a lab's certificate for violation of the rules. Further, fines up to $10,000 for each violation or each day of noncompliance can be levied, and jail sentences of 3 years can be imposed. The law continues to permit, subject to approval by the Secretary, States or private associations to substitute for the Federal accreditation process. Currently, these include at least the College of American Pathologists (CAP), the American Association of Bioanalysts, accrediting agencies in three States, the Joint Commission on Accreditation of Healthcare Organizations, and the American Osteopathic Association.

### Monitoring Laboratory Performance

HCFA will continue using State agencies for onsite monitoring because those agencies have the most experience in inspection activities, have the ongoing responsibilities for assessing laboratory compliance, inspect an entire facility (HCFA agents inspect only specific areas), and make periodic recertification (56 FR 13430).

Beyond onsite monitoring and inspection, proposed HCFA regulations aim to help physicians and patients avoid laboratories that perform poorly by issuing an annual laboratory registry (42 CFR 493.1850). The registry will include, for example, those facilities that have had their CLIA certificates suspended and those that have had their accreditation withdrawn or revoked. The registry is designed to create a national enforcement mechanism that affects virtually every clinical laboratory in the country.

For the first time, CLIA regulates the estimated 98,000 physician office laboratories. In total, HCFA estimates that from 300,000 to 600,000 physician, hospital, and freestanding laboratories in the Nation could potentially come under these provisions, and that the registry will likely change the practice patterns of laboratories across the country. Some laboratories might close because they cannot meet the requirements. Others, out of fear of being sanctioned, might choose not to perform certain tests. Some laboratories will increase their fees to private patients to cover the costs of upgrading facilities to meet CLIA standards and to pay the user fee (57 FR 7188) being imposed to fund the survey and other CLIA requirements. Some laboratories, however, are exempt, including certain State facilities and some performing drug abuse tests (57 FR 7190). Facilities limited to some types of insurance testing could also be exempt (108).

### State Authority Under CLIA

States will be substantially affected by CLIA. On one level, they will probably experience some additional administrative burden if they identify an increased number of noncompliant laboratories. The principal impact, however, will be on the relationship between the Federal Government and the States in the area of direct laboratory regulation. Prior to CLIA's enactment, the Federal Government had no regulatory authority over the numerous intrastate laboratories, including those located in physicians' offices. These were, in many cases, though not all, regulated by the States; such facilities are now subject to CLIA requirements.

As mentioned earlier, however, CLIA does not preclude continued State regulation and licensure (57 FR 7188), although the thrust of States' role is changed. Primary emphasis focuses on licensing personnel and providing information, inspection, and some proficiency testing services. (A later section in this chapter describes specific State initiatives in overseeing clinical laboratories.)

### Proficiency Testing

One issue of critical concern to Congress in passing CLIA was proficiency testing programs. Until CLIA, such programs varied broadly in testing criteria and in grading of test results. Moreover, uniform or minimally acceptable Federal standards did not exist. Now, except under certain circumstances, proficiency testing shall be conducted every 4 months, with uniform criteria for all examinations and procedures. The Secretary shall also establish a system for grading proficiency testing performance. HCFA expects to propose rules on proficiency testing before the end of 1992. None of these rules is expected to apply to DNA-based CF tests (65,185).

### Sanctions

HCFA has moved more quickly on the issue of sanctions against laboratories not meeting Federal requirements (57 FR 7218). Such sanctions can be imposed instead of, or before, suspending, limiting, or revoking the laboratory's certificate and canceling the laboratory's approval to receive Medicare payment for its services.

Prior to CLIA and the Omnibus Budget Reconciliation Act of 1987, the only recourse HCFA had against a noncomplying laboratory was cancellation of its approval to receive Medicare payment for its services. In developing a range of new sanctions, HCFA has attempted to establish consistency between the enforcement approach for Medicare laboratories and for laboratories that do not participate in Medicare. At the direction of Congress, the sanctions include directed plans of correction, civil money penalties, payment for the costs of onsite monitoring by the agency responsible for conducting certification inspections, and suspension of all or part of Federal payments to which the laboratory would otherwise be entitled for services furnished after the effective date of sanction.

HCFA proposes three levels of noncompliance, with graduated severity according to levels of deficiencies: those posing immediate jeopardy to patients, those not posing immediate jeopardy, and those that are minor. HCFA can also impose sanctions in specific categories or subcategories, and thus discourage laboratories from performing tests in which they do not comply with CLIA without discouraging testing in categories in which no deficiencies are identified. CLIA also provides for incarceration and fines for any person convicted of intentionally violating any CLIA requirement. It specifies administrative and judicial review procedures available to a laboratory when HCFA imposes a sanction or suspends, revokes, or limits the facility's CLIA certificate.

### Impact of CLIA on DNA Tests

As with other clinical diagnostic tests, CLIA will affect DNA analysis performed by clinical facilities. Currently HCFA can limit CLIA certificates at the specialty or subspecialty level. No special category exists for DNA tests, but facilities performing such assays clearly fall within CLIA's regulatory rubric. Furthermore, HCFA theoretically could limit CLIA certificates at the level of individual tests rather than at the specialty or subspecialty level. For example, a laboratory could lose its authority to perform CF mutation analyses, while retaining authority to continue performing, and receiving payment for, sickle cell tests. Such detailed oversight, however, would probably strain HCFA's administrative capacities (34).

One aspect of CLIA important to carrier testing and screening for CF will be the development of proficiency testing standards. The legislation is quite detailed in addressing proficiency testing for other clinical tests, but is silent for DNA analyses. Nonetheless, HCFA expects voluntary participation of DNA laboratories in a proficiency testing program (148). As described later, professional societies and nonprofit associations are likely to play the major role in this aspect of quality assurance, although their involvement is neither required nor approved by HCFA.

#### *State Authorities*

Since CLIA, the principal State role in quality assurance for clinical facilities is licensure and certification of personnel. All licensing of medical and clinical personnel is based on State law. State and Federal tort actions to remedy issues related to personnel and service quality are discussed separately in this chapter.

As mentioned earlier, however, CLIA does not prevent States from regulating and licensing facilities within certain guidelines (55 FR 33936). At least one State views CLIA as too broad-based to appropriately address issues raised by DNA tests. California established an expert advisory committee to develop standards and to hire qualified consultants to conduct onsite inspections. After a pilot study using voluntary approvals, the California Department of Health Services (CDHS) intends to ask for specific licensing laws and regulations for DNA and cytogenetic laboratories. CDHS will use any acceptable national proficiency testing program, but will develop its own if those being developed by professional organizations (described in a following section) are not satisfactory (34).

Another State, New York, has regulated clinical laboratories since 1964, prior to enactment of the original Federal legislation in 1967 (184). More important to the issue of quality assurance for CF carrier screening, New York State has established a genetics quality assurance program that includes requirements for licensing personnel, licensing facilities, laboratory performance standards, and DNA-based proficiency testing (box 5-A).

#### *The Role of Professional Societies*

While CLIA clearly expanded the Federal role in clinical laboratory oversight, the law continues to permit, subject to approval by the Secretary, the involvement of other parties in regulating laboratory practices. In particular, private nonprofit associations and professional societies could have the greatest impact in proficiency testing. Of those associations with standing under the past Federal

### Box 5-A—The New York State Genetics Quality Assurance Program

Responding to the development of DNA-based tests for genetic conditions, New York State has created a permit category for genetic tests. Since January 1, 1991, all facilities within the State, or that handle samples from the State, have had to be State licensed. Included among the types of technologies for which a permit is required is DNA analysis for carrier or disease status. To date, 40 facilities—15 within New York and 25 out-of-State—have been accredited.

In the area of personnel qualifications, the New York State regulations detail specific minimum requirements for training and education of the laboratory director, including experience with molecular biology and genetic linkage analysis. To receive a laboratory license, applying facilities must undergo an onsite inspection by the New York State Department of Health. The laboratory also must meet several other requirements, including documenting that it: periodically tests equipment; monitors and performs proper quality control of its reagents and standards; adheres to appropriate confidentiality of records; participates in some form of external quality assurance program (where available); and demonstrates that it has a clear, appropriate, interpretive report format that explains findings for nongeneticist physicians. These reports must also caution the provider about possible inaccuracies and suggest alternative or additional testing if necessary. Finally, to maintain its license, the facility must undergo interlaboratory proficiency testing for DNA analytical methods.

Beginning in August 1992, New York State will administer a quarterly proficiency testing program. Under the program, a single sample will be sent to accredited laboratories. Using five systems of their choosing, the laboratories will analyze and interpret results for the unknown sample and report the findings to the State Department of Health, Clinical Laboratory Evaluation Unit.

SOURCE: Office of Technology Assessment, 1992, based on P.D. Murphy, "New York State Genetics Quality Assurance Program," meeting abstract, Biotechnology and the Diagnosis of Genetic Disease: Forum on the Technical, Regulatory, and Societal Issues, Arlington, VA, Apr. 18-20, 1991.

regulatory structure, CAP is likely to be most important to quality assurance of laboratories doing DNA analysis.

In 1989, CAP established the Molecular Pathology Resource Committee to develop appropriate guidelines for all clinical tests involving DNA probes or other molecular biological techniques. Its scope includes not only DNA genetic diagnostics, but also the use of DNA assays to detect infectious diseases and neoplasms, and for forensic identification. The Committee has administered two DNA-based proficiency testing pilot programs, although their focus was not genetic disorders (66).

Besides CAP, several organizations are poised to facilitate the development of monitoring laboratories through proficiency testing for DNA-based assays. The Council of Regional Networks for Genetic Services (CORN), which receives Federal support, has been active in an array of quality assurance issues for genetic service facilities, including proficiency testing since 1985 (38). The CORN Proficiency Testing Committee sponsored a DNA-based genetic test pilot of 20 laboratories in 1990. The Southeastern region has a regional proficiency testing program, and will be enlarging its planned second survey into a national test, to be completed in 1992; this effort includes CF mutation analysis (100).

The American Society of Human Genetics (ASHG) has recently become active in the area. A joint ASHG/CAP DNA pilot proficiency testing program commenced in 1992. Full proficiency testing is planned by 1994 (5,66,99). ASHG and CORN also have designated liaisons with each other's efforts.

Proficiency testing is widely viewed as an important component of quality assurance. It provides a reliable and identifiable benchmark to assess quality performance; in the past, professional societies' involvements have predominated. Today, each of three principal organizations clearly fills a niche in the evolving area of proficiency testing programs for genetic DNA assays: Historically, CAP has led and administered an array of proficiency testing programs; CORN, with its extensive regional structure and practitioner community emphasis, has long been active in improving education, training, and laboratory quality to improve genetic services delivery; and ASHG has served as the leading national professional society for genetics researchers and service providers. Cooperation among these groups will be essential for the timely development of proficiency testing programs for DNA-based genetic diagnostics. Such cooperation will become increasingly important, since professional programs could affect proficiency testing for CF mutations (and other DNA tests) well before HCFA proposes proficiency testing rules under CLIA (100).

# U.S. FOOD AND DRUG ADMINISTRATION AND MEDICAL DEVICE REGULATION

Today, DNA-based CF tests are done at research laboratories, commercial facilities, public health laboratories, and hospitals. Most attention on ensuring high quality focuses on the institution or individual performing the assay. At some future date, however, DNA-based genetic tests—e.g., for CF mutation analysis—will be marketed widely in the form of kits such as those that exist for pregnancy testing, infectious disease analysis, or forensic DNA identification. At least one U.S. company has begun evaluating a prototype CF mutation test kit in pilot studies (47,48). Cellmark Diagnostics, U.K., is also testing a kit that detects ΔF508 plus three additional mutations (figure 5-1; ch. 10).

The U.S. Food and Drug Administration (FDA) has authority to ensure the safety and efficacy of diagnostic test kits.[1] This section briefly analyzes FDA approval procedures that might apply to new genetic diagnostic kits. A comprehensive analysis of Federal policies and the medical devices industry appears in a 1984 OTA report (168).

## *FDA Authority to Regulate Test Kits*

FDA regulates drugs, devices, and biologics during all phases of their development, testing, production, distribution, and use. Genetic diagnostic kits fall within the definition of a device—i.e., a medical device is a health care product that does not achieve its primary, intended purposes through chemical action in or on the body, or by being metabolized. Thus, the extent to which physicians, genetic counselors, and their clients come to rely on CF mutation analysis kits—or other DNA-based genetic test kits—will depend on FDA regulation of devices.

**Figure 5-1—DNA-Based Test Kit for Cystic Fibrosis Mutations**

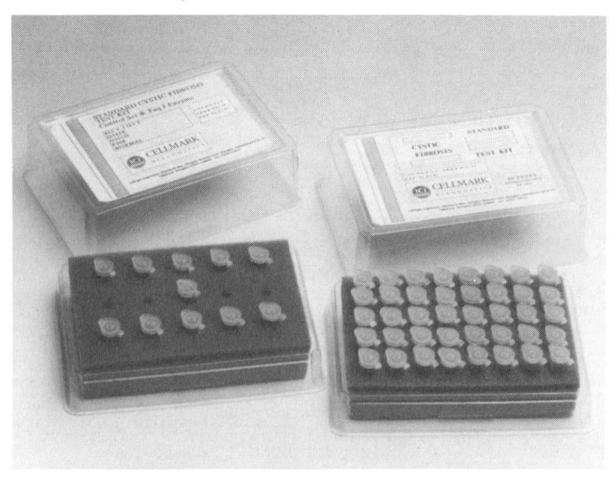

SOURCE: Cellmark Diagnostics (Imperial Chemical Industries PLC), United Kingdom, 1992.

FDA's regulatory options range from registering an item's presence and periodically inspecting facilities to ensure good manufacturing practices, to setting performance and labeling requirements, to premarket review of a device. In addition, the agency may engage in postmarketing surveillance to identify ineffective or dangerous devices; it may ban devices it deems unacceptable.

### The Federal Food, Drug, and Cosmetic Act of 1938

Products such as in vitro DNA diagnostics are regulated under section 351 of the Public Health Service Act (42 U.S.C. 262), but are also subject to the adulteration, misbranding, and registration provisions of the Federal Food, Drug, and Cosmetic Act of 1938 (FFDCA) (21 U.S.C. 301 et seq.). Additionally, "good manufacturing practices" are currently applied to licensed in vitro diagnostics.

### The Medical Device Amendments of 1976 and Safe Medical Devices Act of 1990

The Medical Device Amendments of 1976 (MDA) (Public Law 94-295) and the Safe Medical Devices Act of 1990 (SMDA) (Public Law 101-629) clarified and enlarged the 1938 FFDCA definition of "device" to include items used in diagnosing conditions other than disease (e.g., pregnancy, in vitro diagnostic products), and specific products previously

[1] Though FDA also could have regulated reagents currently used in CF mutation assays, it does not and likely will not. FDA does not regulate reagents unless they are submitted by manufacturers for clearance or approval. Manufacturers of reagents offer them labeled "for investigational use only." Facilities may develop such reagents into analytical procedures, and then offer tests such as CF mutation analysis—and other DNA-based genetic diagnostics—as clinical services. The practices, but not reagents, are regulated under CLIA.

### Box 5-B—FDA Regulation of In Vitro Diagnostic Devices

Under section 510(k) of the Federal Food, Drug, and Cosmetic Act of 1938, manufacturers must notify the U.S. Food and Drug Administration (FDA) 90 days prior to marketing any medical device thought to be substantially equivalent to one legally on the U.S. market. On the basis of this submission, FDA evaluates how similar the new device is to the existing device. (Devices manufactured before passage of the Medical Device Amendments of 1976 (Public Law 94-295) may be exempt from certain regulatory controls.) If FDA finds the proposed device is substantially equivalent, FDA notifies the manufacturer that it can be marketed. Since 1976, 6 percent of new devices underwent stringent premarket review (clinical trials and other demonstrations of safety and effectiveness); 94 percent were reviewed and entered the market on data provided by manufacturers that indicated they were substantially equivalent to existing devices (162).

At present, the majority of biotechnology-based medical devices represent clinical laboratory or in vitro diagnostic applications. In vitro diagnostic devices include reagents, instruments, and systems intended for use in the diagnosis of disease or other conditions—including a determination of the state of health—in order to cure, mitigate, treat, or prevent disease or its sequelae. Manufacturers submit about 1,200 new in vitro diagnostic applications each year to FDA, of which a significant percentage are biotechnology-based.

Although most biotechnology in vitro diagnostic devices submitted are monoclonal antibody-based reagent systems, a number employ DNA technologies, particularly those that detect and identify infectious agents in clinical specimens. The majority of these applications are processed through FDA's 510(k) premarket notification program. Under the 510(k) process, a proposed device may be marketed if it is demonstrated to be substantially equivalent to a legally marketed U.S. product. In many cases, a biotechnology-based in vitro diagnostic device can be shown to be equivalent if the sponsor demonstrates that the new item has essentially equivalent intended use, performance characteristics, and patient risk to an existing product. For example, the first DNA tests for infectious agents were compared to previously cleared 510(k) monoclonal antibody reagents for the same intended uses.

In some instances, comparison to an existing conventional product is not possible and, therefore, introduction raises new types of risk questions that require scientific evaluation of safety and effectiveness through the premarket approval process. In this case, the new product would be classified as Class III, and subject to the regulatory scheme described elsewhere in this chapter. Such was the case for the review of a DNA test for gene rearrangements to assess certain leukemias.

With enactment of the Safe Medical Devices Act of 1990 (Public Law 101-629), manufacturers now introducing a permanently implantable device, a life supporting or life sustaining device, or a device that potentially presents a serious risk to health must conduct postmarket surveillance of the device. FDA may also require any other manufacturer of a device, such as a CF mutation test kit, to conduct postmarket surveillance.

SOURCE: Office of Technology Assessment, 1992, based on K.B Hellman and J.L Hackett, U.S. Food and Drug Administration, Rockville, MD, personal communication, December 1991.

regulated as new drugs (e.g., bone cement, sutures, or soft contact lenses). Based on the 1976 amendments, DNA-based genetic tests would be considered "devices." Box 5-B describes the general regulatory process FDA employs for in vitro diagnostic devices, similar to those under development for CF mutations.

FDA formed the Center for Devices and Radiological Health in 1982 to centralize both the implementation of MDA (and now SMDA) and the development of programs to ensure that unsafe and ineffective medical devices are not sold in the United States. With SMDA, Congress intended that perceived shortcomings in MDA would be addressed. SMDA expands FDA authority to require postmarketing surveillance and to order a temporary or permanent halt to sales of a device in light of postmarketing surveillance results. (FDA's new authority was demonstrated in early 1992 with its consideration of silicone breast implants (73).) SMDA also expands the category of facilities and users required to communicate problems to FDA. MDA/SMDA directs FDA to classify devices into one of three categories, with different levels of regulation applying to each.

***Class I Devices.*** Class I contains devices for which general controls authorized by MDA/SMDA are sufficient to provide a reasonable assurance of safety and effectiveness. Before they can be marketed in the United States, new Class I devices that have not been exempted require premarket notification to FDA demonstrating their substantial equivalence to legally marketed devices. Manufacturers of Class I devices are subject to general controls, meaning they must register their establishments, list the devices with FDA, conform to good manufacturing practices, and submit to periodic inspections (21 U.S.C. 360).

Theoretically, genetic test kits could fall within this first of three classifications. Included in Class I are: chlamydia serological reagents, dye and chemical solution stains, tissue processing equipment, blood bank supplies, and examination gowns. One current Class I product used for genetic diagnosis is the chromosome culture kit, defined as "a device containing the necessary ingredients . . . used to culture tissues for diagnosis of congenital chromosomal abnormalities" (21 CFR 864.2260).

*Class II Devices.* Class II is a regulatory class of devices for which general controls are insufficient to provide a reasonable assurance of safety and effectiveness and for which scientific information is sufficient to establish "special controls" to provide such assurances. The general control provisions of Class I devices also apply to Class II devices, as does the premarket notification requirement. In addition, Class II devices must meet special controls, which can include adherence to performance standards, postmarketing surveillance, establishment of patient registries, and clinical data submission. Older, established genetic test kits not involving DNA methods (e.g., abnormal hemoglobins and alpha-1-antitrypsin assays) have been designated as Class II (21 CFR 862, 864, 868).

In theory, DNA-based genetic diagnostic kits could be classified at this level if FDA determined that general controls, such as good manufacturing practices, are insufficient to give the kits the reliability already exhibited by similar kits classified in Class I. If, for example, FDA considered a DNA-based CF mutation analysis kit similar to abnormal hemoglobin assays, it might classify it as Class II. On the other hand, if FDA finds the reliability of the technologies used in DNA-based diagnostic tests differs substantially, or if the tests raise new issues of safety and effectiveness, FDA could define it as Class III. In fact, concern about the reliability of a DNA-based kit that employs essentially the same methods—PCR and DNA probes—as those that might be developed for CF tests has been raised in criminal court (74).

*Class III Devices.* Devices purported to be "life supporting, life sustaining, or for a use which is of substantial importance in preventing impairment of human health," or "devices which present an unreasonable risk of illness or injury" comprise Class III. In addition to general controls, these products require premarket approval by FDA based on data demonstrating that a device is safe and effective for its intended use. Manufacturers introducing Class III devices since January 1991 have been required to conduct postmarket surveillance. (SMDA additionally empowered FDA to require any other manufacturer of an existing device to conduct postmarket surveillance.)

Examples of Class III devices include a DNA probe to detect the "Philadelphia chromosome" in patients with myelogenous leukemia, tests to detect chromosomal rearrangements in certain immune cells, and maternal serum alpha-fetoprotein (MSAFP) assays for neural tube defects. Class III is an automatic classification level for new devices not yet shown to be substantially equivalent to an existing device on the market—about 2 percent per year. About 5 percent of all medical testing devices are ultimately subject to Class III regulation (51 FR 26342), and it is likely that DNA-based test kits will be categorized as Class III (185,189,190).

*Investigational Device Exemption.* Under the Investigational Device Exemption, FDA may exempt investigational devices from regulatory requirements that might hinder developing scientific data demonstrating safety and effectiveness. In most cases, these clinical studies of medical devices are performed to gather data or to support a premarket notification submission or a premarketing approval application.

### Regulatory Future of Cystic Fibrosis Mutation Test Kits

Experience with other test kits, such as that for MSAFP (box 5-C), could shed light on the regulatory future of CF mutation test kits. On the other hand, congressional concerns about medical device regulation and SMDA have occurred since the debate about MSAFP test kits (152-160,164-166,168), although questions persist about the adequacy of medical device regulation (163). Further,

### *Box 5-C—Maternal Serum Alpha-Fetoprotein Test Kits and the FDA*

In many respects, questions raised in the 1970s and 1980s about screening the serum of pregnant women to determine the concentration of alpha-fetoprotein parallel today's debate about routine carrier screening for cystic fibrosis (CF). (See also ch. 6.) One controversy surrounding maternal serum alpha-fetoprotein (MSAFP) screening involved FDA approval of test kits.

In the 1970s, British medicine had taken the lead in assessing MSAFP screening to detect neural tube defects. Based on a study of 5,800 patients screened for MSAFP in the United Kingdom, the Immunological Panel of the Bureau of Medical Devices, U.S. Food and Drug Administration (FDA), recommended in June 1977 that MSAFP test kits be classified as Class II devices. The panel further recommended that FDA require kits be labeled to indicate that a single positive test did not constitute an accurate diagnosis in and of itself and was insufficient to warrant pregnancy termination, although some panel members (and outside experts) viewed this recommendation as an overextension of FDA authority and an inappropriate attempt to regulate medical practice.

Historically, reagents used in MSAFP screening either qualified under the Investigational Device Exemption (IDE) (21 CFR, part 812) or were not directly regulated because the components were produced within a laboratory for its own diagnostic use. MSAFP test kits, however, were not commercially marketed in the United States prior to the enactment of the Medical Device Amendments of 1976 (MDA) (Public Law 94-295), and thus were subject to MDA. In October 1978, FDA appeared to be on the verge of releasing MSAFP test kits on an unrestricted basis.

Concern about the kits quickly mounted from laboratories, physicians, consumers, and professional societies such as the American College of Obstetricians and Gynecologists and the American Academy of Pediatrics. As with CF carrier screening, objections about the accuracy of the test, the difficulty in interpreting results, and the potential burden from increased caseloads for genetic counselors were raised. Concern was also voiced that anti-abortion groups were influencing FDA to slow approval. Others complained, by contrast, that commercial influences were rushing the move to approve MSAFP kits, which had 1979 sales of $250 million outside the United States. Some contended that FDA's decision to send the device to the Immunological Advisory Group (rather than to a panel of obstetricians and geneticists) resulted in inadequate attention to some of the clinical and programmatic aspects of widespread use of the kits (34).

In February 1979, FDA held a public hearing on MSAFP test kits before two of its advisory committees (the Obstetrics-Gynecologic Device Section and the Immunology Device Section); in August 1979, FDA announced it intended to restrict the sale, distribution, and use of MSAFP test kits. FDA classified MSAFP test kits as Class III, which required premarket approval. Further, premarketing approval applications (PMAs) would not be approved until FDA determined what restrictions, if any, were necessary to ensure the reliability, safety, and efficacy of the kits. Manufacturers were not permitted to distribute MSAFP kits in the United States for investigational use under the IDE. In November 1980, FDA published 13 proposed restrictions for MSAFP test kits (45 FR 74158), and announced public comment would be received at hearings in January 1981.

Despite support for the restrictions—based on concerns about accuracy and efficacy as just described—several objections were raised to FDA's proposed rules. Testimony was offered that 90 percent of hospitals offered MSAFP screening and were using materials not regulated (because they were not kits); that FDA's lack of action on pending PMAs violated due process under MDA; that the proposed regulations extended beyond medical device regulation under MDA into the realm of clinical laboratory regulation (generally the domain of the Health Care Financing Administration and the Centers for Disease Control); and that the proposed regulations thrust FDA into an inappropriate role of regulating private medical practice.

FDA issued revised regulations in 1983, and MSAFP test kits have been widely employed in the United States since that time. Today, the debate is less a matter of the approval of the test kit per se, but on ancillary issues that include the role of State health agencies (35) and whether results generated by small, decentralized laboratories are of lower quality (98, 144).

SOURCE: Office of Technology Assessment, 1992.

because each device is evaluated on a case-by-case basis, the regulatory future of a PCR-based CF (or comparable) test kit remains speculative until one wends its way through the FDA process.

The 1984 medical devices reporting (MDR) regulation and SMDA require a report to FDA of any association between a device and serious injury or death of a patient and could be one level of quality assurance of CF test kits. The regulation, however, is limited to instances of patient death or serious injury. Because use of test kits like those under development for CF is unlikely to result in injury or death to a patient, problems are not likely to be reported on this basis. Kits with poor reliability could, however, lead to unnecessary pregnancy terminations, as well as cause significant emotional harm to patients. The prospect of such pregnancy terminations might prompt FDA to order post-marketing surveillance of CF test kits.

If the MDR regulation and SMDA were to apply to genetic diagnostic kits, they might serve as an early warning system of problems with accuracy and reliability. The congressional General Accounting Office (GAO) found that the MDR regulation generally increased the flow of information about device defects by a factor of 7. Nonetheless, GAO estimated that, prior to SMDA, only one-fourth of manufacturers were in compliance and that FDA was ill-equipped to handle the data flow, data management, and data analysis required (161,162).

In the absence of an actual product, what is the regulatory outlook for test kits like those under development for CF carrier screening? Enhanced postmarketing surveillance under SMDA, coupled with a shift within FDA toward increased regulatory attention to medical devices, might indicate CF mutation analysis kits will be subject to more stringent review than previous non-DNA genetic test kits. FDA recently embarked on a series of measures directed at tightening up regulation and postmarketing surveillance of devices, as well as other items regulated by FDA (73,78). One target of increased regulatory attention, for example, has been monoclonal antibody kits, which are being subjected to increased scrutiny (53). The general change in tone at FDA and accompanying personnel changes have led to consternation among some industry spokespersons (81). Thus, it is difficult to predict how MDA and SMDA will ultimately apply to DNA-based diagnostic test kits.

# GENETIC SERVICES DELIVERY

Delivering high-quality genetic services to clients depends on ensuring a sufficient number of skilled professionals, which in turn demands adequate education and training. Developing and ensuring that high standards are maintained, providing mechanisms to evaluate professional performance, and affording methods for client redress when lapses occur are the subjects of the following section of this chapter. In particular, this section addresses:

- whether primary care physicians (e.g., obstetricians/gynecologists, internists, or family practice specialists) are now expected to discuss CF mutation tests or to provide genetic services related to them as an aspect of routine medical care;
- what all genetic professionals—physicians, genetic counselors, nurses, social workers, or Ph.D. clinical geneticists—are expected to do when counseling individuals about the assays; and
- what remedies exist for consumers harmed by inadequate care.

## *Licensing and Certification*

For genetic tests and information—as for other medical procedures—the quality of care is largely determined by the expertise of the health professionals and the quality of the laboratory services. The expertise and reliability of the providers, in turn, depends on the quality of medical and genetics education (ch. 6) and the quality of State certification, licensure, and discipline of such professionals within its jurisdiction.

Genetics professionals who are physicians are formally licensed by States. The process of medical licensure, making the practice of medicine without a license a criminal offense, both permits individuals to practice medicine and forbids those without a license from competing. As well as providing minimum standards, licensing of physicians provides States with the right to review an individual's practice and to discipline the person. Sanctions range from simple censure to license revocation for failure to follow proper standards in delivering services. As such, licensing can have an impact on the quality of services. A State license is the only one required to practice medicine or any of its specialties. Neither failure to obtain specialty board certification nor failure to maintain membership in a professional medical specialty society in any way limits a physician's legal ability to practice a medical specialty.

Nonetheless, economic and intellectual incentives in the 1930s and 1940s led to the development of certification procedures for specialties, to hospital-based specialty training programs, and finally to the growth of voluntary professional medical specialty societies (143). Genetic counselors and Ph.D. geneticists are not licensed by States, but until 1992 were certified by the American Board of Medical Genetics (ABMG) (as are M.D. geneticists). Beginning in 1993, Ph.D. and M.D. geneticists will be certified by ABMG, but future certification of master's-level counselors is uncertain.

### *Factors Affecting Physician Decisions About Cystic Fibrosis Carrier Screening*

No definitive mechanism exists for determining when physicians should routinely inform people about the availability of tests that could reveal their propensity to have a child with a genetic disorder, such as CF (70). Physician practice may be driven by judgment of what is in a patient's best interest, consumer demand, patient autonomy, liability fears, economic self-interest, or a combination of these factors. CF carrier screening presents a classic instance of the perennial problem of appropriately controlling the evolution of practice standards as a new technology becomes available.

Physicians can now offer individuals with no family history of CF a test that can determine, with 85 to 95 percent sensitivity, whether they are CF carriers. With professional opinion in a state of flux and knowledge of the test's existence continuing to spread among patients, physicians might wonder whether they are obligated to inform patients of its availability, even before patients ask about it. Determining when to routinely inform people about the availability of tests that reveal their propensity for having a child with CF is a contentious issue.

OTA's survey of genetic counselors and nurses in genetics revealed that some consumers are interested in CF carrier screening: about 19 percent of respondents said they were "frequently" or "very frequently" asked by clients about DNA carrier testing or screening for CF (170). On the other hand, some physicians report that consumer willingness to undertake CF carrier screening is modest at present (11,13). This reticence could stem from, in part, resistance to the tests' costs, which patients must generally self-pay.[2] It might also arise from a barrier common to many types of medical screening: lack of interest and reluctance to uncover what might be perceived as potentially unpleasant news (145).

Generally, physicians are obligated to inform patients of the risks and benefits of proposed tests and procedures, so that patients themselves may decide whether to proceed. This obligation extends to diagnostic techniques (150). Where a patient specifically asks about a test, physicians would seem to be obligated to discuss the test, even if they do not recommend that it be taken. Preliminary results from one survey, for example, indicate that up to 90 percent of physicians responding would order a CF carrier test if asked to by a patient (76). Physicians do not appear, however, to be obligated to ask patients about their potential interest in a test or procedure that the physician does not view as warranted by individuals' circumstances (box 5-D) (104), although they are under an obligation to elicit family histories that reveal whether a person is at a particular risk for conceiving a child with a genetic disorder.

A 1989 California appellate court held that a couple, whose family did not appear to have members of an ethnic group at elevated risk for Tay-Sachs disease, had no basis to complain of malpractice when a physician failed to inform them that Tay-Sachs carrier screening is available (104). Expert witnesses advised that the 1 in 167 carrier frequency for Tay-Sachs in the general population was sufficiently low that customary medical practice does not recommend carrier screening for those not at elevated risk—i.e., those who are neither Ashkenazic Jews nor descended from a few other groups with elevated carrier incidence.

For CF, however, the incidence of carriers is more common in the general Caucasian population (1 in 25) than is Tay-Sachs for Ashkenazic Jews (1 in 31) (134). Physicians might ponder whether the 1 in 25 carrier frequency, which results in a 1 in 2,500 incidence of CF among live births in the general population, is sufficiently high that they should inform patients that CF carrier mutation analysis

---

[2] Physicians seeing patients who rely on health insurance to cover part of their costs usually inform them that their coverage generally precludes reimbursement for CF mutation analysis without a family history of the condition (i.e., for screening purposes). OTA recognizes that in the present health care system, and with current reimbursement policies by insurers (ch. 7), the reality is that choosing to be screened usually depends on the ability to self-pay. As mentioned earlier, however, the issue of economic access to CF carrier screening is no different—and inextricably linked—to the broad issue of health care access in the United States (172), a topic beyond the scope of this report. In this report, OTA analyzes the issue in the context of today's health care system, but points out that in the view of some opponents of widespread CF carrier screening, nonuniversal access is an a priori reason for why CF carrier screening should not proceed.

### *Box 5-D—Medical Malpractice and Standards of Care*

Tort law permits individuals to sue those who have negligently caused them harm, achieving financial and emotional compensation for some victims and providing one means of quality assurance in medical practice. Theoretically, making providers responsible for their actions provides an incentive for them to act reasonably and prevent patient harm. In practice, medical malpractice litigation sometimes suffers the shortcoming of juries and judges second guessing past physician practice as a means of stimulating future improvements. In general, tort suits do a better job of enforcing standards after they have been developed. Nevertheless, medical malpractice litigation allows a jury to review the acts of a treating physician, remedy individual grievances, and force development of a good practice standard.

A physician whose treatment complied with the standard of care in the field, i.e., conformed with that offered by the "reasonable prudent physician" (or specialist, if the defendant is a specialist) under the same or similar circumstances, can rarely be found liable for medical malpractice. Statements issued by a relevant professional society are viewed as evidence of what a reasonably prudent physician might have done; so is expert witness testimony (43,51,58,173). Thus, current customs of practice protect physicians. The law assumes, however, that customary medical practice adequately reflects scientific learning and otherwise represents appropriate public policy to be enforced by the courts against individual practitioners (70).

Yet a court can devalue a standard of care by asserting that limited adoption of a practice by some professionals is sufficient to call into question the reasonableness of the defendant's practice—regardless of the extent to which that practice was accepted generally by the profession (40). The plaintiff no longer needs to show a deviation from what the average practitioner would have done. Instead, he or she can establish negligence based only on the defendant's failure to do what some cohort of the same profession was doing (40,188). Even with uniform practice within an industry (147) or profession (75,82,95), conformity with guidelines and customary practice is not an absolute defense because "there are precautions so imperative that even their universal disregard will not excuse their omission" (147).

In the context of medical care, however, only a few courts have followed this reasoning (52,71,96,149). Instead, most courts have deferred to the usual and customary practice of the majority of similarly skilled physicians—sometimes limiting review to local practice standards—when evaluating the actions of a particular physician (37,101,115).

No empirical data exists on current customs of practice about generally informing individuals about CF carrier tests. Physicians are somewhat protected, however, by professional society statements that advise against CF carrier screening for all individuals. On the other hand, because the content of some professional statements are in flux and because the technology changes quickly, a provider might worry that failure to offer the test—or at least to inform couples of the assay's existence—will fall below rapidly evolving customs of care.

Since a variety of professionals provide genetic counseling, another question is whether the same standard of care should apply to all. Generally, each class of health care professionals is held to a separate standard of care (24). But this rule is premised, in part, on the notion that each group performs distinct types of services. Where the service is identical—e.g., CF carrier screening and subsequent counseling about risks by a genetic counselor, nurse, social worker, fertility specialist, obstetrician/gynecologist, internist, or family practitioner—anyone performing the service would be expected to meet at least a common minimum standard of care (24). Where the professional in contact with the patient does not possess the requisite skill, that professional will be under a duty to recognize his or her limitations and refer to the appropriate specialist (90).

SOURCE: Office of Technology Assessment, 1992.

exists. Whether physicians are *obligated* to do so depends, however, on the customary practice of similarly skilled and situated physicians.

With respect to CF carrier screening, customary physician practice might evolve faster than that recommended by physicians' own professional societies (box 5-E), by managed health care facilities or insurance companies, or by government programs. This raises the question of how customary practices develop during a time of diverse opinion. The policy statements of professional societies (6,44,49) and participants at a Government-sponsored workshop (107) all state that CF mutation tests are not recommended for individuals without a family history of CF.

In addition to taking their cues from professional society and government guidelines, physicians might oppose informing patients of the availability of CF

### Box 5-E—Professional Societies and Standards of Care

Professional societies can set voluntary, informal standards for professional behavior, require members to participate in continuing professional education to maintain active membership status, or require periodic examination. They can have codes of ethics governing general behavior, as do the American Medical Association (AMA) and the National Society of Genetic Counselors. A professional organization, such as the new American College of Medical Genetics, can also survey its members and gather data on new techniques. Membership in professional societies is voluntary, as is members' adherence to an organization's code of conduct and standards and participation in membership surveys.

When faced with a complaint about malpractice, courts will generally hold that the customary practice of similarly skilled physicians will be deemed "reasonable" care. To determine what is customary and appropriate, courts often look to guidelines established by the relevant professional societies. Conversely, to protect their members, customary practices are often incorporated in professional statements and guidelines.

Identification of ΔF508 in 1989 resulted in intense speculation about the appropriate standard of care for general population CF carrier screening—speculation that heightens as the assay's capability to detect prevalent mutations improves. At the center of the discussions, professional societies faced the question: Should offering CF carrier screening become the standard of care in medical practice?

While acknowledging that a spectrum of individual opinions exists, the American Society of Human Genetics (ASHG), the largest professional society comprised of members of the human genetics research and clinical communities, issued a statement in 1990 about the timing of widespread carrier screening for CF. ASHG's leadership, based on its own analysis and not a poll of the membership, took the position that routine CF carrier screening is "*NOT* yet the standard of care" (25). The Committee on Obstetrics: Maternal and Fetal Medicine of the American College of Obstetricians and Gynecologists endorsed the ASHG position statement soon thereafter (44), and the AMA has also issued a similar position statement on CF carrier screening (49).

In mid-1992, after extended discussion, ASHG's leadership approved a revised statement that CF mutation analysis "is not recommended at this time" for those without family histories of CF (6). Some argue that the subtle change in language of the 1992 statement retreats from the absoluteness of the 1990 statement. This view holds that the new statement reflects an evolution of debate within the society—that some believe CF carrier screening *may now be offered* to individuals without a family history of CF, although it might not be the "standard of care." Others argue that ASHG's position is unchanged—that the new statement is tantamount to restating that CF carrier screening *should not be offered* to individuals without a family history of CF. In either case, the statement cannot be interpreted to mean that CF carrier screening *should be offered* to all individuals.

Professional statements can exert significant influence beyond helping courts and juries to evaluate malpractice claims. On the basis of the first ASHG statement, at least one commercial facility initially did not promote its CF tests for population screening purposes (56,61), although it appears to do so now. Additionally, OTA's survey of genetic counselors and nurses revealed that 53 percent felt in June 1991 that it was inappropriate to provide CF carrier screening compared to 20.6 percent who believed CF carrier tests for cases of negative family history was appropriate (20.6 percent uncertain); 74 percent of respondents knew of the ASHG statement (versus 31 percent who knew of the NIH statement), and many specifically cited the ASHG statement as influencing their or their institutions' policies.

SOURCE: Office of Technology Assessment, 1992.

carrier screening because they judge that the test is too psychologically risky to be worth any potential benefits to those without a family history of CF. The very existence of prenatal diagnoses can produce stress in potential parents (89). For some patients, tests' availability sharpens otherwise low-level, diffuse concerns that surface only "on bad days," and turns them into real and dreaded possibilities (89). Even with accurate delivery of statistical information concerning the incidence of CF, people can become worried about their carrier status—even if they never worried about it when the test was unavailable. The effects of this concern can be significant, ranging from sleepless nights to hesitation about conceiving or bearing a child (131). Physicians might also decline to screen patients because a third-party payor or managed care provider judges the test to be too expensive for expected benefits.

Opponents of CF carrier screening also argue that inappropriate financial incentives drive the practice—

327-046 - 92 - 5 : QL 3

that physicians paid on a fee-for-service basis find CF mutation tests profitable, as has been the case for other diagnostic procedures (72), or contend that physicians' recommendations might be influenced by laboratories marketing their tests in the same way that pharmaceutical companies currently market drugs to doctors (32,64). Some opponents also express concern that increased CF carrier screening will pressure third-party payors and managed care facilities to provide reimbursement for the test's cost, thereby necessitating a rise in premiums or discontinuation of coverage for other tests that these opponents view as more important.

Some physicians, however, have already chosen to incorporate CF carrier screening into their practices because they disagree with the existing guidelines. They believe the assays are sufficiently sensitive for general use, and that even patients with unknown risk of conceiving a child with CF should now have the information to exercise choice in managing their health care. Still other physicians might be offering the assay out of concern that failing to could subject them to charges of medical malpractice if a couple has a child with CF and a court subsequently finds that CF carrier screening had indeed become the standard of care, despite professional statements to the contrary (19). They may worry and practice "defensive medicine" (171), afraid that the growing practice of offering the test to self-paying patients—those who have specifically asked about and those who have not—sets a de facto (and therefore de jure) standard of care for all individuals (box 5-F).

Concerns about defensive medicine are especially important because, although courts look to professional society statements for evidence of practice standards, in the end it is the actual practice of similarly skilled professionals that tends to set the minimum threshold for reasonable care. Defensive medicine has been blamed for the proliferation of many other medical tests and procedures of limited value to certain populations. The problem is particularly acute with regard to procedures performed in the context of reproductive medicine, since the birth of a baby with severe medical problems can result in substantial damage awards to cover medical expenses of the child's projected lifespan (79). CF carrier screening seems to fall squarely within this concern (57).

As of mid-1992, customary medical practice has not evolved to routine CF carrier screening. Nor has any court had occasion to consider whether the standard of care for good medicine requires CF carrier screening. To date, the statements of professional societies have slowed the adoption of such a standard of care by signaling to physicians, third-party payors, and courts that CF carrier screening is not necessary to meet definitions of reasonable care. On the other hand, while no empirical data exist on current customs of practice about informing individuals in a clinical setting about the availability CF carrier tests, trends in the number of assays performed suggest increasing numbers of providers are informing individuals about their availability (ch. 2). Whether such practices will be sustained—and hence become the standard of care—is unclear. But if doubts about the appropriateness of CF carrier assays fade, an obligation to offer them to all individuals is likely to heighten (128).

Clearly, a balance among professional guidelines, physician views, and patient demand must be struck with regard to CF carrier screening. Overall, physicians acting on behalf of individuals will establish customs of care. Nevertheless, standard setting in the area of medical practice is diffuse and generally unregulated. In the end it might be up to courts and juries to determine, on a case-by-case basis through retrospective review in the context of medical malpractice litigation, what level of care is owed to patients.

### *Duties of Care for Genetic Counseling*

Genetic counseling requires professionals to educate patients about the availability of genetic services, to elicit enough information to determine whether patients are in particular need of genetic tests, to help patients decide whether genetic information would be useful to them, especially in light of their personal and religious values, and to assist patients in obtaining quality genetic analysis if desired.

A decision to offer information about tests for CF carrier screening—or to provide the assay itself—raises questions: What constitutes quality genetic counseling? What about confidentiality of information obtained in the course of counseling?

### *Box 5-F—The Maine Medical Liability Demonstration Project: An Alternative Approach to Set Practice Standards?*

In response to concerns about defensive medicine, Maine enacted legislation in 1990 that creates a demonstration project designed to ensure high quality medical care, to reduce costs associated with medical malpractice litigation, and to decrease incentives to practice defensive medicine. The project hopes to accomplish these goals by having groups of physicians work with representatives of patients and insurers to form consensus opinions on practice standards in defined areas of medical care. These practice standards are then available to participating physicians in the form of professional education. If a participating physician complies with the practice standards, then he or she is largely protected from claims of medical malpractice.

Advisory committees in a particular area of practice will be composed of experts relevant to the area, as well as public members. For example, the Medical Specialty Advisory Committee on Obstetrics and Gynecology consists of nine members, including six physicians representing diverse interests (e.g., a tertiary hospital, mid-sized hospital, and rural practice) and three public members (one representing the interests of payors of medical costs, one representing consumers, and a representative of allied health professionals).

Each medical specialty advisory committee shall develop practice parameters and risk management protocols in the area relating to that committee. Practice parameters must define appropriate clinical indications and methods of treatment within that specialty. Risk management protocols must establish standards of practice designed to avoid malpractice claims and increase the defensibility of those that are pursued. Once the medical specialty advisory committee recommends a set of actions, the Board of Registration in Medicine and the Board of Osteopathic Examination and Registration shall review and approve the parameters and protocols for each medical specialty area, and adopt them as rules under the Maine Administrative Procedure Act. Each medical specialty advisory committee shall also provide a report to the Legislature setting forth the parameters and protocols that have been adopted, and describe the extent to which the risk management protocols reduce the practice of defensive medicine.

For claims of professional negligence against a participating physician (or the employer of a participating physician) that allege a violation of the standard of care, only the participating physician (or the physician's employer) may introduce into evidence, as an affirmative defense, the existence of the practice parameters and risk management protocols developed and adopted for that medical specialty area. Unless independently developed from a source other than the demonstration project, the practice parameters and risk management protocols are not admissible in evidence in a lawsuit against any physician who is not a participant in the demonstration project.

For malpractice policies beginning on or after July 1, 1990, the State superintendent of insurance shall determine the amount of the savings in professional liability insurance claims and claim settlement costs to insurers anticipated in each 12-month period as a result of the project. A portion of the savings could be subject to an assessment that would be used to address other health care needs of the State.

The Maine project represents an innovative approach to questions raised about appropriate medical practice standards. It formalizes the role of professional societies in establishing standards of care, giving them statutory authority and protection. It also expands the decisionmaking process to explicitly include members of the public. As the project progresses, it could provide interesting perspectives and results in the area of standards of care and medical liability to policymakers.

SOURCE: Office of Technology Assessment, 1992, based on Title 24. Insurance, Chapter 21. Maine Health Security Act, Subchapter IX. Medical Liability Demonstration Project, 24 M.R.S. 2971 (1990).

## Pretesting and Prescreening

To meet standards of responsible care, a genetics professional must understand enough about the patient's health and his or her reproductive plans. The provider must also be aware of what technologies are available to take an appropriate family history and proceed with necessary analyses. For a nonspecialist, it might be enough to recognize the need for a referral (24,106). These tasks have become more difficult as the timing for genetic counseling has moved back from after the birth of an affected child to prior to a patient's first conception. Thus, the usual signal of a patient at risk (i.e., the birth of an affected child) would not be present.

Today, providers might elicit information concerning a patient's plans with regard to children and family history for a wide variety of detectable disorders, some of which are quite rare, during general checkups and annual gynecological visits. This would then be followed by client education to help individuals decide whether they wish to pursue further counseling or tests. Less than that could give patients grounds to complain of a false assurance of safety.

### Counseling

Having elicited information and obtained test results, the professional's next task is to communicate the results in a meaningful way. An important aspect of this task involves explaining the reproductive risks clients face. Because statistical information can be difficult to understand, this responsibility is more complex than merely stating the odds that a child will be born with a genetic condition.

All professionals who do genetic counseling—genetic counselors, nurses, doctors, social workers, and clinical geneticists—realize that translating technically accurate information into understandable information is essential and difficult (63,91, 114,124,130,137,179). Making information meaningful to nonprofessionals includes :

- supplying an accurate, or at least tentative, diagnosis;
- explaining the pattern of inheritance along with known uncertainties;
- recognizing and understanding psychosocial and ethnocultural issues;
- presenting the range of therapeutic options for treatment and management of the disorder;
- offering options for further diagnostic tests, if available; and
- counseling on medical options for preventing the birth of affected children, if desired by the patient (1,7,24,111,123,178).

Judging whether information has been delivered in a sufficiently comprehensible way is not simple. People interpret information about genetic risk in a highly personal manner (93,183), and a counselee can misperceive, misunderstand, or distort information. Such an effect could have significant emotional impact that affects the individual's decisionmaking or adjustment to the circumstance (131,176,179). Some consumers could perceive that a negative result from the use of "cutting edge" DNA technology means no risk, thus mistakenly interpreting the assay's resolution. Still others might believe that administration of the test itself conveys protection from risk.

In one study on risk communication and patient interpretation, over one-quarter of women surveyed could not correctly explain the meaning of 1 out of 1,000. Of those who gave the correct answer, 16 percent said the defect occurred "often or occasionally," versus "rarely or very rarely." Thirty-one percent of those who incorrectly answered the question judged the defect occurred "often or occasionally" (26). Another study of clients showed that those perceiving their numerical risk as higher than others who were at the same risk were more likely to ask the genetic counselor about having another child. At the same time, patients tend to interpret a given mathematical risk as "low" more often than do the counselors describing it (183). Finally, leaving the mathematical ability of patients aside, parents' perceptions of uncertainty in genetic counseling significantly affects qualitative decisions they make (94).

Photo credit: Beth Fine

A genetic counselor discusses results with clients. Genetic counseling can help individuals and families understand the implications of positive and negative test outcomes.

Given the nuances of information delivery and reception, and differences in situations encountered, is there a standard for genetic counseling? One commentator argues in favor of a standard for genetic counseling based on what patients would want to know (modeled after informed consent requirements) because there is no fixed professional norm as an alternative, and because adequacy of the information conveyed turns more on the values of the patient being counseled than on professional norms: "It seems proper for a counselor to aim at informing the counselees about everything the latter would find material to the decision they have to make as determined on the basis of a lay, not expert, standard'' (24). The more prevalent standard, however, appears to be based on a review of what most professionals do, rather than what an individual patient wants (29,87). The problem with relying solely on professional custom is that standards are still evolving, and there are distinctive schools of thought about methods of counseling (78,137,142).

In light of the 1991 Supreme Court decision in *Rust* v. *Sullivan* (132), it is unlikely that patients at a clinic receiving Federal funds from Title X of the Public Health Service Act could easily receive information about the option of choosing to terminate a pregnancy that is at risk of resulting in the birth of a child with a genetic disorder. This judicial decision runs counter to beliefs of many in the genetics community: ASHG members overwhelmingly assert that genetic counseling about all reproductive options is imperative (4). *Rust* upheld Federal regulation requiring clinics receiving these monies to respond to all inquiries concerning abortion by stating that abortion is not an appropriate family planning option. In the context of genetic counseling, many argued that the rule is inconsistent with the standards of medical care required by State malpractice statutes and cases—i.e., that abortion is within the range of options that physicians and health care providers are expected to disclose to their patients (23). In March 1992, the rule was reinterpreted to permit physicians to discuss abortion, but it does not permit them to counsel where it can be obtained. Nurses and counselors may not discuss abortion. As the vast majority of interactions at clinics receiving Title X funds are between patients and nurses, the reinterpretation will have limited effect on counseling practices.

Depending on the condition, pregnancy termination is chosen by 57 to 97 percent of parents who learn that a fetus will be born with a genetic disorder (16,41,50,62). A 1990 general survey of the American public found 32 percent of respondents would undergo an abortion if a genetic test proved positive; an additional 18 percent reported that having an abortion would depend on the nature of the defect (140). With respect to CF specifically, a recent sampling in New England of parents who had a child with CF found 80 percent would continue a pregnancy even if prenatal testing determined that the fetus was affected, although the majority said abortion should be a legally available option for those who utilize such testing (181,182). Another study of families who had children with CF in the Rochester, NY area found 56 percent approved that a woman should have the option of terminating CF-affected pregnancies, but the question did not ask whether they, themselves, would (102). No data on general population attitudes (no family history) toward CF carrier screening, prenatal diagnosis, and abortion exist, but data are accumulating from a limited number of experiences in CF carrier screening pilots. Indications are that, compared to couples with family histories of CF, fewer individuals from the general population would continue an affected pregnancy (14,21).

The 1992 U.S. Supreme Court decision in *Planned Parenthood of Southeastern Pennsylvania* v. *Casey*[3] means abortion now turns largely on State law, although it also appears that absolute obstacles to such a choice will be held unconstitutional. At issue was the 1973 decision in *Roe* v. *Wade* (129), which held a woman's liberty of conscience and bodily integrity may not be sacrificed to State interests in protecting fetal life. The 1992 decision announced such liberty remains protected from State efforts to prohibit abortion, but States now may make a woman's choice to terminate a pregnancy following prenatal diagnosis more financially, medically, or emotionally difficult unless the restriction is a substantial obstacle to choosing abortion. This new standard, the ''undue burden'' test, represents a retreat from the standard in *Roe*, which held that abortion was a fundamental right. As a fundamental right, abortion was protected from all State impediments except those based on a compelling need and implemented in the least restrictive manner possible—the ''strict scrutiny'' test. The new standard requires empirical data as to the effect of State regulations on women's ability to choose pregnancy terminations. Thus, State regulations restricting or shaping genetic counseling might be evaluated to determine if they pose a ''substantial'' obstacle to a woman's choice of pregnancy termination. The 1992 opinion appears to tolerate a potentially wide range of State laws that might be enacted to discourage women from using prenatal testing or aborting affected fetuses. The decision explicitly upholds the constitutionality of State preferences for childbirth over abortion throughout pregnancy, and not merely following viability.

[3] *Planned Parenthood of Southeastern Pennsylvania* v. *Casey*, __S.Ct.__ (1992), 60 U.S.L.W. 4795.

## Keeping Genetic Information Confidential

Both ethics and law obligate professionals with information on the carrier or affected status of a patient to keep that information confidential absent a few, specific exceptions (3,8,60,88,187). Professional codes of ethics guide practitioners (3,109), as do State statutes or case law in States with no specific statutory authority (27,67,68,77,110,139). Not all genetic information must remain confidential, however.

A patient is presumed to have consented to disclosure to his or her partner, for example, if the individual comes in for genetic counseling as part of a couple and the initial history is performed on both partners. Similarly, consent to disclosure is generally assumed when test results are raised as an issue in a malpractice suit or in an appeal for benefits denied by a government agency (126). State statutes also require that certain medical findings be reported, such as certain birth defects or communicable diseases, sexual and physical abuse of children, gunshot wounds, and drug abuse (8,60). At least 12 States require that birth defects be reported to a registry (Florida, Indiana, Iowa, Louisiana, Maryland, Michigan, Minnesota, New Jersey, Virginia, Washington, West Virginia, and Wisconsin).[4]

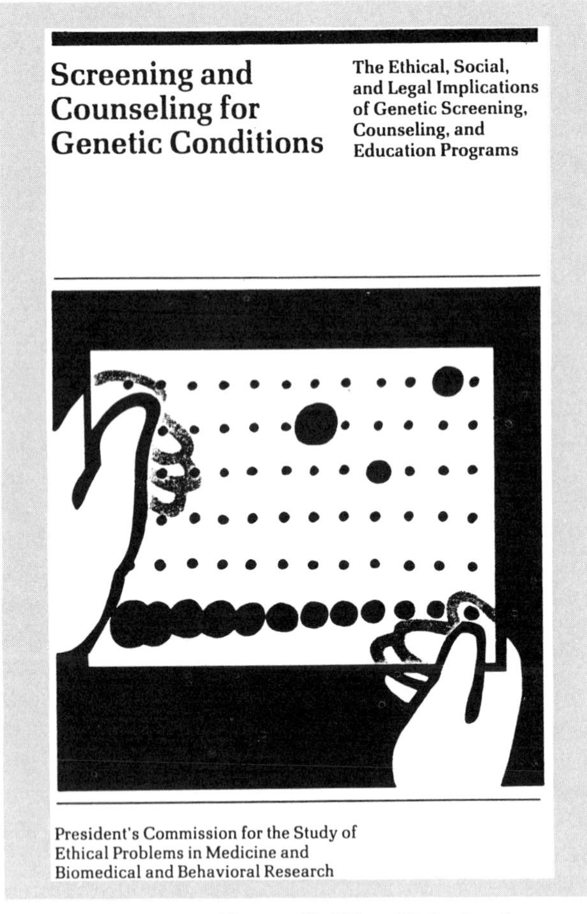


Photo credit: Office of Technology Assessment

A 1983 report of the President's Commission for the Study of Ethical Problems in Medicine and Biomedical and Behavioral Research.

The closest analogy to a situation in which a provider might wish to reveal genetic information to third parties without a patient's permission appears to be the case of disclosure of communicable diseases. As with disclosing a risk of infection, the motivation would be to spare third parties the ill effects of the disorder. In the case of CF, this would entail a wish to help relatives of the patient be aware that they, too, could be at higher than average risk of conceiving a child with CF. If, after a patient has been advised to inform relatives that they could carry a CF mutation, the provider is persuaded that the relatives will not be notified, he or she may want to breach confidentiality. The impulse to breach confidentiality could be legal as well as humanitarian. Genetics providers might be concerned that they have a legal duty to protect third parties from intentional, foreseeable harm when they know that one of their patients poses a threat (146), although courts still narrowly construe such a duty.

From an ethical perspective, the 1983 President's Commission delineated four conditions that should be satisfied before overriding confidentiality with regards to genetic information. First, reasonable efforts to elicit voluntary consent for disclosure have failed. Second, a high probability exists that harm will occur if the information is withheld and that the disclosed information will actually be used to avert harm. Third, the harm that identifiable individuals would suffer would be serious. Fourth, appropriate precautions are taken to ensure that only the genetic

[4] Florida Stat. Ann. Sec. 411.203(9)(b) (West 1986); Florida Admin. Code. Ann. R. 10J-8.007 (1990) and Guideline VII(b) (1981); Indiana Stat. Ann. Sec. 16-4-10-1 (Michie 1990); Iowa Code Ann. Sec. 136A.6 (West 1989); Louisiana Department of Health and Human Services, Guidelines: Neonatal Screening, Sec. III(F) (1988); Maryland Regs. Code Tit. 10, Sec. 10.38.11 (1975); Michigan Comp. Laws 333.5721 (1990); Minnesota Reg. 7.1.172(c)(2)(b) (1979); New Jersey Stat. Sec. 26:8-40.22 (1989); Virginia Code Ann. 32.1-69.1 (1990); Wash. Rev. Code Sec. 70.58.320 (1990); West Virginia Code Sec. 16-5-12a (Michie 1990); Wisconsin Stat. Sec. 146.028 (1987-88).

information needed for diagnosis or treatment of the condition in question is disclosed (119).

Evaluating CF carrier screening in light of the President's Commission's criteria reveals that:

- The patient is not the source of the danger to the third party, as is the case with communicable disease. Rather, the patient's carrier status is merely an indication that the mutation is present in the family and that each blood relative is at increased risk of being a carrier.
- Even if the relative is a carrier, it poses no threat to his or her person because only those carrying two copies of the mutation exhibit any ill effects. Being a carrier carries no personal threat of illness.
- Being a carrier does not pose a problem in reproductive planning unless the relative's partner is also a carrier.

Overall, then, the risk to the third party from nondisclosure must be balanced against the benefit of maintaining the expected confidentiality of the provider-patient setting. A provider contemplating breach of confidentiality and disclosure to a patient's spouse must weigh the patient's own confidentiality against the spouse's interest in sharing decisions concerning conception, abortion, or preparation for the birth of a child with extraordinary medical needs. For CF, the chances of harm also must be evaluated: The spouse must also be a CF carrier for the probability of having a child with CF to rise to 25 percent.

In actual practice, a recent international survey of M.D. and Ph.D. geneticists revealed that 60 percent of respondents in the United States (and 66 percent from 17 other countries) said they would disclose a child's diagnosis of hemophilia A to interested maternal relatives who might be at risk for conceiving children with hemophilia A—against the wishes of the client. Twenty-four percent of respondents would seek out relatives and tell them even if they did not ask for the information (180). (Hemophilia A is an X-linked, recessive disorder, and hence carriers are only female. For CF, both paternal and maternal relatives could be carriers. Hence, the situation is not strictly equivalent, but is illustrative for CF carrier screening.)

Finally, as important as maintaining confidentiality of CF carrier status within families, is confidentiality with respect to third parties. At least one life insurance group acknowledges that existing mechanisms to maintain confidentiality of genetic test information might not be appropriate, and that special protection might be necessary (2). Recent evidence indicates that the Cystic Fibrosis Foundation (CFF), as part of its annual survey of CF centers, has begun to request information on mutation status (31,174). CFF releases only aggregate information for research purposes, but because data are delivered by centers to CFF without express consent in a fashion that is identifiable to individuals and their families, questions about the advisability of this practice have arisen. A sampling of institutions reveals that institutional review boards have not reviewed such practices nor been consulted about releasing such information (31).

## *Compensation for Inadequate Genetic Counseling*

Practitioners who provide inadequate genetic counseling, including failing to prescribe needed tests or failing to keep results confidential, might be subject to sanctions by a regulatory body or a professional society. As with other areas of civil law enforcement, sanctions can range from a mild reprimand to revocation of applicable licenses to practice. Courts also can issue an injunction to prevent a practitioner from disclosing certain information, on pain of being found in contempt of court. Finally, courts (and juries) can award monetary compensation for out-of-pocket expenses and mental or physical suffering to those harmed by poor counseling. In particularly outrageous cases, punitive damages may be assessed against a defendant.

### Failure to Adequately Test or Counsel

People are human and mistakes are made. But what happens when the birth of an affected child occurs because a health professional breaches a duty to adequately test or counsel a client? Increasingly, courts have become arbiters of whether a health care provider has met his or her professional obligations to a patient, which has increased the impact of judicial decisionmaking on quality assurance of professional services.

Inadequate counseling for genetic tests can result in a number of outcomes. First, patients might forego conception or terminate a pregnancy when correct information would have reassured them. Second, people might choose to conceive children when they otherwise would have practiced contraception, they might fail to investigate using donor gametes that are free of the genetic trait they wish to avoid, or they might lose the opportunity to choose to terminate a pregnancy. The latter situations could result in the unwanted birth of an affected child.

Medical personnel have a duty to provide information to expectant parents so that they can be fully informed about any reproductive decision they choose to make (17,20,24,39,86,105,125,127). Courts have also occasionally recognized a duty for health care providers (and even parents) to prevent persons not yet conceived from being conceived in a manner that would result in their birth under conditions where they would suffer serious genetic or congenital disorders (36,69).

***Wrongful Birth and Wrongful Life.*** Wrongful birth and wrongful life are the terms used to describe the forms of malpractice claims that arise from the birth of an affected child. In a wrongful birth claim, parents assert that failure to receive timely, accurate information robbed them of the opportunity to avoid conception or birth of an affected child. Wrongful birth claims can result in special damages (usually medical expenses and other special costs associated with the care of an affected child) and general damages (those encompassing all the ordinary costs of raising the child). Since the 1973 *Roe* v. *Wade* decision (129), courts tend to award special damages when a case has merit.[5] Most courts remain reluctant to award general damages.

Although some courts have rejected the wrongful birth claim altogether (10,86), most jurisdictions allow compensation to parents for the negligent failure to inform or to provide correct information in time for them to either prevent conception or to decide whether to terminate a pregnancy if a fetus shows evidence of a genetic disorder. Rarely have State legislatures acted to categorically deny parents access to wrongful birth claims,[6] except to forbid claims based on allegations that the parents would have terminated the pregnancy had they been adequately counseled and tested.[7] Such State statutes limit parental claims for wrongful birth to cases of preconception counseling and testing that result in parental loss of opportunity to forego conception in favor of adoption or the use of donor gametes.

With regard to CF, at least one court has ruled that parents might collect the extra medical costs associated with managing the condition. In this case, the couple complained that they would have avoided conceiving a second child had their physicians accurately diagnosed CF in their first child and thus realized both parents were carriers (136).

In wrongful life claims, the child asserts he or she was harmed by the failure to give the parents an opportunity to avoid conception or birth, because never having existed would be better than to exist with severe disabilities. U.S. courts, however, have been reluctant to allow damages because most have been uncomfortable with any decision that hints nonexistence might be preferable to life, even when that life includes pain and suffering.

As of 1991, courts in only three States recognized a cause of action for wrongful life: California (151), New Jersey (120), and Washington (69). The status of the cause of action is unclear in Louisiana and Indiana, whose courts have held that physicians do have a duty to advise parents prior to conception that they have an elevated risk of giving birth to a severely afflicted child and that a wrongful life action might be an appropriate remedy for the child (33,118).[8] In contrast, 21 States have judicially rejected a common law cause of action for wrongful

[5] Should *Roe* v. *Wade* ultimately be overturned, wrongful birth and wrongful life cases would again turn largely on State abortion law. Where abortion becomes illegal, State courts could conclude that failure to inform a woman of significant fetal abnormalities does not deprive her of the choice to terminate the pregnancy, as that choice is foreclosed by State law. Since it is unlikely, however, that all States would outlaw abortion, the ability to travel to a jurisdiction where abortion remained legal could lead courts to conclude that an opportunity nonetheless had been lost due to a faulty diagnosis of genetic impairment. Thus, while wrongful life and wrongful birth claims might be weakened by overturning *Roe* v. *Wade*, they would not be eliminated. The ASHG recently endorsed model statutory language designed to protect the reproductive options of women at risk for bearing children with serious genetic or congenital disorders (5).

[6] Colorado Stat. Sec. 13-64-502.

[7] Minnesota Stat. Ann. Sec. 145; Missouri Ann. Stat. Sec. 188.130; South Dakota Stat. Sec. 21-55-2; Utah Stat. Sec. 78-11-24 (30).

[8] In the Indiana case, the court held that the State statute prohibiting wrongful life suits applied only when the child asserted he or she should have been aborted. By its terms, the statute did not prohibit suits that claim the child should never have been conceived (33).

[9] Alabama (45); Arizona (177); Colorado (92); Delaware (59); Florida (103); Georgia (9,55); Idaho (18); Illinois (138); Kansas (22); Kentucky (135); Massachusetts (175); Michigan (121); Missouri (186); New Hampshire (141); New York (15); North Carolina (10); Pennsylvania (46); South Carolina (117); Texas (112); West Virginia (80); Wisconsin (42).

life.[9] At least eight State statutes prohibit a cause of action against a physician for wrongful life.[10]

Overall, then, parents can sue successfully for extraordinary medical costs associated with the birth of a child with a disability whom they would not have conceived or carried to term if they had received timely, accurate information about the risks the pregnancy posed to the affected child and to their own emotional and financial stability. Ordinary costs of raising the infant, however, usually are not reimbursed. Children suing on their own behalf for wrongful life are far less successful; most courts are unable to conclude that they have been harmed by living with severe disabilities when the only alternative is never to have lived.

### Breach of Confidentiality

At least 21 State statutes explicitly protect patient information regarding medical conditions and treatment.[11] Offending physicians can have their licenses revoked or be subject to other disciplinary action. Four of these States—Illinois, New Mexico, North Dakota, and Oregon—punish both negligent and willful disclosures. Idaho and Michigan do not differentiate between the two types of disclosure, and the remainder punish only willful breaches of confidentiality (8).

Patients whose confidential records have been revealed can also bring civil suit against the physician or facility for tortious public disclosure of private facts (122). This is not the same as a suit for defamation, which requires that the information divulged be false; it merely requires that the disclosure offend community standards of decency or expectations of privacy (12,68). Like defamation, however, the plaintiff must demonstrate an actual harm before compensation can be awarded.

Other civil suits a patient could bring for breach of confidentiality include a breach of contract action. While not common, such suits have been recognized as legitimate by State courts (68,77,97). They are premised on the notion that the provider-patient relationship is contractual, and that breach of contract litigation may be used to enforce the implied contract of confidentiality—for example through an injunction or, alternatively, to obtain financial redress following an unauthorized disclosure (8,113). Actions brought under breach of contract would also be possible against employees and nonphysician health care workers, either because these individuals are party to the contractual relationship (e.g., clerks at a medical facility) or under a theory of respondent superior (24).

Some suits for unauthorized disclosure could be premised on Federal and State guarantees of a right to privacy, thus limiting the ability of government agencies, such as health departments, to obtain medical records (116,133). In addition, several States have statutes that protect the confidentiality of medical records, independent from State licensure and discipline legislation.[12]

Health care professionals who release genetic or other medical information about a patient, however, would not be legally liable to that patient or subject to disciplinary action if there were a valid defense to the action. Such defenses would include the consent of the patient, waiver of the right to object to disclosure, the need to comply with a valid State or Federal law, or, at times, the need to prevent physical harm to a third party, as discussed earlier.

## SUMMARY AND CONCLUSIONS

Quality assurance for CF carrier screening means ensuring the safety and efficacy of the tests themselves, whether they are performed de novo in

[10] Indiana Code Ann. Sec. 34-1-1-11 (Burns Supp. 1989); Maine Rev. Stat. Ann. Tit. 24, Sec. 2931 (Supp. 1989); Minnesota Stat. Ann. Sec. 145.424 (West 1990); Missouri Ann. Stat. Sec. 188.130-1 (Vernon Supp. 1990); North Dakota Cent. Code Sec. 32-03-43 (Supp. 1989); Pennsylvania 42 C.S. Sec. 8305 (Supp. 1991); South Dakota Codified Laws Sec. 21-55-1 (1987); Utah Code Ann. Sec. 78-11-24 (Supp. 1989).

[11] Arizona, Rev. Stat. Ann. Secs. 32-1401(20)(b), -1451 (Supp. 1990); Arkansas, Stat. Ann. Sec. 17-93-409(15) (1987); California, Bus. and Prof. Code Secs. 2227, 2228 (West Supp. 1991) and 2263 (West Supp. 1990); Delaware, Code Ann. Tit. 24, Sec. 173(a), (b)(12) (1987); Idaho, Code Sec. 54-1814(13) (1988); Illinois, Ann. Stat. ch. 111, para. 4400-22(A)(30) (Smith-Hurd Supp. 1990); Kansas, Stat. Ann. Sec. 65.2386(c) (Supp. 1989); Kentucky, Rev. Stat. Ann. Sec. 311-595 (Baldwin Supp. 1990); Maine, Rev. Stat. Ann. Tit. 32, Sec. 3282-A(2) (Supp. 1990); Michigan, Comp. Laws Ann. Sec. 333.16221(e)(ii) (West Supp. 1990); Minnesota, Stat. Ann. Sec. 147-091(1)(m) (West 1989); Nebraska, Rev. Stat. Secs. 71-147(10), 71-148(9) (1990); Nevada, Rev. Stat. Ann. Sec. 630-3065(1) (1989); New Mexico, Stat. Ann. Sec. 61-6-15(D)(5) (1989); North Dakota, Cent. Code Sec. 43-17-31(13) (Supp. 1990); Ohio, Rev. Code Ann. Sec. 4731.22 (B)(4) (Baldwin 1987); Oklahoma, Stat. Ann. Tit. 59, Secs. 503, 509(4) (West 1989); Oregon, Rev. Stat. Sec. 677.190(5) (1989); South Dakota, Codified Laws Ann. Secs. 36-4-29 (1986) and 36-4-30(4) (Supp. 1990); Tennessee, Code Ann. Sec. 63-1-120(15) (1990); Utah, Code Ann. Secs. 58-12-35(1)(a), 56-12-36(3) (1990).

[12] California Civil Code Sec. 56.10 (West 1988); Montana Code Ann. Sec. 50-15-525 (1987); Rhode Island Gen'l Laws Secs. 5-37.3-1 to .3-11 (1987); Wisconsin Stat. Ann. Sec. 146.82 (West 1987).

clinical diagnostic laboratories or via test kits. It also encompasses guarantees for accurate interpretation of the test results by health care professionals. Ensuring that consumers receive high-quality technical and professional service for DNA-based CF carrier tests is the responsibility of providers, under the shared oversight of the Federal Government, State and local governments, private entities, including professional societies, and the courts.

Quality assurance to assess clinical laboratory performance is still in flux, in large measure because the 1967 legislation governing regulation of clinical testing facilities was overhauled by Congress in 1988. Rulemaking by the executive branch is under way for some aspects of clinical laboratory regulation, but not others. The Health Care Finance Administration hopes to propose rules, for example, on proficiency testing, a key quality assurance component, by the end of 1992.

The Clinical Laboratory Improvement Amendments of 1988 clearly encompass facilities performing DNA-based genetic analyses. But, while CLIA details particular performance standards for several types of clinical diagnostic procedures, it does not specifically address DNA-based tests. This lack of detailed directives for DNA-based diagnostics could be a strength in the short-term, since the field is rapidly changing. Whereas a predominant Federal role appears the likely result for certain clinical laboratory protocols, multiple stakeholders might ultimately share oversight of DNA-based genetic assays (e.g., CF carrier tests). For example, the efforts of New York State in the area of genetic test laboratory certification and proficiency testing could influence the Federal approach to regulating genetic analyses. Similarly, the impact of professional organizations (e.g., the College of American Pathologists, the Council of Regional Networks for Genetic Services, and the American Society of Human Genetics) on proficiency testing will be important.

To a certain extent, truly broad dissemination of CF carrier screening depends on the availability of test kits now under development. Such kits will be subject to regulation by the U.S. Food and Drug Administration. Since no DNA-based genetic diagnostic test kit comparable to that being developed for CF carrier assays exists, it is difficult to predict what regulatory status will evolve for such kits. Two events could, however, serve as a gauge: The enactment of the Safe Medical Devices Act of 1990, which reflected congressional response to concern about FDA's oversight of medical devices, and indications that FDA is increasing medical device regulation and postmarketing surveillance. If congressional intent is served and increased FDA scrutiny is extended to DNA-based diagnostic test kits, developers can expect more stringent regulation of their products than previous non-DNA genetic tests (e.g., assays for abnormal hemoglobinopathies). Increased regulation could, in turn, slow the implementation of widespread CF carrier screening, since the availability of an easy, quick kit—similar to what exists for maternal serum alpha-fetoprotein screening—would otherwise facilitate screening individuals in primary care settings.

Finally, quality assurance for CF carrier screening ultimately depends on the interaction of the health care professional with the client. Customs of care are still evolving regarding the obligation of physicians, genetic counselors, and other health professionals to inform individuals about the availability of CF carrier screening. Although professional societies and government advisory bodies currently state that CF mutation assays are too imperfect to be used in the general population, physicians are nonetheless free to offer information and screening. Absent consistent resistance on the part of insurers to reimburse for the assays, it would appear that practitioner interest, patient demand, and the perceived threat of medical malpractice litigation will encourage some physicians and genetic counselors to offer information about CF carrier screening to a larger population than that recommended by their own professional societies. The increase in information concerning patients' genetic backgrounds can be expected to increase the number of situations in which health professionals will need to balance confidentiality of patient information against demand from relatives and other third parties for access to that information.

## CHAPTER 5 REFERENCES

1. Allen, C.C., Murray, R.F., Quinton, B.A., et al., "Adapting Genetic Counseling to Meet the Needs of Black Americans in the District of Columbia," *Journal of Genetic Counseling* 1:93-94, 1992.
2. American Council of Life Insurance, *Genetic Test Information and Insurance: Confidentiality Concerns and Recommendations* (Washington, DC: American Council of Life Insurance, 1992).

3. American Medical Association, Current Opinions of the Judicial Council of the American Medical Association (1984).
4. American Society of Human Genetics, "American Society of Human Genetics Statement on Clinical Genetics and Freedom of Choice," *American Journal of Human Genetics* 48:1011, 1991.
5. American Society of Human Genetics, "Genetic Services Committee Update: Report on ASHG/CAP Joint Proficiency Testing" (Rockville, MD: American Society of Human Genetics, 1991).
6. American Society of Human Genetics, "Statement of the American Society of Human Genetics on Cystic Fibrosis Carrier Screening," personal communication, June 1992.
7. American Society of Human Genetics, Ad Hoc Committee on Genetic Counseling, "Genetic Counseling," *American Journal of Human Genetics* 27:240-241, 1975.
8. Andrews, L., and Jaeger, A., "Confidentiality of Genetic Information in the Workplace," *American Journal of Law and Medicine* 17:75-108, 1991.
9. *Atlanta Obstetrics and Gynecology Group* v. *Abelson*, 955-2-C (S.Ct. Ga. 12/5/90).
10. *Azzolino* v. *Dingfelder*, 315 N.C. 103, 337 S.E. 2d 528 (1985), cert. denied 479 U.S. 835 (1986).
11. Barathur, R., Roche Biomedical Laboratories, Raritan, NJ, personal communication, September 1991.
12. *Bazemore* v. *Savannah Hospital*, 171 Ga. 257, 155 S.E. 194 (1930).
13. Beaudet, A.L., Howard Hughes Medical Institute, Houston, TX, remarks at the 8th International Congress of Human Genetics, Washington, DC, October 1991.
14. Beaduet, A.L., Howard Hughes Medical Institute, Houston, TX, personal communication, March 1992.
15. *Becker* v. *Schwartz*, 46 N.Y. 2d 401, 386 N.E. 2d 807, 413 N.Y.S. 2d 895 (1978).
16. Benn, P.A., Hsu, L.Y.F., Carlson, A., et al., "The Centralized Prenatal Genetics Screening Program of New York City: III, The First 7,000 Cases," *American Journal of Medical Genetics* 20:369-384, 1985.
17. *Bergstresser* v. *Mitchell*, 577 F.2d 22 (8th Cir. 1978).
18. *Blake* v. *Cruz*, 108 Idaho 253, 698 P.2d 315 (1984).
19. Blumenthal, D., and Zeckhauser, R., "Genetic Diagnosis: Implications for Medical Practice," *International Journal of Technology Assessment in Health Care* 5:579-600, 1989.
20. *Bonbrest* v. *Kotz*, 65 F. Supp. 138 (D.D.C. 1946).
21. Botkin, J.R., and Alemagno, S., "Carrier Screening for Cystic Fibrosis: A Pilot Study of the Attitudes of Pregnant Women," *American Journal of Public Health* 82:723-725, 1992.
22. *Bruggeman* v. *Schimke*, 239 Kan. 245, 718 P.2d 635 (1986).
23. *Canterbury* v. *Spence*, 464 F.2d 772 (D.C. Cir.), cert. denied, 409 U.S. 1064 (1972).
24. Capron, A.M., "Tort Liability in Genetic Counseling," *Columbia Law Review* 79:618-684, 1979.
25. Caskey, C.T., Kaback, M.M., and Beaudet, A.L., "The American Society of Human Genetics Statement on Cystic Fibrosis Screening," *American Journal of Human Genetics* 46:393, 1990.
26. Chase, G.A., Faden, R.R., Holtzman, N.A., et al., "Assessment of Risk by Pregnant Women: Implications for Genetic Counseling and Education," *Social Biology* 33:57-64, 1986.
27. *Clark* v. *Geraci*, 29 Misc. 2d 791, 208 N.Y.S. 2d 564 (Sup. Ct. 1960).
28. Clarke, A., "Is Non-Directive Genetic Counselling Possible?," *Lancet* 338:998-1001, 1991.
29. Clayton, E.W., The Vanderbilt Clinic, Nashville, TN, personal communication, December 1991.
30. Clayton, E.W., "Reproductive Genetic Testing: Regulatory and Liability Issues," *Fetal Diagnosis and Therapy* in press, 1992.
31. Collins, D., University of Kansas Medical Center, Kansas City, KS, personal communication, February 1992.
32. *Consumer Reports*, "Pushing Drugs to Doctors," February 1992, pp. 87-94.
33. *Cowe* v. *Forum Group, Inc.*, 541 N.E. 2d 962 (Ind. App. 1989).
34. Cunningham, G.C., Genetic Disease Branch, California Department of Health Services, Berkeley, CA, personal communication, December 1991.
35. Cunningham, G.C., and Kizer, K.W., "Maternal Serum Alpha-Fetoprotein Screening Activities of State Health Agencies: A Survey," *American Journal of Human Genetics* 47:899-903, 1990.
36. *Curlender* v. *Bioscience Laboratories*, 106 Cal. App. 3d 811, 165 Cal. Rptr. 477 (1980).
37. Danzon, P., *Medical Malpractice: Theory, Evidence, and Public Policy* (Cambridge, MA: Harvard University Press, 1985).
38. Davis, J.G., New York Hospital, New York, NY, personal communication, October 1991.
39. DeWalt, T., "Wrongful Life and Wrongful Birth: The Current Status of Negligent Genetic Counseling," *Cooley Law Review* 3:175-191, 1985.
40. *Doe* v. *University Hospital of the New York University Medical Center*, slip op. (N.Y. Sup. Ct., 7/31/90).
41. Drugan, A., Greb, A., Johnson, M.P., et al., "Determinants of Parental Decisions to Abort for Chromosome Abnormalities," *Prenatal Diagnosis* 10:483-490, 1990.
42. *Dumer* v. *St. Michael's Hospital*, 69 Wis. 2d 766, 233 N.W. 2d 372 (1975).

43. *Edwards* v. *United States*, 519 F.2d 1137 (5th Cir. 1975).
44. Elias, S., Annas, G.J., and Simpson, J.L., "Carrier Screening for Cystic Fibrosis: Implications for Obstetric and Gynecologic Practice," *American Journal of Obstetrics and Gynecology* 164:1077-1083, 1991.
45. *Elliot* v. *Brown*, 361 So. 2d 546 (Ala. 1978).
46. *Ellis* v. *Sherman*, 512 Pa. 14, 515 A.2d 1327 (1986).
47. Erlich, H.A., Roche Molecular Systems, Inc., Emeryville, CA, personal communication, September 1991.
48. Erlich, H.A., Gelfand, D., and Sninsky, J.J., "Recent Advances in the Polymerase Chain Reaction," *Science* 252:1643-1651, 1991.
49. Evans, R.M., "Report of the Council on Scientific Affairs: Carrier Screening for Cystic Fibrosis," *Proceedings of the American Medical Association House of Delegates Annual Meeting*, June 1991, pp. 330-337.
50. Faden, R.R., Chwalow, A.J., Quaid, K., et al., "Prenatal Screening and Pregnant Women's Attitudes Toward the Abortion of Defective Fetuses," *American Journal of Public Health* 77:288-290, 1987.
51. *Faircloth* v. *Lamb-Grays Harbor Co.*, 467 F.2d 685 (5th Cir. 1972).
52. *Favalora* v. *Aetna Casualty & Surety Co.*, 144 So. 2d 544 (La. App. 1962).
53. Fleisher, L.D., Sidley & Austin, Chicago, IL, personal communication, September 1991.
54. Fost, N., University of Wisconsin School of Medicine, Madison, WI, remarks at U.S. Congress, Office of Technology Assessment workshop, "Cystic Fibrosis, Genetic Screening, and Insurance," Feb. 1, 1991.
55. *Fulton-Dekalb Hospital Authority* v. *Graves*, 252 Ga. 441, 314 S.E. 2d 653 (1984).
56. Funk, R.L., GeneScreen, Dallas, TX, personal communication, March 1991.
57. Furrow, B.R., Widener University School of Law, Wilmington, DE, personal communication, December 1991.
58. *Gardner* v. *General Motors Corp.*, 507 F.2d 525 (10th Cir. 1974).
59. *Garrison* v. *Medical Center of Delaware, Inc.*, No. 193, 1988 (Sup. Ct. Del. 12/12/89).
60. Gellman, R.M., "Prescribing Privacy: The Uncertain Role of the Physician in Protecting Privacy," *North Carolina Law Review* 62:255-295, 1984.
61. GeneScreen, "Two Years Later . . . Seven Mutations in the CF Panel," *In the Genome*, Addendum, summer 1991.
62. Golbus, M.S., Loughman, W.D., Epstein, C.J., et al., "Prenatal Diagnosis in 3,000 Amniocenteses," *New England Journal of Medicine* 300:157-163, 1979.
63. Gordon, H., "Genetic Counseling: Considerations for Talking to Parents and Prospective Parents," *Journal of the American Medical Association* 217: 1215-1225, 1971.
64. Gorman, C., "Can Drug Firms Be Trusted?," *Time* 139:42-46, Feb. 10, 1992.
65. Greenstein, R.M., University of Connecticut Health Center, Farmington, CT, personal communication, February 1992.
66. Grody, W.W., "Status of the CAP Program in Molecular Diagnosis," meeting abstract, Biotechnology and the Diagnosis of Genetic Disease: Forum on the Technical, Regulatory, and Societal Issues, Arlington, VA, Apr. 18-20, 1991.
67. *Hague* v. *Williams*, 37 N.J. 328, 181 A.2d 345 (1962).
68. *Hammonds* v. *Aetna Casualty & Surety Co.*, 243 F. Supp. 793 (N.D. Ohio 1966).
69. *Harbeson* v. *Parke-Davis*, 98 Wash. 2d. 460, 656 P.2d 483 (1983).
70. Havighurst, C.C., Duke University School of Law, Durham, NC, personal communication, March 1992.
71. *Helling* v. *Carey*, 83 Wash. 2d 514, 519 P.2d 981 S.Ct. Wash. (1974).
72. Hillman, B.J., Joseph, C.A., Mabry, M.R., et al., "Frequency and Costs of Diagnostic Imaging in Office Practice—A Comparison of Self-Referring and Radiologist-Referring Physicians," *New England Journal of Medicine* 363:1604-1608, 1990.
73. Hilts, P.J., "U.S. Cracks Down on Health Devices Made Before 1976," *New York Times*, Feb. 24, 1992.
74. Hodel, M.B., "California Physician Says DNA Test Kits Unreliable," *Associated Press*, June 27, 1991.
75. *Hoemke* v. *New York Blood Center*, 912 F.2d 550 (2d Cir. 1990).
76. Holtzman, N.A., Johns Hopkins University Hospital, personal communication, December 1991.
77. *Horne* v. *Patton*, 291 Ala. 701, 287 So. 2d 824 (1973).
78. Ingersoll, B., "Changes Vowed for Reviewing Medical Devices: FDA Says It Will Improve Its Regulation Program After Charges of Lapses," *Wall Street Journal*, Mar. 26, 1992.
79. Institute of Medicine, *Medical Professional Liability and the Delivery of Obstetrical Care (Vol. I)* (Washington, DC: National Academy Press, 1989).
80. *James G.* v. *Caserta*, 332 S.E. 2d 872 (W.Va. 1985).
81. Kaplan, S., "Agency Door Revolves: As the FDA Gets Going, Food-and-Drug Bar Gets Rich," *Legal Times*, July 29, 1991, p. 1.
82. *Kemmerlin* v. *Wingate*, 274 S.C. 62, 261 S.E. 2d 50 (1979).

83. Kilshaw, D., "Quality Assurance. 1. Philosophy and Basic Principles," *Medical Laboratory Sciences* 43:377-381, 1986.
84. Kilshaw, D., "Quality Assurance. 2. Internal Quality Control," *Medical Laboratory Sciences* 44:73-83, 1987.
85. Kilshaw, D., "Quality Assurance. 3. External Quality Assessment," *Medical Laboratory Sciences* 44:178-186, 1987.
86. King, P.A., "The Juridical Status of the Fetus: A Proposal for the Legal Protection of the Unborn," *Michigan Law Review* 77:1647-1687, 1979.
87. Knoppers, B.M., University of Montreal School of Law, Montreal, Canada, personal communication, January 1992.
88. Kobrin, J.A., "Confidentiality of Genetic Information," *University of California—Los Angeles Law Review* 30:1283-1315, 1983.
89. Kolker, A., "Advances in Prenatal Diagnosis: Social Psychological and Policy Issues, *International Journal of Technology Assessment in Health Care* 5:601-617, 1989.
90. *Larsen* v. *Yelle*, 246 N.W. 2d 841 (Minn. 1976).
91. Leonard, C.O., Chase, G.A., and Childs, B., "Genetic Counseling: A Consumers' View," *New England Journal of Medicine* 287:433-439, 1972.
92. *Lininger* v. *Eisenbaum*, 764 P.2d 1202 (Colo. 1988).
93. Lippman-Hand, A., and Fraser, F.C., "Genetic Counseling: Provision and Reception of Information," *American Journal of Medical Genetics* 3:113-127, 1979.
94. Lippman-Hand, A., and Fraser, F.C., "Genetic Counseling—The Post Counseling Period: I. Parents' Perceptions of Uncertainty," *American Journal of Medical Genetics* 4:51-71, 1979.
95. *Longoria* v. *McAllen Methodist Hospital*, 772 S.W. 2d 663 (Tex. App. 1989).
96. *Lundahl* v. *Rockford Memorial Hospital Association*, 93 Ill. App. 2d 461, 235 N.E. 2d 671 (1968).
97. *MacDonald* v. *Clinger*, 84 A.2d 482, 444 N.Y.S. 2d 801 (1982).
98. Macri, J.N., Kasturi, R.V., Krantz, D.A., et al., "Maternal Serum Alpha-Fetoprotein Screening. II. Pitfalls in Low-Volume Decentralized Laboratory Performance," *American Journal of Obstetrics and Gynecology* 156:533-534, 1987.
99. Matteson, K.J., "DNA Based Genetic Testing: Current Proficiency Testing Activities," meeting abstract, Biotechnology and the Diagnosis of Genetic Disease: Forum on the Technical, Regulatory, and Societal Issues, Arlington, VA, Apr. 18-20, 1991.
100. Matteson, K.J., University of Tennessee Medical Center-Knoxville, Knoxville, TN, personal communication, December 1991.
101. McCoid, A.H., "The Care Required of Medical Practitioners," *Vanderbilt Law Review* 12:549-632, 1959.
102. Miller, S.R., and Schwartz, R.H., "Attitudes Toward Genetic Testing of Amish, Mennonite, and Hutterite Families With Cystic Fibrosis," *American Journal of Public Health* 82:236-242, 1992.
103. *Moores* v. *Lucas*, 405 So. 2d 1022 (Fl. Dist. Ct. App. 1981).
104. *Munro* v. *Regents of the University of California*, 215 Cal. App. 3d 977, 263 Cal. Rptr. 878 (1989).
105. Myers, J.E.B., "Abuse and Neglect of the Unborn: Can the State Intervene?," *Duquesne Law Review* 23:1-76, 1984.
106. *Naccarato* v. *Grob*, 384 Mich. 248 (1970).
107. National Institutes of Health, Workshop on Population Screening for the Cystic Fibrosis Gene, "Statement From the National Institutes of Health Workshop on Population Screening for the Cystic Fibrosis Gene," *New England Journal of Medicine* 323:70-71, 1990.
108. *National Intelligence Report* 13(2), Oct. 28, 1991.
109. National Society of Genetic Counselors, "National Society of Genetic Counselors Code of Ethics," *Journal of Genetic Counseling* 1:41-43, 1992.
110. *Nationwide Mutual Insurance Co.* v. *Jackson*, 10 Ohio App. 2d 137, 226 N.E. 2d 760 (1967).
111. Naveed, M., Phadke, S.R., Sharma, A., et al., "Sociocultural Problems in Genetic Counselling," *Journal of Medical Genetics* 29:140, 1992.
112. *Nelson* v. *Krusen*, 678 S.W. 2d 918 (Tex. 1984).
113. Newman, S., "Privacy in Personal Medical Information: A Diagnosis," *University of Florida Law Review* 33:394-424, 1981.
114. Note, "Father and Mother Know Best: Defining the Liability of Physicians for Inadequate Genetic Counseling," *Yale Law Journal* 87:1488-1524, 1978.
115. Pearson, R.N., "The Role of Custom in Medical Malpractice Cases," *Indiana Law Journal* 51:528-557, 1976.
116. *Peninsula Counseling Center* v. *Rahm*, 105 Wash. 2d 929, 719 P.2d 926 (1986).
117. *Phillips* v. *United States*, 508 F. Supp. 537 (D.S.C. 1980).
118. *Pitre* v. *Opelousas General Hospital*, 530 So. 2d 1151 (La. 1988).
119. President's Commission for the Study of Ethical Problems in Medicine and Biomedical and Behavioral Research, *Screening and Counseling for Genetic Conditions: The Ethical, Social, and Legal Implications of Genetic Screening, Counseling, and Education Programs* (Washington, DC: U.S. Government Printing Office, 1983).
120. *Procanik by Procanik* v. *Cillon*, 97 N.J. 339, 478 A.2d 755 (1984).

121. *Proffitt* v. *Bartolo*, 162 Mich. App. 35, 412 N.W. 2d 232 (1987).
122. Prosser, W., and Keeton, W., "The Law of Torts," 5th ed. (St. Paul, MN: West Publishing, 1984).
123. Rapp, R., "Chromosomes and Communication: The Discourse of Genetic Counseling," *Medical Anthropology Quarterly* 2:143-157, 1988.
124. Reilly, P., "Genetic Counseling and the Law," *Houston Law Review* 12:640-675, 1975.
125. *Renslow* v. *Mennonite Hospital*, 67 Ill. 2d 348, 10 Ill. Dec. 484, 367 N.E. 2d 1250 (1977).
126. Riskin, L., and Reilly, P., "Remedies for Improper Disclosure by Genetic Data Banks," *Rutgers-Camden Law Review* 8:480-506, 1977.
127. Robertson, H.B., "Toward Rational Boundaries of Tort Liability for Injury to the Unborn: Prenatal Injuries, Preconception Injuries and Wrongful Life," *Duke Law Journal* 6:1401-1457, 1978.
128. Robertson, J.A., "Legal Issues in Genetic Testing," *The Genome, Ethics, and the Law: Issues in Genetic Testing* (Washington, DC: American Association for the Advancement of Science, 1992).
129. *Roe* v. *Wade*, 410 U.S. 113 (1973).
130. Rosenstock, I.M., "The Evaluation of Genetic Counseling: A Committee Report," *Public Health Reporter* 92:332-335, 1977.
131. Rothman, B.K., *The Tentative Pregnancy* (New York, NY: Viking Penguin, 1986).
132. *Rust* v. *Sullivan*, 111 S.Ct. 1759 (1991).
133. *Sanderson* v. *Bryan*, 361 Pa. Super. 491, 522 A.2d 1138 (1987).
134. Sandhoff, K., Conzelmann, E., Neufeld, E.F., et al., "The $G_{M2}$ Gangliosidoses," The Metabolic Basis of Inherited Disease, C.R. Scriver, A.L. Beaudet, W.S. Sly, et al., (eds.) (New York, NY: McGraw Hill, 1989).
135. *Schork* v. *Huber*, 648 S.W. 2d 861 (Ky. 1983).
136. *Schroeder* v. *Perkel*, 87 N.J. 53, 432A A.2d 834 (1981).
137. Shiloh, S., and Sagi, M., "Effect of Framing on the Perception of Genetic Recurrence Risks," *American Journal of Medical Genetics* 33:130-135, 1989.
138. *Siemieniec* v. *Lutheran General Hospital*, 117 Ill. 2d 230, 111 Ill. Dec. 302, 512 N.E. 2d 691 (1987).
139. *Simonson* v. *Swenson*, 104 Neb. 224, 177 N.W. 831 (1920).
140. Singer, E., "Public Attitudes Toward Genetic Testing," *Population Research and Policy Review* 10:235-255, 1991.
141. *Smith* v. *Cote*, 128 N.H. 231, 513 A.2d 341 (1986).
142. Sorenson, J.R., Swazey, J.P., and Scotch, N.A., *Reproductive Pasts, Reproductive Futures: Genetic Counseling and Its Effectiveness* (New York, NY: Alan R. Liss, Inc., 1981).
143. Starr, P., *The Social Transformation of Medicine* (New York, NY: Basic Books, 1982).
144. Strickland, D.M., Butzin, C.A., and Wians, F.H., Jr., "Maternal Serum Alpha-Fetoprotein Screening: Further Consideration of Low-Volume Testing," *American Journal of Obstetrics and Gynecology* 164:711-714, 1991.
145. Swee, D.E., and Micek-Galinat, L., "Part I: Basic Approach to Health Screening," *New Jersey Medicine* 88:421-426, 1991.
146. *Tarasoff* v. *Regents of the University of California*, 17 Cal. 3d 425 131 Cal. Rptr. 14, 551 P.2d 334, 83 A.L.R. 3d 1166 (1976).
147. *The T.J. Hooper*, 60 F.3d 737 (2d Cir. 1932).
148. Tirone, A.J., Office of Survey and Certification, Health Standards, and Quality Bureau, Health Care Financing Administration, letter to M.S. Watson, Cytogenetics Laboratory, St. Louis Children's Hospital, Oct. 24, 1991.
149. *Toth* v. *Community Hospital at Glen Cove*, 22 N.Y. 2d 255, 292 N.Y.S. 2d 440, 239 N.E.2d 368 (1968).
150. *Truman* v. *Thomas*, 27 Cal. 3d 285, 165 Cal. Rptr. 308, 611 P.2d 902 (S.C. Cal. 1980)
151. *Turpin* v. *Sortini*, 31 Cal. 3d 220, 182 Cal. Rptr. 337, 643 P.2d 954 (1982).
152. U.S. Congress, General Accounting Office, *Federal Regulation of Medical Devices—Problems Still To Be Overcome*, GAO/HRD-83-53 (Gaithersburg, MD: General Accounting Office, September 1983).
153. U.S. Congress, General Accounting Office, *Medical Devices: Early Warning of Problems is Hampered by Severe Underreporting*, GAO/PEMD-87-1 (Gaithersburg, MD: U.S. General Accounting Office, December 1986).
154. U.S. Congress, General Accounting Office, *Medical Devices: Early Warning of Problems is Hampered by Severe Underreporting*, Statement of Eleanor Chelimsky, Assistant Comptroller General, GAO/PEMD-87-4 (Gaithersburg, MD: U.S. General Accounting Office, May 1987).
155. U.S. Congress, General Accounting Office, *Medical Devices: FDA's Forecasts of Problem Reports and FTEs Under H.R. 4640*, GAO/PEMD-88-30 (Gaithersburg, MD: U.S. General Accounting Office, July 1988).
156. U.S. Congress, General Accounting Office, *Medical Devices: FDA's 501(k) Operations Could Be Improved*, GAO/PEMD-88-14 (Gaithersburg, MD: U.S. General Accounting Office, August 1988).
157. U.S. Congress, General Accounting Office, *Medical Devices: FDA's Implementation of the Medical Device Reporting Regulation*, GAO/PEMD-89-10 (Gaithersburg, MD: U.S. General Accounting Office, February 1989).
158. U.S. Congress, General Accounting Office, *Medical Device Recalls: An Overview and Analysis 1983-88*, GAO/PEMD-89-15BR (Gaithersburg, MD: U.S. General Accounting Office, August 1989).

159. U.S. Congress, General Accounting Office, *FDA Resources: Comprehensive Assessment of Staffing, Facilities, and Equipment Needed*, GAO/HRD-89-142 (Gaithersburg, MD: U.S. General Accounting Office, September 1989).
160. U.S. Congress, General Accounting Office, *Medical Device Recalls: Examination of Selected Cases*, GAO/PEMD-90-6 (Gaithersburg, MD: U.S. General Accounting Office, October 1989).
161. U.S. Congress, General Accounting Office, *Medical Devices: The Public Health at Risk*, statement of Charles Bowsher, Comptroller General, GAO/T-PEMD-90-2 (Gaithersburg, MD: U.S. General Accounting Office, November 1989).
162. U.S. Congress, General Accounting Office, *Medical Devices: Underreporting of Problems, Backlogged Systems, and Weak Statutory Support*, Statement of Eleanor Chelimsky, Assistant Comptroller General, GAO/T-PEMD-90-3 (Gaithersburg, MD: U.S. General Accounting Office, November 1989).
163. U.S. Congress, General Accounting Office, *Medical Technology: Quality Assurance Needs Stronger Management Emphasis and Higher Priority*, GAO/PEMD-92-10 (Gaithersburg, MD: U.S. General Accounting Office, February 1992).
164. U.S. Congress, House of Representatives, Committee on Energy and Commerce, Subcommittee on Health and the Environment, *Health and the Environment Miscellaneous, Part 2, Oversight Hearing on the Medical Device Amendments of 1976*, hearing Feb. 24, 1984, serial No. 98-108 (Washington, DC, U.S. Government Printing Office, 1984).
165. U.S. Congress, House of Representatives, Committee on Energy and Commerce, Subcommittee on Oversight and Investigations, *FDA Oversight: Medical Devices*, hearing July 16, 1982, serial No. 97-144 (Washington, DC: U.S. Government Printing Office, 1982).
166. U.S. Congress, House of Representatives, Committee on Energy and Commerce, Subcommittee on Oversight and Investigations, *Medical Device Regulation: The FDA's Neglected Child*, Committee Print 98-F (Washington, DC: U.S. Government Printing Office, May 1983).
167. U.S Congress, House of Representatives, Committee on Energy and Commerce, Subcommittee on Oversight and Investigations, "CLIA Oversight," hearing, May 2, 1991.
168. U.S. Congress, Office of Technology Assessment, *Federal Policies and the Medical Devices Industry*, OTA-H-229 (Springfield, VA: National Technical Information Service, October 1984).
169. U.S. Congress, Office of Technology Assessment, *Genetic Witness: Forensic Uses of DNA Tests*, OTA-BA-438 (Washington, DC: U.S. Government Printing Office, July 1990).
170. U.S. Congress, Office of Technology Assessment, *Cystic Fibrosis and Genetic Screening: Policies, Practices, and Attitudes of Genetic Counselors—Results of a Survey*, OTA-BP-BA-97 (Washington, DC: U.S. Government Printing Office, forthcoming 1992).
171. U.S. Congress, Office of Technology Assessment, *Defensive Medicine and the Use of Medical Technology*, forthcoming 1993.
172. U.S. Congress, Office of Technology Assessment, *Technology, Insurance, and the Health Care System*, forthcoming 1993.
173. *United States ex rel Fear* v. *Rundle*, 506 F.2d 331 (3rd Cir. 1974), cert. denied 421 U.S. 1012 (1975).
174. Valverde, K.D., The Cystic Fibrosis Center, St. Vincent's Hospital and Medical Center, New York, NY, personal communication, February 1992.
175. *Viccaro* v. *Milunsky*, 406 Mass. 777, 551 N.E. 2d 8 (Mar. 1, 1990).
176. Vlek, C., "Risk Assessment, Risk Perception and Decision Making About Courses of Action Involving Genetic Risk: An Overview of Concepts and Methods," *Decision Making Under Uncertainty*, R. Scholz (ed.) (Amsterdam, The Netherlands: North Holland Publishing Co., 1983).
177. *Walker* v. *Mart*, 790 P.2d 735 (1990).
178. Wang, V., and Marsh, F.H., "Ethical Principles and Cultural Integrity in Health Care Delivery: Asian Ethnocultural Perspectives in Genetic Services," *Journal of Genetic Counseling* 1:81-92, 1992.
179. Weil, J., "Mothers' Postcounseling Beliefs About the Causes of Their Children's Genetic Disorders," *American Journal of Human Genetics* 48:145-153, 1991.
180. Wertz, D.C., Fletcher, J.C., and Mulvihill, J.J., "Medical Geneticists Confront Ethical Dilemmas: Cross-Cultural Comparisons Among 18 Nations," *American Journal of Human Genetics* 45:1200-1213, 1990.
181. Wertz, D.C., Janes, S.R., Rosenfield, J.M., et al., "Attitudes Toward the Prenatal Diagnosis of Cystic Fibrosis in Decision Making Among Affected Families, *American Journal of Human Genetics* 50:1077-1085, 1992.
182. Wertz, D.C., Rosenfield, J.M., Janes, S.R., et al., "Attitudes Toward Abortion Among Parents of Children With Cystic Fibrosis," *American Journal of Public Health* 81:922-996, 1991.
183. Wertz, D.C., Sorenson, J.R., and Heeren, T.C., "Genetic Counseling and Reproductive Uncertainty," *American Journal of Medical Genetics* 18:79-88, 1984.
184. Willey, A.M., "Genetics Laboratory Quality Assurance—Overview," meeting abstract, Biotechnology and the Diagnosis of Genetic Disease: Forum

on the Technical, Regulatory, and Societal Issues, Arlington, VA, Apr. 18-20, 1991.
185. Willey, A.M., New York State Department of Health, Albany, NY, personal communication, January 1992.
186. *Wilson* v. *Kuenzi*, 751 S.W. 2d 741 (Mo. 1988).
187. Winslade, W.J., "Confidentiality of Medical Records," *Journal of Legal Medicine* 3:497-533, 1982.
188. Wolf, E.L., "Blood Collector Liability and the Professional Standard of Care," *Courts, Health Science & the Law* 2:184-188, 1991.
189. Yoder, F., U.S. Food and Drug Administration, Rockville, MD, remarks at the 8th International Congress of Human Genetics, Washington, DC, October 1991.
190. Yoder, F., U.S. Food and Drug Administration, Rockville, MD, personal communication, February 1992.

Chapter 6

# Education and Counseling

# Contents



## *Boxes*



## *Figures*



## *Tables*

# Chapter 6
# Education and Counseling

Major questions raised by the implementation of routine CF carrier screening include:

- Can pretest education and post-test counseling adequately inform participants and prevent harm?
- What is the role of public education as a source of pretest information?
- Who should be offered screening and in what setting?
- Can confidentiality be assured?
- Is quality assurance possible?

The answers to these questions will surface as time passes and the results of several pilot projects in the United Kingdom (ch. 10) and the United States become available. How education for cystic fibrosis (CF) carrier screening is conducted will likely serve as a prototype for screening for other autosomal recessive disorders. This chapter examines the educational and counseling issues that will need addressing as CF carrier screening becomes more widespread, as well as the roles of various health professionals.

There are no mandatory genetic screening programs of adult populations in the United States. In this regard, OTA finds it highly unlikely that CF carrier screening will set a precedent. In a 1991 OTA survey of 431 genetic counselors and nurses in genetics, 99 percent of those who responded said CF carrier screening should be voluntary (71). Thus, this chapter assumes voluntary screening and examines factors in delivering CF carrier screening to a large group of Americans. Topics discussed are: the complexity of pretest education and post-test counseling for CF carrier status; the capacity of the professional clinical genetics community to assume the primary role in the provision of counseling; the most appropriate sites, facilities, and resources for screening; and contributions to be made by CF carrier screening pilot projects.

## THE NEED FOR SUFFICIENT AND APPROPRIATE EDUCATION AND COUNSELING

For some individuals, even considering whether to undergo genetic screening or testing constitutes a potential life crisis because of the possible outcomes. If the results are positive, the crisis is exacerbated. How results affect an individual has much to do with the person's own frame of reference, but also with the implications of the condition and its prognosis.

Psychological issues permeate every aspect of genetic consultation. Information received can be ego-threatening, or even life-threatening, as individuals find they are "flawed," "imperfect," "defective," "inadequate," or "abnormal," and could potentially transmit these "flaws" to their progeny (37). How the information is obtained, communicated, retained, and eventually used by the person being tested or screened involves a "series of complex, multidimensional processes with major rational and nonrational components" (37).

Beyond the intrapsychic consequences of receiving genetic information are the potential impacts on family. Genetic information affects not only the individual, but also the partner, parents, grandparents, siblings, and children of the individual being tested or screened. Social and psychological stress, as well as future financial and emotional burdens, can strain family functioning (61).

The psychological impact of a positive diagnosis varies with its severity, treatability, and potential for different families to react uniquely to similar situations. Support, counseling, and followup can assist individuals and their families in coping with positive test results. The knowledge and skills of a properly trained counselor can help the individual understand the diagnosis, risk, prognosis, and relevant preventive and therapeutic measures, and also aid in communicating important information to other family members. When these goals are accomplished, genetic counseling is usually perceived as a valuable experience by the counselee and the professional (50).

### *Pretest Education*

In routine genetic counseling, the genetics professional elicits the reasons for testing or screening and discusses the implications of possible outcomes. The counselor prepares the individual for both positive and negative test results. It is also the time

Photo credit: Peter T. Rowley, University of Rochester School of Medicine

Educational materials, such as this pamphlet developed at the University of Rochester School of Medicine, Rochester, NY, can be useful for pretest education.

to discuss risk reduction strategies, if relevant, and the nature and severity of the disorder for which the test is being done. Counseling for CF carrier screening, if conducted in the typical genetic services setting, is no different from routine counseling for other disorders for which tests have less-than-perfect sensitivity.

Most studies of genetic counseling have focused on cases where the client already has an affected child or relative and is familiar with the disorder. Experience with Tay-Sachs, sickle cell anemia, and β-thalassemia screening provides some information about the effects of education prior to screening people with no previous family history of the condition for which they are being screened (app. B). Clearly, without pretest education—and in some cases even after such education—misperceptions can be great. Pretest education is imperative in cases where there has been no family history—i.e., screening.

### Understanding Risk

One task of the genetics professional is to communicate risks to the client—a job not easily performed (31). A decision to be screened or tested will be influenced by a person's perception of the chance that the test will be positive. The interpretation of numerical risk varies depending on:

- prior perception of the magnitude of the risk,
- anxiety of the client at time of test,
- familiarity with the outcome (whether there is an affected relative or friend),
- how treatable the condition is, and
- belief that the outcome with which the individual is familiar is representative of all such outcomes (34).

Cultural differences need to be understood, as they also influence interpretation of risk (46).

The perception of risk is a more important determinant of decisionmaking than the actual risk. The way risks are posed can influence the client's choices. When confronted with the risk of genetic disease in their offspring and when making reproductive decisions, people tend to place greater weight on their ability to cope with a disabled or fatally ill child than on precise numerical risks. For some couples, for example, a risk of 10 percent could be perceived no differently from a risk of 50 percent if they believe that they cannot cope with the situation. In prenatal counseling, regardless of actual risk, parents might perceive the chance of occurrence as either 0 or 100 percent—it either will or will not happen. By processing risks in this way, individuals simplify probabilistic information and shift their focus to the implications of being at risk and the potential impact of what could occur (40).

In addition to subjective factors that influence the interpretation of risk, many individuals have difficulty understanding risk in arithmetic terms. Comprehension of the concepts of probability and risk influence a client's understanding of the information provided by genetic tests. In a Maryland study of 190 predominantly Caucasian, middle-class women, over one-fifth thought that "1 out of 1,000" meant 10 percent, and 6 percent thought it meant greater than 10 percent (12). In addition, interpretation of risks varies according to whether they are presented as a single figure or in comparison to a variety of genetic risks (65).

Even for individuals who are familiar with a disease because they have an affected family member, comprehending risks can be elusive. In a study of 190 subjects from 100 families with a family member with adult polycystic kidney disease (APKD) —a kidney disease inherited in an autosomal dominant fashion (each child of an affected individual has a 50 percent chance of inheriting the disease)—most tested poorly on questions reflecting their knowledge of the genetics of APKD. Twenty-nine percent had a genetic knowledge score considered good or excellent, while 46 percent had poor or absent knowledge (30).

How risks are framed also influences understanding and choices (45). Deciding to have a genetic test can be different if the risk is presented as a 25 percent chance of having an affected child rather than a 75 percent chance of having an unaffected child (31). Risk perceptions vary among individuals and among counselees and counselors: Patients tend to interpret the same level of numerical risk as lower than do geneticists (79).

Presenting risks in personal terms can improve the chance that the information will be understood. For example, one genetics center that conducts routine carrier screening for hemoglobinopathies in at-risk populations uses a videotaped drama to present information about screening during pregnancy and has demonstrated effective and lasting transmission of information (41). Videos of patients are particularly useful in demonstrating the gestalt of a condition, its natural history at various ages, and variability among affected individuals (4).

A recent survey of public attitudes regarding genetic tests revealed that belief in the technology's accuracy is one of the strongest predictors of favorable attitudes toward its use (66). Thus, an individual's decision to undergo screening or testing is likely influenced by the manner in which the accuracy of the test is presented. Even under the best circumstances, however, counseling might not be satisfactory. In a survey of 1,369 counseling cases at 47 clinics—after counseling sessions that typically lasted 45 to 60 minutes—both the counselor and the patient were aware of what topic the other party had most wanted to discuss in only 26 percent of the sessions. In almost half the sessions, neither party ever became aware of what the other had most wanted to discuss (80). Many of the undesirable psychological effects of screening might be avoided or reduced by careful attention to patients' needs at each stage of the screening process.

Maternal serum alpha-fetoprotein (MSAFP) screening is instructive in this regard. Women who are notified that their MSAFP test is abnormally low or high must await further testing, either through ultrasound or amniocentesis, to rule out the possibility of a fetus with Down syndrome or a neural tube defect, but a small percentage of neural tube defects will still be missed (box 6-A). (This situation parallels that of a pregnant woman who finds she is a carrier for CF and must sometimes await the results of her partner's test and subsequent fetal tests.) The mental anguish and apprehension that women encounter when faced with the prospect of having a child with a condition such as Down Syndrome is well documented (20,69). Although increased anxiety among women might persist until a definite chromosomal diagnosis by amniocentesis rules out the condition, initial anxiety can be reduced by comprehensive genetic counseling (36).

The importance of informed consent, careful presentation of counseling, and confidentiality have long been recognized as essential components of genetic testing and screening (29). Geneticists, perhaps more than any other medical specialists, have advocated a nondirective approach to counseling and have a strong commitment to patient autonomy (6). Further, a history of concern exists about genetic information being delivered by health professionals used to a more directive approach (13). This concern has been played out in the debate over MSAFP and is a factor in the reluctance of the clinical genetics community to rush toward wide-

### *Box 6-A—Maternal Serum Alpha-Fetoprotein Screening*

Alpha-fetoprotein (AFP) is a glycoprotein that is synthesized first by the embryonic yolk sac and later by the fetal gastrointestinal tract and liver. Its concentration in fetal serum is 150 times that found in amniotic fluid; an even smaller amount enters maternal blood, most likely through the placenta. Although its function in the fetus is unclear, the association of an abnormally increased AFP concentration in maternal serum with a variety of fetal abnormalities and adverse pregnancy outcomes has been known for 20 years. Elevated maternal serum AFP (MSAFP) can indicate an open neural tube defect (NTD) or an abdominal wall defect in the fetus. Abnormally low MSAFP levels are associated with fetal chromosomal abnormalities such as Down syndrome.

As with some other screening tests, the incidence of false positive MSAFP tests is high, and about 4 to 5 percent of women have abnormally high initial MSAFP values. More than half these women will have abnormal results when tests are repeated and will need to be assessed by ultrasonography. Because MSAFP levels must be correlated precisely with gestational age and the number of fetuses, ultrasonography provides an explanation for the elevated MSAFP in about half these women because of either incorrect dating of the last menstrual period or multifetal gestation. Thus, 1 to 2 percent of all women screened will be candidates for diagnostic amniocentesis to assess the AFP concentration of their amniotic fluid, and about 1 in 10 will ultimately be found to have a fetus with a NTD.

At the other end of the curve, another 5 percent of all women screened will have abnormally low MSAFP. Ultrasound eliminates half of these women from further evaluation because of inaccurate gestational dating, and the other half will be advised to undergo amniocentesis for karyotyping; 1 in 40 will have a chromosomally abnormal fetus.

Because of the number of false results at both ends of MSAFP screening, the cost-effectiveness of universal screening was debated widely during the early years of the assay's development. Moreover, there was heightened concern about the possibility that the test could fall into the hands of practitioners not familiar with the complexities of the required followup. Of further concern was the reproducibility of results with the available assay kits, as well as the lack of standardized controls for the specific population studies. In 1978, the U.S. Food and Drug Administration (FDA) was on the verge of releasing AFP test kit reagents on a totally unrestricted basis. The American College of Obstetricians and Gynecologists (ACOG) and the American Academy of Pediatrics (AAP) formed a task force on the issue that concluded:

> There is no precedent in the United States for a mass disease detection program that requires the coordination of so many disparate elements of the health care system as in AFP screening. Maternal serum AFP screening should be implemented initially only where it can be done within a coordinated system of care which contains the requisite resources and facilities to provide proper safeguards and potential for follow through.

AAP and the American Society of Human Genetics (ASHG) asked FDA to restrict MSAFP testing to programs that can provide qualified personnel and equipment, rather than making the test available to every pregnant woman regardless of capability for followup. In 1983, the FDA adopted unrestricted regulations approving MSAFP test kits, making expanded use of the test possible.

A properly designed MSAFP program demands public education, assurance of quality laboratory work, accurate interpretation of test results, and public health tracking to assure appropriate followup and testing of positive results. Misinterpretation or improper followup could result in either the unwanted birth of a seriously impaired child or the termination of a healthy fetus. MSAFP screening also depends on the availability of ancillary services (counseling, ultrasound, amniocentesis) for further diagnostic evaluation, particularly in underserved areas. At the time the test kit was introduced, there was concern that sufficiently equipped facilities and trained personnel were not yet available to handle the large number of potential cases.

In 1986, ACOG recommended to physicians that MSAFP screening be offered to all women, and, at least legally, such screening became the de facto standard of care. In the same year, ASHG developed a policy statement detailing the conditions necessary to provide for appropriate use of this test. Today, numerous MSAFP screening programs operate in the United States. By its very nature, MSAFP screening is the kind of program that lends itself to a State or regional public health approach since there is no clearly definable high-risk group. It is standard practice to refer women with positive test results to specialized diagnostic centers for confirmation. It is unclear, however, whether the counseling needs of this high-anxiety group are being met.

SOURCE: Office of Technology Assessment, based on G.C. Cunningham and L.C. Gilstrap, "Maternal Serum Alpha-Fetoprotein Screening," *New England Journal of Medicine* 325:55-57, 1991; and G.C. Cunningham and K.W. Kizer, "Maternal Serum Alpha-Fetoprotein Screening Activities of State Health Agencies: A Survey," *American Journal of Human Genetics* 47:899-903, 1990.

spread screening for any disease. Nearly 82 percent of the respondents surveyed by OTA said the human genetics community should be the primary organizer of CF carrier screening programs. Also mentioned were State or local health departments (59 percent) and primary care givers (27 percent). Over 89 percent felt CF population screening should be provided in genetics centers, but 59 percent felt that carrier screening could also be provided in the primary care setting or organized, community-wide programs (54 percent). Concern about the sometimes difficult nature of communicating risk information regarding CF (as described in the following section)—even for experienced genetic centers—has led some in the clinical genetics community to warn against rapid movement to routine CF carrier screening (5).

## THE SCENARIO FOR CYSTIC FIBROSIS COUNSELING

As mentioned earlier, while the CF mutation assay poses probabilistic uncertainties, similar circumstances exist for other tests, such as MSAFP. Education and counseling for CF carrier screening need not be viewed in a vacuum.

OTA's survey found genetic counselors and nurses in genetics estimate that, on average, 1 hour is needed to obtain a three-generational family history and to discuss carrier assays and recurrence risks, regardless of family history (71). The majority (70 percent) of genetic counselors and nurses in genetics surveyed by OTA feel that widespread screening should be withheld until a sensitivity of 95 percent or more is attained (71). This section focuses on some common scenarios likely to be encountered in routine CF carrier screening.

### *A Priori Risks*

Before providing DNA analysis for CF carrier status, it is important to understand and explain to the client his or her a priori risk—that is, an individual's risk prior to any test result. Two factors significantly influence an individual's a priori risk of being a CF carrier: ethnicity and presence or absence of a family history of CF. In the case of a positive history, an individual's chance of being a carrier depends on his or her relationship to the individual with CF. With a negative family history for CF, an individual's ethnic background is most important in defining a priori risk. Table 6-1 presents the a priori probabilities of carrier status for individuals with negative and positive family histories of CF.

**Table 6-1—A Priori Carrier Risks for Cystic Fibrosis**

| | |
|---|---|
| **Negative family history** | |
| Caucasian | 1 in 25 (4%) |
| African American | 1 in 60 to 65 (1.5 to 1.7%) |
| Asian American | 1 in 150 (0.7%) |
| Hispanic American | 1 in 46 (2.2%) |
| **Positive family history** | |
| Parent of child with CF | 1 in 1 (100%) |
| Sibling with CF | 2 in 3 (67%) |
| Aunt or uncle with CF[a] | 1 in 3 (33%) |
| First cousin with CF | 1 in 4 (25%) |
| Niece/nephew with CF[a] | 1 in 2 (50%) |

[a] Consanguineous.

SOURCE: Office of Technology Assessment, 1992.

Prior to any CF carrier screening, patient prescreening education is imperative. It is important potential screenees understand information regarding a priori risk, types of tests available, and uncertainties in risk assessment based on screening results.

### *Testing To Determine Risks for Relatives of Affected Individuals*

Currently, it is accepted practice to offer CF carrier tests to individuals who have a positive family history of CF (71). OTA's survey of genetic counselors and nurses revealed 86 percent of respondents' clinics have policies stating CF carrier tests should be routinely offered to individuals with a positive family history of CF but not to those with a negative family history.

OTA's finding that clinics' policies offer CF carrier assays to those with positive family histories is not surprising. An unaffected sibling of an individual with CF has a 2 in 3 likelihood of being a CF carrier. A consanguineous uncle or aunt of an individual with CF has a 1 in 2 likelihood of being a carrier. A first cousin of an individual with CF has a 1 in 4 likelihood of being a carrier (table 6-1).

In families where CF is due to the ΔF508 or another common mutation, carrier tests are relatively straightforward. If an undescribed CF mutation is involved, however, carrier detection might not be possible via direct mutational analysis. In these cases, indirect methods must be used, and there is a possibility that several family members will require analysis to arrive at an answer regarding carrier status. For at-risk families (those with a consanguineous relative who has or had CF), the use

**Table 6-2—Adjusted Carrier Risks After Negative Test Results at Various Detection Rates**

| Relationship of affected | a priori | Percent detection 75 | 80 | 85 | 90 | 95 |
|---|---|---|---|---|---|---|
| Sibling | 67% | 34% | 29% | 23% | 17% | 9% |
| Niece or Nephew[a] | 50% | 20% | 17% | 13% | 9% | 5% |
| Aunt or Uncle[a] | 33% | 11% | 9% | 7% | 5% | 2% |
| First cousin | 25% | 8% | 6% | 5% | 3% | 1% |

[a]Consanguineous.

SOURCE: Office of Technology Assessment, 1992.

of restriction fragment length polymorphism analysis in conjunction with direct detection of ΔF508 and other CF mutations can improve the tests' sensitivity to nearly 100 percent (ch. 4). The individual whose test reveals a CF mutation is definitely a carrier (the test is 100 percent predictive). However, if no mutation is found (a negative test), carrier status cannot be ruled out. Thus, CF tests, at best, can provide a definitive positive answer. At worst, they alter or modify the assessment of individuals' or couples' risk from their a priori risk. CF mutation analysis does not alter the actual risk, it merely reduces the uncertainty about what that actual risk is.

CF mutation analysis can adjust the assessment of risk downward: Table 6-2 displays how those risks are adjusted after analysis at several test sensitivities. As the sensitivity improves, the uncertainty—i.e., possibility of a false negative result—diminishes. For example, without any testing, a sibling of an individual with CF has an a priori risk of 2 in 3, or 67 percent, of being a carrier. If the sibling is tested for CF carrier status using a test that is 85 percent sensitive and is found to be negative, the adjusted risk of being a carrier is 23 percent. At 90 percent sensitivity, the carrier risk drops to 17 percent.

### *Screening To Determine Risks of Those With No Family History*

To date, most genetic counselors do not offer unsolicited CF mutation assays to individuals with a negative family history (71). The American Society of Human Genetics (ASHG) and a workshop of the National Institutes of Health (NIH) published policy documents in 1990 discouraging CF carrier screening (11,53); a 1992 ASHG statement reaffirms

**Table 6-3—Potential for Detecting Couples at Risk for a Child With Cystic Fibrosis**

| Percent of cystic fibrosis mutations detectable | Percent of carrier couples detectable |
|---|---|
| 75 | 56.3 |
| 85 | 72.3 |
| 90 | 81.0 |
| 95 | 90.3 |

SOURCE: W.K. Lemna, G.L. Feldman, B.-S. Kerem, et al., "Mutation Analysis for Heterozygote Detection and the Prenatal Diagnosis of Cystic Fibrosis," *New England Journal of Medicine* 322:291-296, 1990.

that position (50). OTA's survey found 21 percent of respondents felt that CF carrier tests should be offered to individuals with no family history of CF.

CF mutation analysis can, however, yield information about the carrier status or risk for individuals who have no family history of CF. In the recent past, the sensitivity of the carrier test was limited to the frequency of the ΔF508 mutation. Today, however, most commercial and university laboratories examine ΔF508 and 6 to 12 additional mutations. Taken together, these mutations comprise 85 to 90 percent of CF mutations in U.S. Caucasians (95 percent in Ashkenazic Jews). Counselors report an almost even split between commercial and university-based laboratories as the facility performing their CF mutation assays (45 percent and 48 percent, respectively) (71).

As with individuals who have affected relatives, determining risks in persons with no family history of CF also depends on test sensitivity. At 85 percent test sensitivity, about 72 percent (0.85 x 0.85) of at-risk couples will be identified. As the sensitivity of the assay improves, a greater proportion of +/+ couples can be identified (table 6-3). Simlarly, at 85 percent sensitivity, the probability that a couple from the general population would bear a child with CF is reduced from 1 in 2,500 to about 1 in 16,100 if one partner is screened and found to be negative; risk is reduced to 1 in 103,700 if both have negative test results. In roughly 7 percent of couples screened for CF mutations, one partner will be found to carry a mutation and the other will not (+/-). Such couples constitute the most difficult counseling situation because, assuming 85 to 90 percent sensitivity, the risk of having an affected child remains 1 in 644 (table 6-4). Computerized programs have been developed to incorporate factors such as a priori risk and mutation frequency when determining risk. In addition, a "slash sheet," or pedigree flow chart,

**Table 6-4—Cystic Fibrosis-Related Risks After Mutation Analysis for Carrier Status[a]**

| | Risk of cystic fibrosis in offspring | | | |
|---|---|---|---|---|
| Percent of cystic fibrosis mutations detectable | Carrier risk for person with negative test | One screened; negative result | One parent positive, one parent negative | Both parents negative |
| 0 | 1 in 25 | 1 in 2,500 | NA[b] | 1 in 2,500 |
| 75 | 1 in 97 | 1 in 9,700 | 1 in 388 | 1 in 37,600 |
| 85 | 1 in 161 | 1 in 16,100 | 1 in 644 | 1 in 103,700 |
| 90 | 1 in 241 | 1 in 24,100 | 1 in 964 | 1 in 232,300 |
| 95 | 1 in 481 | 1 in 48,100 | 1 in 1,924 | 1 in 925,400 |

[a] Assumes carrier frequency of 1 in 25.
[b] NA = Not applicable.

SOURCE: Office of Technology Assessment, based on W.K. Lemna, G.L. Feldman, B.-S. Kerem, et al., "Mutation Analysis for Heterozygote Detection and the Prenatal Diagnosis of Cystic Fibrosis," *New England Journal of Medicine* 322:291-296, 1990.

was developed to assist in the estimation of risk (21). The value of either of these tools in the clinical setting is yet to be determined.

## *Prenatal Diagnosis of Cystic Fibrosis*

In addition to CF carrier screening of adults, prenatal diagnosis of CF can be performed. The type of test used to establish carrier status of the parents determines the DNA protocol for the fetus. The choice of technique to obtain a sample for DNA analysis involves consideration of timing, procedural risk, access to procedures, cost, and the presence or absence of other indications for prenatal diagnosis. Cells can be obtained via chorionic villus sampling (CVS) (performed at 9 to 12 weeks of pregnancy), amniocentesis (performed at 16 to 18 weeks), or through percutaneous umbilical blood sampling, also called cordocentesis (performed at 20 weeks). When CF mutation analysis is unavailable or inconclusive, microvillar intestinal enzyme levels can sometimes be measured in the amniotic fluid at 17 to 18 weeks, but this method suffers from a high false-positive and false-negative rate.

A new, experimental procedure, called blastomere analysis before implantation, or BABI, has been used to diagnose the CF status of an in vitro fertilized embryo before implanting it in the woman's uterus. The technique involves extracting a single cell from an embryo at about the eight-cell stage and analyzing its CF mutation status. Recently, an

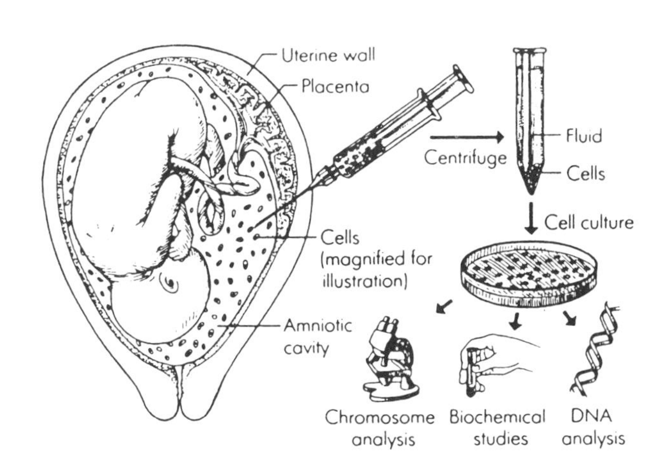


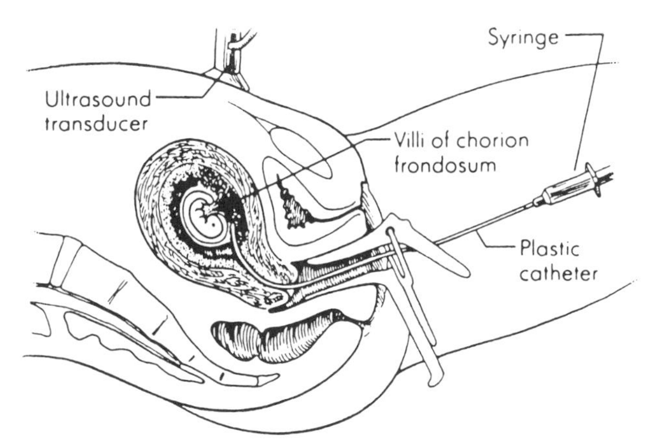


*Photo credit: National Institute of General Medical Sciences*

**Left:** Amniocentesis—the most widely used technique for prenatal diagnosis, generally at 16 to 18 weeks of a pregnancy. Cells shed by the developing fetus are extracted from a sample of amniotic fluid that has been withdrawn from the expectant mother's uterus by a hypodermic needle. The cells are cultured and then can be analyzed for chromosomal defects, such as Down syndrome. DNA analysis can also be performed (e.g., for CF mutation status).

**Right:** Chorionic villus sampling—a method of prenatal diagnosis that provides results as early as the 9th week of pregnancy. Fetal cells from the chorionic villi (protrusions of a membrane called the chorion that surrounds the fetus during its early development) are suctioned out through the uterine cervix and their DNA is analyzed. Preliminary results of this process can be obtained within a day.

unaffected baby that had been tested with BABI was born to a couple in London at risk for CF (70).

For some families, prenatal diagnosis for CF is inconclusive. As in screening for carrier status, couples receive test results that yield percentages rather than certainties (54). There is speculation that this "restructuring of uncertainty" versus a "yes/no answer" will be confusing to individuals and cause undue stress; pilot studies to examine levels of anxiety in test populations before and after screening are under way. Assessments of anxiety levels of participants in other carrier screening programs are limited in what they offer to CF carrier screening because the levels of test uncertainty are not as great. Women offered hemoglobinopathy carrier screening, for example, were found to manifest appropriate levels of concern rather than undue anxiety, but the test results were less ambiguous because the test is more sensitive (59).

### *Informing Relatives and Inductive Screening*

In any type of genetic screening, a real possibility exists that test results will affect other family members. In the usual genetic counseling setting, the person being screened (the proband) is routinely advised of risks to other family members. If, for example, an individual is found to be a carrier for an autosomal recessive disorder such as CF, the genetics professional informs the client that siblings also each have a 50 percent chance of being carriers. In most cases, the suggestion is made that the individual contact his or her siblings, and that they consult with their personal physician or come to the same clinic. The genetics professional does not usually confirm that the proband has informed relevant family members due to limits of confidentiality in the counselor-client relationship. Some argue that genetics providers should be legally permitted to disclose such relevant information to relatives at risk (76).

As discussed in chapter 5, breaching confidentiality to disclose medical information to relatives raises legal and ethical issues. Not all families are emotionally and psychologically secure. Sibling relationships could impede full disclosure. Sharing highly personal medical information that involves reproductive and health futures can cause personal embarrassment or emotional stress for family members.

From the point of view of identifying the largest number of carriers by screening the fewest number of individuals, however, encouraging carriers to notify relatives provides economic and pragmatic benefits. Encouraging known carriers to refer siblings and first cousins for testing can detect a larger percentage of at-risk couples. This approach, also known as "inductive screening," improves the efficiency of screening by 10 to 15 percent (15). Testing those known to be at higher risk because of family history is more effective than screening those with unknown risk. In reality, complex psychological factors enter when family members of individuals with CF contemplate screening, and it cannot be assumed that all will want to be tested (63).

OTA's 1991 survey of genetic counselors and nurse geneticists revealed that most families who have a child with CF are not routinely seen in genetics service settings, and few counselors have routine contact with CF families (71). For inductive screening to work, those providing health care and counseling to CF families will have to actively participate in referrals of relatives to genetics centers. Less than 10 percent of respondents reported contacting previously identified CF families regarding the availability of CF mutation analysis.

### *Post-Test Counseling*

When attending a genetics clinic for reasons other than prenatal genetic screening, people typically come because they have or had an affected relative, usually a child. These individuals tend to be aware of the disorder, and the affected relative, rather than a test, serves as the indicator of potential disease for other offspring. As the number of genetic tests administered to healthy individuals with no apparent family history of genetic disease increases, genetics providers might need to spend more time describing the disorder to those with positive results.

Results are best reported in person, by the same person who provided the pretest education—although this is not always possible or practical (16). If the test results are positive, prior contact might have alerted the professional as to who else should be informed, whose help might be needed on behalf of the client (e.g., for emotional support), and important information about the client's lifestyle and family (as well as financial and insurance information).

Followup counseling and support are also strongly advised for persons with positive results. News of a positive result can impede a person's ability to receive information on both emotional and practical levels. Faced with positive results, most individuals are unable to act until they overcome the shock and possible denial that their fate or their children's fate could suddenly shift in a negative direction. People's perceptions of their own health can worsen, for example, as occurred for some when they were made aware of their carrier status for Tay-Sachs disease (42,43); such anxiety can be prolonged (83). Knowledge of carrier status can also have an impact on reproductive intentions or behavior, including decisions relating to marriage or choice of marriage partner (49,68).

Even in the best of all worlds, where consistent counseling has been provided throughout the process, the effectiveness of counseling is sometimes questionable. An analysis of nine studies on counseling published since 1970 concluded, "many parents of children with a genetic disorder have an inadequate understanding of the genetic implications of the disease, even after one or more genetic counseling sessions" (23). One survey found that more than half of the 87 percent of people who came to a genetic counseling center with inaccurate knowledge of risk were still misinformed after counseling (67). It is not clear whether they were incrementally better informed.

The task of communicating genetic information is formidable. Genetic counseling programs continually try to improve the process (74). A major impediment to satisfactory genetic counseling has been a profound lack of understanding of basic genetics by almost everyone in the general public. Anyone administering tests necessarily takes on the role of educator as well as practitioner.

### *The Need for Better Public Education*

Whether CF carrier screening programs are offered to prenatal or preconception populations, public education efforts aimed at better understanding of genetic conditions and inheritance will be increasingly essential. The need for better scientific literacy has been a topic of great concern in recent years. Even by the 12th grade, fewer than half of students can use data in tables or graphs (72). A recent survey of 1,006 Americans regarding attitudes about genetic testing revealed that fewer than half were able to correctly answer 4 of 5 technical questions regarding genetic testing (66).

*Photo credit: University of South Carolina School of Medicine*

Genetic consultations can require longer office-visit time because of the need for gathering detailed information about the client and for followup counseling and support. This genetic counselor is discussing the client's chromosome profile with her.

Most counselors and nurses responding to the OTA survey, however, indicated they spend little to no time on general public education in schools and communities (71). Thus, most people will rely on their primary care provider for preliminary genetic information. Survey respondents indicated that they think primary care providers and public health departments should play an active role in educating the public about DNA tests for CF carrier status (71).

Public education programs for genetic diseases have been nearly nonexistent since the programs established under the National Genetic Diseases Act (Public Law 94-278) were phased out in 1981. In terms of public education, the National Science Foundation has supported a teacher training program in genetics for school teachers in Kansas, but there is no similar program through NSF at the national level (14). In the Kansas program, lead teachers were trained to teach peer teachers in genetics. Teachers who participated in the program showed a 3-fold increase in genetics instruction at the high-school level and a 22-fold increase at the elementary school level (28). More recently, a project to prepare 50 selected science teachers per year for 3 years to become State resource teachers in human genetics received funds though the Department of Energy's Ethical, Legal, and Social Issues Program. Nevertheless, fewer than half the Nation's elementary schools and about one-third of high schools make science education a curriculum priority (72).

The importance of a period of community education before the implementation of genetics programs has been demonstrated in sickle cell and Tay-Sachs screening programs in the United States, and in thalassemia screening programs in Sardinia and Cyprus (app. B). More public awareness about genetic diseases and tests could result in less time needed for individual counseling. Experiences with β-thalassemia carrier screening in Sardinia and Cyprus demonstrate the impact of public education. Aggressive public education campaigns orchestrated in these countries placed information on the disease and carrier screening on television and in large department stores and factories, marriage registry offices, general practitioners' offices, and family planning clinics. In Sardinia, the birth rate of thalassemia-affected newborns fell from 1 in 250 to 1 in 1,200 (10). In Cyprus, genetics is taught in school and screening is encouraged before marriage. As a result, the time needed for counseling has decreased as public education has increased (2).

In an ideal world, better education in the schools would make individuals more aware of genetic risks before they are confronted with genetic screening programs. And if pregnant women are informed, they might initiate genetics discussion with their obstetricians rather than waiting to be informed (8).

## STRATEGIES FOR SCREENING VARIOUS POPULATIONS

Two key considerations in deciding how best to implement routine CF carrier screening are the clinical settings in which it will take place and the target populations. Delineation of a target group (or groups) determines other elements such as location, educational approach and tools, time, format, types of counseling, facilities, and publicity.

The NIH statement on CF carrier screening emphasized the importance of preconceptional screening (53). Most pilot projects in the United Kingdom are directed at preconceptional populations (ch. 10). Pilot studies under way in the United States are both prenatal and preconceptional (see following section). One program in Canada targets high school students (35).

### *Newborn Screening*

Numerous newborn screening programs exist for genetic disorders such as sickle cell anemia and phenylketonuria. These are programs intended to screen for the presence of disease, although some can also detect the carrier status of the newborn. Wisconsin has performed statewide neonatal screening for CF disease since 1985, using the immunoreactive trypsin assay. Primary care physicians have been cooperative in referring screened patients to designated CF centers (47). But even newborn screening for CF disease is not without controversy. Evidence of heightened anxiety and disrupted maternal-infant bonding have been reported in cases of false-positive diagnoses (7).

For at least two reasons, many believe that newborn screening is an inappropriate and inefficient mechanism for carrier detection. First, newborns determined to be carriers must be tracked through their reproductive years to ensure they are aware of their carrier status. Second, detection of newborn carriers might unnecessarily raise the anxiety level of parents. Thus, newborn screening for CF carrier status is not generally viewed as acceptable (48). OTA's survey of genetic counselors and nurses bears this out; a minority of respondents (33 percent) felt the newborn population would be an appropriate target group for widespread CF carrier screening (71).

### *Adolescent Preconceptional Screening*

Some geneticists advocate screening at the high-school level (35). A recent nationwide survey of American attitudes about, and knowledge of, genetic tests showed better knowledge and more positive attitudes in younger populations (66). Studies of pregnant women known to be carriers of a hemoglobinopathy gene have shown that age is a predictor of postcounseling knowledge—younger women (and adolescents as young as 12 years old) are more likely to understand genetic information (41). While not routinely done in the United States, high-school screening programs have been conducted in Canada for some time (app. B). For any disease where screening is done in childhood or adolescence, however, the benefits of such screening, including savings in inconvenience, resources, and anxiety, must be balanced against the potential problems, such as the possibility that an adolescent will be falsely assigned to a low-risk group because of poor test sensitivity (thereby obviating further screening), or the possibility of psychosocial harm to the child as a result of identified carrier status (29).

Adolescents were not considered to be an appropriate target by the genetic counselors and nurses surveyed by OTA (71). Less than 18 percent felt individuals ages 13 to 18 years should be screened. Only 6 percent felt that children ages 2 through 12 years should be screened.

### *Adults—Preconceptional or Prenatal?*

Current debate surrounding CF carrier screening focuses on whether the goals are best accomplished by targeting preconceptional adults or pregnant women. These approaches are not necessarily mutually exclusive. Many feel, however, that receipt of troubling information during pregnancy is not desirable, and that it would be better for individuals to know their risks before getting pregnant (39). Others argue that individuals not facing a pregnancy are not motivated to seek or use information on their carrier status, but will wait until they are either planning a family or starting a family before viewing such information as useful (9).

CF carrier screening offered as part of primary health care rather than prenatal care is likely to encourage preconceptional CF carrier screening. For most individuals, however, the first real opportunity for carrier screening takes place post-conception (22). It could well be that the primary responsibility for providing CF carrier screening will reside with the obstetrician, as has happened with MSAFP screening. Sixty-six percent of respondents to OTA's survey identified pregnant women or couples as the appropriate target population for CF carrier screening. Yet 88 percent generally identified adults in their reproductive years as the appropriate target group (71). While most respondents state that the *ideal* population to target for carrier screening is the preconceptional adult (71), in reality, the first target population is likely to be the prenatal population because it has been the traditional entry point into genetic services for many people.

Pregnant women often have established a relationship with an obstetrician/gynecologist or prenatal clinic staff, who can provide information about CF carrier screening. Ideally, individuals should receive potentially emotionally loaded information when they have the most latitude for reproductive choice (figure 6-1). Women in the early stages of pregnancy have the choice to continue the pregnancy or electively terminate it; attitudes regarding abortion of CF-affected fetuses indicate that prenatal screening would only modestly reduce the incidence of CF, as many couples with CF-affected children speculate that they would not elect termination of an affected fetus (1,77,78). Studies of pregnant women screened for hemoglobinopathy carrier status have shown little evidence that screening raised anxiety (41). Pregnant women with fetuses at risk for hemoglobinopathies are highly receptive to genetic information (60) because they seek reassurance that fetuses are not affected.

The increasing availability of genetic tests might shift genetic services from specialized clinics, where they are now usually located, to primary care settings. This is likely to be especially true for tests for disorders like CF. First, the possibility exists for the clinical genetics community to become overwhelmed by the volume of tests and counseling. If this occurs, genetic specialists will need to rely on primary care providers and community and public health institutions to bear some of the workload. Two examples of public health models are State-sponsored MSAFP screening in California and newborn screening programs for hemoglobinopathies and phenylketonuria. Second, some aspects of medical genetics—specifically routine screening of those with no family history of genetic disease—could increasingly be considered less of a medical specialty (tertiary care) and more a part of primary care. Thus, genetics education must reach primary caregivers, yet the average 4-year medical school curriculum includes only 21.6 hours of genetics instruction (57). Third, placing CF carrier screening within the realm of obstetric care might decrease out-of-pocket expenses to the client. As part of routine care, insurers might be more likely to reimburse CF carrier screening in the prenatal population.

Overall, then, if carrier screening is to become routine, it is likely that it will be offered as part of family planning or reproductive health, and the medical specialty most likely to offer the test will be obstetrics—a prospect some genetic specialists find unsettling. Experience with MSAFP has shown that despite the development of practice guidelines, obstetricians often perform tests on pregnant women without obtaining their consent (31). In addition, obstetricians who do not screen for certain genetic conditions in individuals at high risk could be at risk

**Figure 6-1—Decisionmaking in Premarital Carrier Screening**

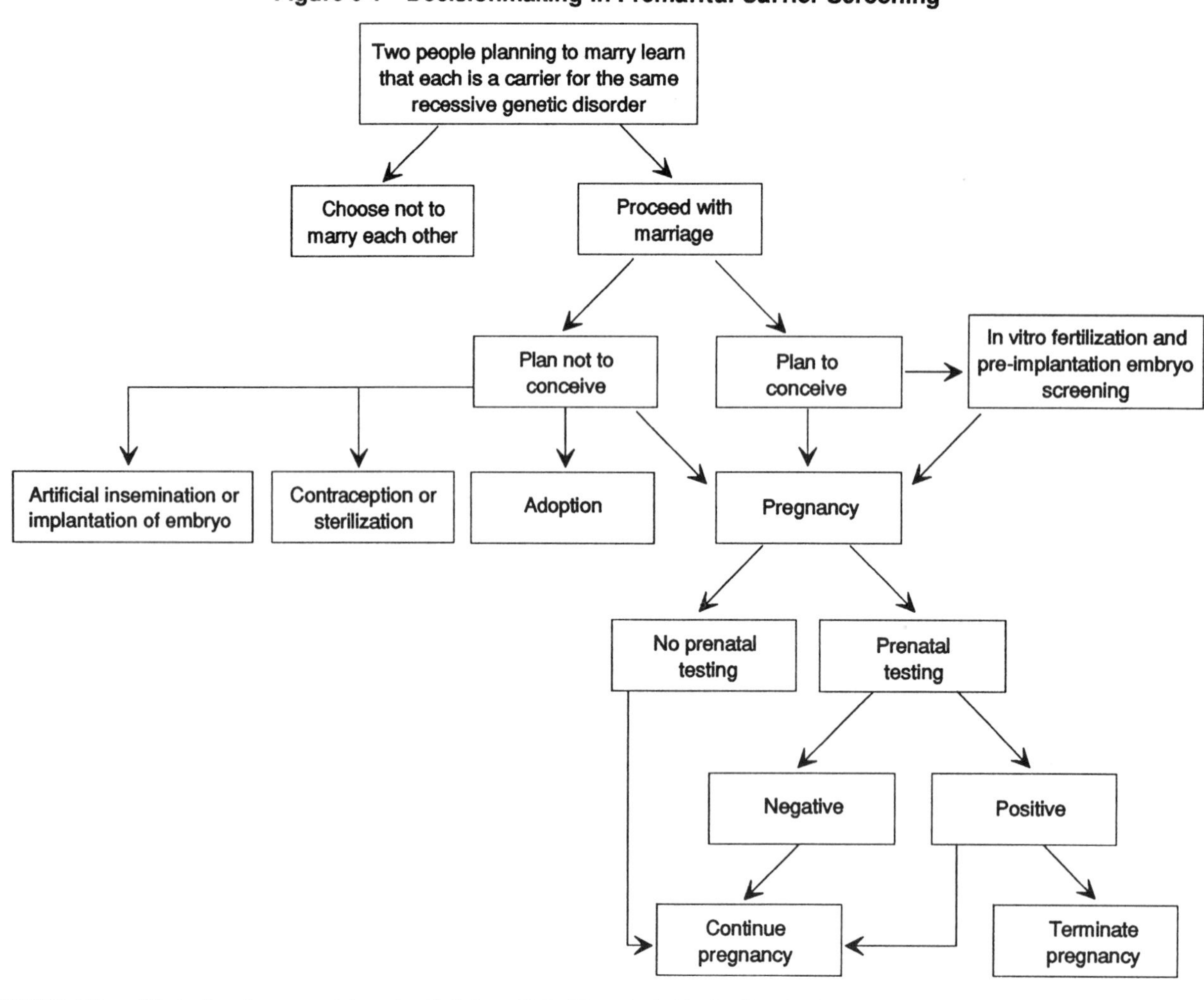


SOURCE: Office of Technology Assessment, based on F. Cohen, *Clinical Genetics in Nursing Practice* (Philadelphia, PA: Lippincott, 1984).

for malpractice or wrongful birth suits (64). This perceived tension—lack of informed consent coupled with pressure to screen—leads many in the clinical genetics community to express concern about premature widespread CF carrier screening before adequate professional education is in place. With regard to CF carrier screening, concern exists that layers of uncertainty will inhibit informed consent and that, ultimately, more harm than good might be done.

The sickle cell and Tay-Sachs carrier screening programs provide valuable information on the importance of prescreening education, understanding the culture and values of the population being screened, and optimizing the setting in which screening occurs. Mistakes were made and lessons learned (app. B). While CF carrier screening also involves identification of carriers and high-risk couples, it differs from these experiences because of the sensitivity of the test and the larger number of couples at risk. There is little experience in the delivery of such complex information to large populations (11). Initial experiences with MSAFP screening revealed some confusion and concern on the part of patients because of a high false positive rate, limited test sensitivity, and apparent lack of understanding within the obstetrics community about the screening procedure (box 6-A).

# PROFESSIONAL CAPACITY

As mentioned earlier, OTA reserves the term ***genetic counselor*** to specifically describe master's-level individuals certified as genetic counselors by the American Board of Medical Genetics (ABMG)

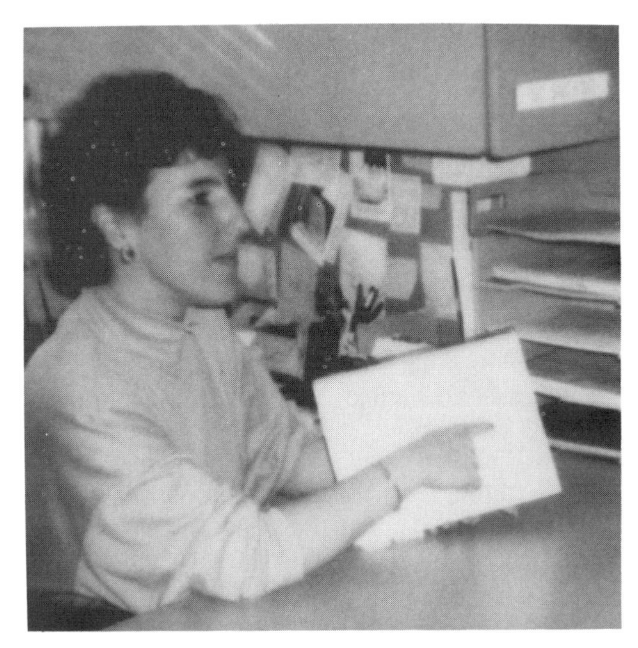

Photo credit: Diane Baker

Genetic counseling prepares the individual for both positive and negative test results.

(or board-eligible) because legal distinctions in licensing and third-party reimbursement exist among the different types of practitioners. OTA uses the term ***genetic counseling*** generically to refer to the information delivery process that is performed by genetic specialists, including physicians, Ph.D. clinical geneticists, genetic counselors, nurses, and social workers. Overall, approximately 630 individuals from the range of types of training are certified by the ABMG to perform genetic counseling. (The exams are given every 3 years so many practitioners are board-eligible but have not yet taken or passed the exam.)

At issue in considering widespread carrier screening for CF is whether there are enough adequately trained health professionals to handle the volume of tests. One study estimated that a minimum of 651,000 counseling hours would be required annually if their maximum estimate of 6 to 8 million preconceptional couples are screened for CF carrier status (81). Considering the current number of board-certified genetic counselors practicing in the United States today, this translates to 17 weeks per year from each genetic counselor to serve CF-related clients (81). On the other hand, one estimate suggests the supply of genetic specialists could absorb routine carrier screening for CF, sickle cell anemia, hemophilia, and Duchenne muscular dystrophy, assuming that obstetricians or other primary care physicians perform the screening on pregnant women, with referral of those with positive results to genetics professionals (31).

The counselors and nurses surveyed by OTA estimate that pretest counseling for CF carrier status, regardless of family history, would take, on average, 1 hour (71). It is unclear to what extent increased demand for CF carrier screening would strain the current system. Current estimates undercount the number of health care professionals who practice genetic counseling and assume that counseling would always be provided in a clinical genetics setting by board-certified or board-eligible counselors. Such estimates also ignore the role that aggressive public education can play in improving pretest knowledge (2,10). Improvements in public education could result in dramatically less time required in formal counseling as could reliance on health professionals not formally trained in genetics.

The following section addresses the traditional roles played by master's-level genetic counselors and presents nontraditional sources as possible options for handling an increasing caseload.

### *Master's-Level Counselors*

The master's-level genetic counselor is a relatively new addition to the health care system. In 1971, 10 graduates of the first such program entered the workforce; in 1979, the National Society of Genetic Counselors (NSGC) was incorporated as a professional organization. Today, there are approximately 1,000 master's-level genetic counselors practicing in the United States.

Master's-level genetic counselors receive specialized multidisciplinary training and experience to prepare them for counseling related to a wide variety of genetic disorders and birth defects. They are typically graduates from a 2-year master's degree program, during which time they receive didactic course work in the principles and application of human genetics, clinical and medical genetics, genetic laboratory methods, and interviewing and counseling. Genetic counselors are also trained in social, ethical, legal, and cultural issues relating to genetic diseases, principles of public health and health care delivery systems, and education for the lay and professional community (73). Over the past 20 years, master's-level graduate programs in genetic counseling have increased to 15; combined, they produce approximately 75 graduates each year (54).

Genetic counselors receive a minimum of 400 hours of supervised clinical training in at least three clinical settings, including a general genetics clinic, a prenatal diagnosis clinic, and a speciality disease clinic. Until 1992, graduates were eligible to sit for the certification examination in genetic counseling by the ABMG, but the continuing certification of these individuals by this body is unclear. In the past, counselors were required to submit their credentials and a logbook of 50 cases obtained in a clinically accredited training site before taking the exam (54). Genetic counselors typically work in university medical centers or private hospitals in metropolitan areas, and tend to be female, Caucasian, and married (71). The mean gross salary in 1990 was $33,879 (56). The majority are board-certified or board-eligible by the ABMG and have been in clinical practice for at least 6 years (71).

Training support for master's-level genetic counselors has been minimal. No financial support is supplied by the U.S. Department of Health and Human Services (DHHS) for the training of genetic counselors or for improving genetics education in medical schools (31). Through support to the Council of Regional Networks for Genetic Services (CORN), the DHHS Bureau of Maternal and Child Health and Resources Development provides support for some continuing professional education programs for physicians and postdoctoral students, but not for master's-level counselors.

As the profession has developed, master's-level counselors have begun to consider taking the role of trainer of other health professionals. This role with respect to "single-gene counselors," discussed in the following section, could serve as an example of how an increase in CF-related counseling could be handled.

### *Non-Master's-Level Counselors*

In some clinical settings, a role has been created for a non-master's-level individual to meet the demand for patient education related to one diagnostic category of disease. In other settings, such individuals assist genetic counselors in overcoming cultural, linguistic, geographic, or economic barriers: The OTA survey of genetic counselors and nurses in genetics, for example, revealed that only 14 percent were fluent in a language other than English (71). Individuals who assist genetic counselors, often called "single-gene counselors," or "non-master's-level counselors," do not have the same training as master's-level genetic counselors and have not been eligible for ABMG certification. With the growth of genetic services and increasing demands on the time and resources of traditionally trained counselors, use of these individuals has raised debate.

Advocates of single-gene counselors cite the current shortage of genetic counselors—the NSGC maintains a jobs hotline, and has consistently over the last 3 years posted at least 35 unfilled positions at a given time. Single-gene counselors could also improve the quality of service in underserved, culturally diverse populations that are disproportionately affected by a particular genetic disease (54).

Those opposed to single-gene counselors express concern about what they view as a lack of genetics training. Some view them as a possible threat to the professional status of genetic counselors. There is also concern about whether single-gene counselors have a broad enough view of clinical genetics to identify complex and obscure risks of other genetic disorders in their patients. Since taking a family history often exposes previously unknown or undiagnosed genetic disorders or predispositions, individuals who focus on one category of disease might not recognize the need to further investigate peripheral information.

An NSGC task force has recommended that the society:

- acknowledge the current and predicted personnel needs for genetic counselors as well as the shortage of master's-level genetic counselors;
- recognize the existing use of non-master's-level counselors and the benefits they offer;
- educate the NSGC membership regarding the potential use of these individuals;
- support the use of non-master's-level counselors in specific settings where genetic counselors can be involved in training, evaluating, and supervising these individuals; and
- establish a committee to collaborate with other organizations.

Genetic counselors have been involved in developing and conducting training programs as well as supervising hemoglobin trait counselors in several

States. Programs in California and Massachusetts, for example, employ non-master's-level professionals to conduct genetic counseling for sickle cell disease (26).

### *Other Health Professionals*

Integration of other health professionals, such as nurses, nurse practitioners, social workers, dietitians, psychologists, and physicians, into the existing genetics network will supplement the skills of the traditional genetic counselor. Similarly, the involvement of other health care professionals will be important to increasing public awareness and education. If CF carrier screening were to become routine, clinical geneticists would have to become more involved in the delivery of community (public health) genetic services, the education and training of other health care professionals, and public education.

#### Nurses in Genetics

There are nearly 2 million registered professional nurses in the United States, many involved in maternal and child health nursing. These professionals provide a unique potential to contribute to the effective delivery of genetic services. Efforts are under way to encourage the incorporation of clinical genetics into the curricula of schools of nursing at both the graduate and undergraduate level (32). The need for better genetics education in nursing stems from the recognition that genetics is generally within the realm of tertiary care; thus, genetic specialists are not always in the position to screen every individual needing genetics referral (32). That is, individuals who need genetic services must first be identified by the primary health care professional, and in some settings—such as community, occupational, or school health—nurses are the only link with the health care system (27). Thus, nurses can assist in the identification, education and counseling, and followup of patients (25,32). Yet while nurses can be a valuable part of genetic services, to date they are a largely untapped resource (27).

Opportunities for clinical genetics experience in nursing programs vary. Genetics is generally a part of the nursing school curriculum, but again, variability exists among programs (27). Four of the 200 universities in the United States that offer graduate degrees in nursing have established programs providing a master's-level genetics major (27). A small number of nurses, particularly those in maternal and child health nursing, have focused on genetics in order to sit for the genetic counseling examination given by the ABMG (27,33). There are over 100 nurses who are employed in genetics, according to the International Society of Nurses in Genetics. Governmental support of genetics education for nurses has been through the Health Resources and Services Administration, Bureau of Health Care Delivery and Assistance, Division of Maternal and Child Health (26).

#### Social Workers and Public Health Professionals

Social workers can play an important role in genetic services delivery, particularly in underserved communities. Nevertheless, only 9 of almost 100 accredited social work graduate programs in the United States offer special courses on genetic topics (28).

Similarly, public education in genetics requires increased commitment at the public health level. This requires educating public health professionals about pertinent issues related to medical genetics and changing the attitudes and staffing patterns of key State agencies (17,18). Yet a survey of curricula at member schools of the Association of Schools of Public Health indicated a decrease in the number of schools offering human genetics as a major area of study (28). Few schools of public health offer genetics as part of their curriculum, and in none is it required (62).

## THE FEDERAL ROLE IN DECIDING THESE ISSUES

In 1990, NIH convened a consensus workshop on "Population Screening for the Cystic Fibrosis Gene." Participants concluded that tests should be offered to all individuals and couples with a family history of CF. However, the group did not recommend population-based screening for individuals and couples with a negative family history, because:

- With a sensitivity [at the time] of 70 to 75 percent, only half of the couples at risk can be identified.
- The frequency of the disease and the different mutations vary according to racial and ethnic background, so that important laboratory and counseling modifications would be required in different populations.

327-046 - 92 - 6 : QL 3

- There are substantial limitations on the ability to educate people regarding the use of an imperfect test.
- Without more definitive tests, about 1 in 15 couples—those in which one partner has a positive test and the other has a negative test—would be left at increased risk (approximately 1 in 500) of bearing a child with CF (53).

The workshop concluded that these difficulties would be substantially reduced if the test were improved to 90 to 95 percent accuracy. Population-based CF carrier screening was considered appropriate if a 95 percent level of carrier detection were achieved. Additional screening guidelines were developed, including:

- Screening should be voluntary and confidentiality assured.
- Informed consent must be obtained via pretest education.
- Providers of screening have an obligation to provide adequate education and counseling.
- Quality control of all aspects of laboratory testing, including systematic proficiency testing, is required and should be implemented as soon as possible.
- There should be equal access to counseling.

The NIH group felt that legislative action regarding CF carrier screening was not required unless it became evident that individuals identified as carriers were suffering from discrimination, either through employment or insurance. It described the most appropriate group for population-based carrier screening as individuals of reproductive age, preferably preconception. Furthermore, the NIH group agreed that the optimal setting for carrier screening is through primary health care providers or via community-based screening. It concluded that newborn or childhood screening would be inappropriate. The NIH statement stressed the importance of providing nondirective genetic counseling for individuals determined to be carriers. Finally, the group called on the Federal Government to fund pilot projects to investigate research questions in the delivery of population-based screening. The pilots were envisioned to address the effectiveness of educational materials, the level of use of screening, laboratory aspects, counseling issues, costs, and beneficial and deleterious effects of screening (53).

### *NIH Clinical Assessments*

Responding to increasing calls for a Federal initiative to evaluate population carrier screening (58), the Ethical, Legal, and Social Issues Working Group of the National Center for Human Genome Research (NCHGR) hosted a workshop in September 1990 to discuss an appropriate role for NIH. This workshop concurred with earlier statements that clinical evaluations of alternative approaches to genetic education, testing, and counseling were needed to establish the professional practices that should govern widespread CF carrier screening. In stressing the importance of setting professional standards as early as possible, CF mutation analysis was viewed as a prototype for future DNA-based genetic tests.

In January 1991, NCHGR, the National Institute of Child Health and Human Development (NICHD) and the National Institute of Diabetes and Digestive and Kidney Diseases (NIDDK) invited a group of consultants to advise NIH on the appropriate issues to be addressed through pilot studies. They arrived at six questions:

- What are the levels of understanding of, and interest in, CF carrier screening among different patient populations?
- What are the optimum forms and levels of pretest education for different patient populations?
- What are the accuracy and cost effectiveness of various types of tests?
- What are the best approaches to post-test counseling, in terms of patient understanding and psychological health?
- What are the optimum settings for providing CF carrier screening services?
- What record-keeping and reporting policies best protect against breaches of confidentiality, stigmatization, and discrimination?

The consultants recommended that NIH develop a consortium of multiple studies, each addressing some subset of the overall agenda. Such an approach would allow for standardization across the participating groups in terms of evaluation measures and tools, cost accounting, laboratory quality control, and human subjects protection. It was also recommended that NIH provide support to underwrite the current laboratory costs of the assay during these clinical studies, to improve access to the pilots by all interested persons.

In April 1991, NCHGR, the National Center for Nursing Research, NICHD, and NIDDK issued a request for applications (RFA) for clinical evaluations related to CF carrier screening. The grant competition was open to nonprofit and for-profit organizations, including universities, public health departments, and voluntary organizations. The award period is for up to 3 years, and is renewable.

Originally, the RFA specifically excluded laboratory costs of the assays as eligible for grant support, since they were considered part of the clinical care of the individuals involved in the studies. Applicants were urged to obtain additional institutional and corporate support for these costs. (After the grants were awarded, the exclusion of test cost was rescinded for those studies involving subjects without a family history of CF.) Another requirement was that minorities and women were to be sufficiently represented in study populations.

In September 1991, NIH awarded eight grants to seven research teams around the country (box 6-B). A consortium approach to the pilot projects has been adopted: Grantees meet in workshops to coordinate and share information. "The underlying goal of these studies is to help determine whether CF mutation analysis should remain focused on members of families already known to be at risk, or whether it is feasible to offer the test more widely in an ethically acceptable manner" (52).

For fiscal year 1993, NCHGR announced that it intends to collaborate with the National Cancer Institute to begin pilot projects to help health care

### Box 6-B—Clinical Studies of Testing, Education, and Counseling for Cystic Fibrosis Mutations, National Institutes of Health

In October 1991, three components of the National Institutes of Health—the National Center for Human Genome Research, the National Institute of Child Health and Human Development, and the National Center for Nursing Research—launched a 3-year research initiative to optimize parameters for educating and counseling individuals who want to be screened for CF mutations. The research teams supported under this initiative coordinate their efforts. Where appropriate, some features of the research, such as evaluation measures and tools, cost assessment, laboratory quality control procedures, and human subjects protections have been standardized across sites. U.S. research teams at seven sites will conduct eight studies:

*University of North Carolina, Chapel Hill, NC, "An Evaluation of Testing and Counseling for CF Carriers" ($231,916).* Close relatives of CF patients will receive pretest education either from a pamphlet in a private physician's office or in a traditional genetic counseling setting. The investigators will also assess the effectiveness of a pre-genetic-counseling video for CF carrier screening clients. Both before and after receiving the results of CF carrier tests, subjects will be assessed to determine genetic and medical knowledge, psychological status, and selected health behaviors.

*Children's Hospital Oakland Research Institute, Oakland, CA, "Perception of Carrier Status by Cystic Fibrosis Siblings" ($73,196).* By interviewing the adult siblings of CF patients and the CF siblings' spouses, the investigators will identify factors motivating or interfering with the pursuit of CF carrier testing in siblings, and assess their spouses' level of interest in screening. In addition to examining interest in testing, this study aims to assess the levels of understanding of test results and knowledge of medical aspects of CF, as well as to assess psychological functioning of CF siblings and spouses following testing.

*Vanderbilt University, Nashville, TN, "Cystic Fibrosis Screening: An Alternative Paradigm" ($206,513).* This study aims to determine the feasibility of a CF carrier screening program that incorporates pre- and post-test education for people with negative screening tests and provides personal counseling primarily for those who test positive for CF carrier status. Written and video materials will be developed. The investigators will examine various settings for provision of carrier screening, determine the factors that affect a couple's decision whether or not to be screened for CF carrier status, and determine general acceptance of population screening.

*University of Rochester, Rochester, NY, "Testing and Counseling for Cystic Fibrosis Mutations" ($274,110).* CF carrier tests will be offered to women of reproductive age to determine what proportion desire it, what proportion of women who are tested adequately comprehend the significance of the results, and what proportion of partners

*(Continued on next page)*

**Box 6-B—Clinical Studies of Testing, Education, and Counseling for Cystic Fibrosis Mutations, National Institutes of Health—Continued**

of the screened women decide to be screened themselves. Anxiety, lack of comprehension, requests for prenatal diagnosis despite low risk, and the costs of the program will be assessed.

*UCLA School of Medicine, Los Angeles, CA, "Cystic Fibrosis Mutations Screening and Counseling" ($179,067).* Women of reproductive age and the partners of those who test positive will be screened. The target population includes large numbers of Hispanic and Asian Americans, two groups that have not been studied extensively for either their CF mutation frequencies or their response to screening and counseling. Pre- and post-test questionnaires will be used to determine level of understanding of CF, predictors of consent to screening, and emotional responses to implications of the test results in the various ethnic and socioeconomic subgroups. Strategies of pre- and post-test counseling will be compared for their effectiveness.

*Johns Hopkins University, Baltimore, MD, "Ethical and Policy Issues in Cystic Fibrosis Screening" ($314,449).* This project focuses on families and individuals receiving care from a health maintenance organization. It seeks to determine the level of interest in learning more about CF and factors that distinguish those who are interested in participating in a CF education program from those who are not. The focus consists of three elements: education of the study population, determination of the characteristics that distinguish those who agree to have the CF carrier test from those who decide not to be screened, and comparison of the responses of individuals identified as CF carriers and those identified as probable noncarriers, with emphasis on the extent to which these responses are influenced by marital status, or carrier status of the partner. All participants who test positive for CF carrier status and a sample of those who test negative will be followed for 1 year.

*University of Pennsylvania, Philadelphia, PA, "Prescriptive Decision Modeling for Cystic Fibrosis Screening" ($197,634 and $180,201).* Decision theory and economic techniques will be used to model decisionmaking about CF carrier screening that addresses the following issues: who should be offered carrier screening and the best method for screening couples; the best course and sequence of further screening and treatment following initial results; rescreening individuals who have been screened in the past for CF mutations as more mutations are uncovered; the anticipated impact of future technologic innovation on CF carrier screening and treatment; tradeoffs between monetary and nonmonetary effects that the alternative answers to these questions imply; and differences in responses of various groups (i.e., patients, health care providers, and insurance companies, which have varying financial, psychological, and moral perspectives).

In addition, a team will conduct a clinical study, *How Much Information About the Risk of Cystic Fibrosis Do Couples Want?*, to complement the theoretical work. This project will analyze the decisionmaking processes of preconceptional and prenatal couples who are offered CF carrier screening one partner at a time, and, in the event of a negative result for the first partner, whether or not the couple chooses to have the second partner screened. The appropriate timing of CF carrier screening, as well as the amount that should be performed, will be investigated.

SOURCE: National Center for Human Genome Research, National Institutes of Health, October 1991.

professionals understand the best way to educate and deliver genetic tests to patients who ask for them, specifically genetic tests related to colon and breast cancer (75).

### *Funding for Genetic Services*

Recent Federal support for genetic services, and thus salaries, has been minimal. Prior to 1981, genetic programs could apply through their State for funds under the National Genetic Diseases Act. The Omnibus Reconciliation Act of 1981 (Public Law 97-35) replaced the National Genetic Diseases Act and amended Title V of the Social Security Act to create the Maternal and Child Health (MCH) Block Grant (ch. 2). This resulted in a drastic reduction of direct Federal support for genetic services not related to newborn screening (38). States that had received Federal support in the past now had to rely on the discretion of their State agencies and compete with other public health initiatives for diminishing dollars.

In fiscal year 1990, Federal and State funds for genetic services other than newborn screening totaled about $34 million, of which the Federal share was approximately $12 million (table 6-5). Other sources of funds that States used for genetic services

## Table 6-5—Total Funding for Genetic Services by State, Fiscal Year 1990[a]

| State | Total funds | Maternal and Child Health block grant | Other Federal | State | Other |
|---|---|---|---|---|---|
| AL. | $ 228,000 | $24,000 | 0 | $60,000 | $144,000[b,c] |
| AK. | 103,000 | 103,000 | 0 | 0 | NA |
| AZ. | 524,600 | 0 | $218,600 | 306,000 | NA |
| AR. | 984,200 | 0 | 0 | 574,600[d] | 409,600[b] |
| CA. | 16,673,300 | 0 | 470,200 | 2,163,800 | 14,039,300[e] |
| CO. | 403,000 | 0 | 0 | 27,000 | 376,000[f] |
| CT. | 371,400 | 0 | 0 | 371,400 | NA |
| DE. | 145,800 | 0 | 0 | 145,800 | NA |
| DC. | 481,200 | 40,000 | 220,700 | 220,500 | NA |
| FL. | 1,070,300 | 0 | 0 | 1,070,300 | NA |
| GA. | 1,567,000 | 162,700 | 0 | 1,404,300 | NA |
| HI. | 318,900 | 102,000 | 216,900 | 0 | NA |
| ID. | 246,900 | 123,400 | 0 | 0 | 123,500[b] |
| IL. | 292,000 | 262,800 | 0 | 0 | 29,200[f] |
| IN. | 843,100 | 254,200 | 0 | 149,300 | 439,600[c,g] |
| IA. | 935,000 | 0 | 0 | 860,200 | 74,800[b] |
| KS. | 50,000 | 0 | 0 | 50,000 | NA |
| KY. | 291,000 | 171,700 | 0 | 0 | 119,300[b,g] |
| LA. | 388,600 | 0 | 0 | 388,600 | NA |
| ME. | 293,400 | 60,000 | 0 | 233,400 | NA |
| MD. | 798,300 | 400,300 | 0 | 398,000 | NA |
| MA. | 716,100 | 0 | 632,500 | 83,600 | NA |
| MI. | 725,000 | 0 | 0 | 100,000 | 625,000[f] |
| MN. | 256,400 | 51,900 | 135,800 | 68,700 | NA |
| MS. | 333,300 | 0 | 173,300 | 0 | 160,000[f] |
| MO. | 1,617,000 | 180,000 | 0 | 1,437,000 | NA |
| MT. | 423,400 | 0 | 59,400 | 0 | 364,000[h] |
| NE. | 202,000 | 167,700 | 0 | 34,300 | NA |
| NV. | 582,100 | 54,400 | 80,000 | 217,500 | 230,200[e,g] |
| NH. | 152,300 | 38,100 | 0 | 91,400 | 22,800[b] |
| NJ. | 500,700 | 197,200 | 160,700 | 142,800 | NA |
| NM. | 635,400 | 31,000 | 0 | 604,400 | NA |
| NY. | 26,654,200 | 1,755,300 | 849,800 | 1,260,000 | 22,789,100[b,c,e,g] |
| NC. | 2,152,900 | 344,500 | 0 | 1,808,400 | NA |
| ND. | 99,900 | 15,000 | 0 | 0 | 84,900[g] |
| OH. | 3,558,500 | 282,200 | 0 | 1,552,400 | 1,723,900[b,e,g] |
| OK. | 386,400 | 236,400 | 150,000 | 0 | NA |
| OR. | 599,000 | 35,200 | 0 | 121,500 | 442,300[b] |
| PA. | 779,800 | 649,000 | 130,800 | 0 | NA |
| PR. | 107,500 | 0 | 0 | 25,000 | 82,500[e,g] |
| RI. | 1,355,000 | 0 | 155,000 | 1,200,000[i] | NA |
| SC. | 300,000 | 0 | 0 | 0 | 300,000[g] |
| SD. | 112,600 | 64,200 | 0 | 48,400 | NA |
| TN. | 2,332,900 | 291,600 | 0 | 1,808,000 | 233,300[b,j] |
| TX. | 4,311,900 | 415,600 | 201,800 | 2,300,000 | 1,394,500[b,c,e,g] |
| UT. | 197,000 | 197,000 | 0 | 0 | NA |
| VT. | 235,000 | 103,700 | 0 | 99,400 | 31,900[b,c,e] |
| VA. | 755,000 | 513,400 | 0 | 241,600 | NA |
| WA. | 1,300,000 | 650,000 | 0 | 0 | 650,000[b,c,e,g] |
| WV. | 375,000 | 155,000 | 0 | 0 | 220,000[e,g] |
| WI. | 540,300 | 47,300 | 0 | 0 | 493,000[e,g] |
| WY. | 125,000 | 0 | 125,000 | 0 | NA |
| TOTAL | $79,430,600 | $8,179,800 | $3,980,500 | $21,667,600 | $45,602,700 |

[a]Excluding newborn screening..
[b]Funds derived from third-party reimbursement.
[c]Funds derived from grants and contracts.
[d]Funds derived from a one-time grant of 18 months.
[e]Funds derived from provider service charges.
[f]Funds derived from newborn screening fee.
[g]Funds derived from provider in-kind services.
[h]Funds derived from surcharge to health insurers.
[i]Funds reported are for an integrated, tertiary care program that includes a genetic service component.
[j]Funds derived from mental health and mental retardation funds.
NA = Not available.
SOURCE: F.J. Meaney, "CORN Report on Funding of State Genetic Services Programs in the United States, 1990," contract document prepared for the U.S. Congress, Office of Technology Assessment, April 1992.

include provider-in-kind services, third-party reimbursement, and user fees. These funding mechanisms provided an additional $45.6 million for at least 26 States (44).

Support for education, training, and services of master's-level genetic counselors and other genetics personnel also comes chiefly through the MCH block grant and has declined precipitously. MCH genetics laboratory training grants totaled just under $1 million in 1991 spread among 9 States through Special Projects of Regional and National Significance (SPRANS) monies, down from $2.6 million in 1981. In real purchasing power, this decrease represents a decline of about 76 percent. Total SPRANS funding, not just that devoted to training, but to provide seed money for services, education, and technical assistance demonstration projects has also declined in real purchasing power since 1981 (figure 6-2). In addition to Federal funding, at least 25 States devote State monies to education, technical assistance, and training (table 6-6) (44).

**Figure 6-2—Federal Support of Genetic Services Through the Special Projects of Regional and National Significance (SPRANS), 1978-91**

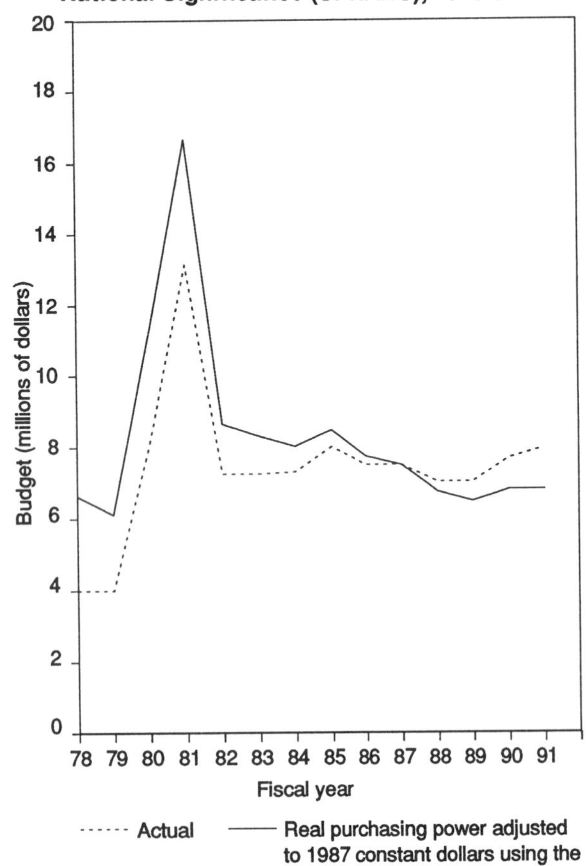


SOURCE: Office of Technology Assessment, based on E. Duffy, Maternal and Child Health Bureau, U.S. Department of Health and Human Services, Rockville, MD, personal communication, February 1992.

# RESULTS OF NONFEDERAL PILOTS

In the absence of federally sponsored pilot projects to evaluate CF carrier screening, several public and private institutions began to systematically screen subsets of the population—pregnant women and their partners, preconceptional teenagers and adults, and fetuses. While most are collecting data on the incidence of carrier status and mutation frequencies, some are also following psychosocial issues, such as levels of anxiety and retention of information. The various populations targeted for screening and the strategies used reflect the lack of consensus on the best approach to CF carrier screening. Some of these privately funded pilot projects are described in the following section. Because most were initiated at least one year before

**Table 6-6—Funding for Genetics Education, Technical Assistance, and Training, Fiscal Year 1990[a]**

| State | Funding level | Percent of total[b] |
|---|---|---|
| AR. | $14,000 | 13.1 |
| DE. | 6,800 | 4.7 |
| GA. | 20,400 | 1.3 |
| HI. | 10,200 | 3.2 |
| ID. | 19,000 | 7.7 |
| IL. | 29,200 | 10.0 |
| IA. | 467,500 | 50.0 |
| LA. | 16,900 | 4.4 |
| MD. | 98,000 | 12.3 |
| MA. | 372,900 | 52.1 |
| MI. | 320,000 | 44.1 |
| MS. | 166,700 | 50.0 |
| MO. | 80,800 | 5.0 |
| NV. | 27,400 | 7.8 |
| NM. | 195,200 | 30.7 |
| NC. | 215,300 | 10.0 |
| OK. | 19,500 | 5.0 |
| OR. | 221,500 | 37.0 |
| SD. | 69,800 | 62.0 |
| TN. | 58,300 | 2.5 |
| UT. | 14,800 | 7.5 |
| WA. | 260,000 | 20.0 |
| WV. | 33,400 | 8.9 |
| WI. | 216,400 | 40.0 |
| WY. | 10,000 | 8.0 |

[a] Figures not available for States not listed.
[b] Calculated as a percentage of genetic services funding (excluding newborn screening) from State, Federal, and other sources.

SOURCE: F.J. Meaney, "CORN Report on Funding of State Genetic Services Programs in the United States, 1990," contract document prepared for the U.S. Congress, Office of Technology Assessment, April 1992.

the NIH studies commenced, these efforts have more data.

### *Baylor College of Medicine: Prenatal and Preconceptional Carrier Screening of Couples*

From 1990 through 1991, Baylor College of Medicine (Houston, TX) processed more than 1,800 samples for CF carrier screening and testing, using six mutations at 84.5 percent sensitivity. Baylor employs a two-step approach. First, both partners are concurrently screened for ΔF508+5. Anxiety of pregnant women who test positive and must wait for the results of their partner's test is reduced if both samples are processed simultaneously, rather than sequentially. Second, partners of identified carriers are subsequently analyzed for 12 additional mutations at no extra charge (24).

The original Baylor population was a mix of prenatal and preconceptional couples, many related to affected individuals. Of the high-risk group, 64 at-risk pregnancies were diagnosed. Of these 64, 14 affected fetuses were found; half of the pregnancies were electively terminated. Sixteen carrier fetuses were identified. Of those couples found to be +/-, no pregnancies were terminated and there did not appear to be undue anxiety. None of these +/-couples requested prenatal fetal diagnosis. Six couples in 1991 were identified as +/+ prior to conception (24).

Starting in September 1991, screening has been offered to all couples of reproductive age who have contact for any reason with Baylor's genetic services. Again, couples are screened, rather than individuals, at a charge of $100 per couple. Identified carriers are encouraged to refer their relatives for testing.

### *Cornell University Medical College: Prenatal Carrier Screening*

Since April 1990, Cornell University Medical College has offered CF carrier screening to couples with a negative family history for CF who are enrolled in the prenatal diagnosis program (primarily for advanced maternal age). In 1992, screening has been extended to all couples of reproductive age coming to the genetic service, whether or not pregnancy is involved.

Initially, one partner was screened for ΔF508 mutation only; the W1282X mutation was added later because 30 percent of the Cornell couples are Ashkenazic Jews. If the partner is positive, followup testing of the other partner is done using six mutations. More than 500 couples have been screened to date using mouth rinse specimens. At a charge of $100 per couple, about 33 percent choose to participate. Those who choose to participate cite an interest in learning about the health of the fetus. Those who choose not to participate primarily cite a perceived low carrier risk and the fact that the patient's referring physicians had not specifically recommended the test.

All those who participate in the screening are informed (and in followup questionnaires, acknowledge) that the assays will miss some at-risk couples. Virtually all agree that the screening should continue and not be slowed until a greater proportion of CF carriers can be detected, or limited to those ethnic groups in which the detection rates are the highest (19).

### *Genetics & IVF Institute: Elective Fetal Screening*

The Genetics & IVF Institute (Fairfax, VA) is a clinical and laboratory facility that provides integrated outpatient services in the areas of human genetics and infertility. In 1990, the Institute began offering CF carrier screening of fetal samples to an unselected Caucasian population undergoing amniocentesis or CVS primarily for advanced maternal age.

As of August 1991, 4,782 consecutive Caucasian patients undergoing a prenatal procedure were offered concurrent CF carrier screening of their fetal sample, on a self-paying basis. Initially, screening only detected ΔF508, but for some time has included ΔF508 and six other mutations. Of 3,013 CVS patients, 1,327 (44 percent) elected screening. Of 1,769 amniocentesis patients, 370 (20.9 percent) chose CF carrier screening. Three carrier fetuses were found in patients with a family history of CF, and 48 carrier fetuses, including a set of twins, were found in patients with a negative family history. Of these 50 couples, 12 declined further testing. In one couple requesting further testing, both partners were found to be ΔF508 carriers. In the other couples who had further testing, only one partner in each carried a mutation; all carried the ΔF508 mutation except for one who carried the G542X mutation. No couples chose to terminate the pregnancy based on these results.

Patients are called by a counselor or physician after delivery of the baby to determine the pregnancy outcome and apparent health of the child, to determine if a sweat test was performed, and to discuss retrospective attitudes toward CF carrier screening (63).

### *Roche Biomedical Laboratories: Prenatal Couples Screening*

Roche Diagnostic Genetics, a national, full-service commercial genetics laboratory, launched a nationwide collaborative research study of CF carrier screening in July 1991. Using a reverse dot blot/ploymerase chain reaction method (ch. 4), Roche intends to screen 20,000 couples in the United States. Originally intended to last 6 months, participation has been less than expected, so the study has been expanded to 1 year. Assays are performed on buccal cell samples obtained at home using a buccal brush, collected in tubes, and mailed to Roche Biomedical Laboratories (Research Triangle Park, NC). Roche believes the ideal patients are those who are 15 to 16 weeks pregnant and are undergoing MSAFP screening. The sample can be collected as early as the first prenatal visit, week 8 of the pregnancy. This "captive" population, believe Roche officials, is more likely to volunteer as research participants (3). A solicitation letter was sent to 100 obstetricians around the country introducing the program.

Roche employs the two-step approach, although samples will be collected simultaneously for both partners. The woman's sample will be screened first for ΔF508, G551D, G542X, and R553X. These mutations collectively account for about 85 percent of all CF mutations, according to Roche. If the woman's sample tests negative for these mutations, analysis is not performed on her partner's sample. The couple is then informed that they are at diminished risk. If the woman tests positive for one of the four mutations, then her partner's sample is tested for the same four mutations. If his sample is negative, the couple is told they are at reduced risk, and the woman is informed that she is a carrier. If the sample is positive, the couple is referred to a genetic counselor and advised of opportunities for prenatal diagnosis.

Roche officials believe this approach avoids undue anxiety on the part of a pregnant woman found to be positive. Roche reports that this approach will detect 72 percent of at-risk couples and 85 percent of carrier females. As of the fall of 1991, the subscription rate was 50 percent. There are no plans to offer the screen to preconception individuals (3).

Hypothetically, if Roche screens 20,000 Caucasian couples, and assuming a carrier frequency of 1 in 25, some 800 women will be identified as carriers. If the partners of all 800 women are tested, 32 men will be found to be carriers. Thus, out of 20,000 couples, at a detection rate of 72 percent, 23 at-risk pregnancies will be identified. Because Roche will not screen the male samples unless the female sample is positive, however, the opportunity to identify 768 male carriers is lost.

### *McGill University: High School Carrier Screening*

In Montreal, Canada, carrier screening for genetic diseases, such as Tay-Sachs, is a common practice in some high schools. In May 1990, nine students and four biology teachers at four schools in the Montreal area conducted a pilot study of attitudes in persons tested for the ΔF508 mutation. Forty percent of the nearly 600 students invited to participate in the project did so. Of these, two carriers were found. The carriers and their families were interviewed and found to hold positive views about their new awareness. Additional family members have been tested at their own request. Followup questionnaires showed that participants who received negative test results were found to be reasonably well informed about the CF clinical phenotype, its inheritance, and its distribution. Most understood that a negative test did not rule out carrier status and were satisfied they had taken the test (35).

### *Permanente Medical Group, Inc. —Vivigen— Integrated Genetics: Carrier Screening of Pregnant Women*

In November 1991, the Kaiser Permanente Health Care System of Northern California undertook screening of 5,000 pregnant Caucasian and Hispanic women—with a negative family history only—for CF carrier status. The analysis and cost of running the samples is equally divided between Vivigen, Inc. (Santa Fe, NM) and Integrated Genetics (Framingham, MA). The samples are screened for six mutations, at a sensitivity of about 85 percent.

Women are screened first. If positive, their partner's specimen is obtained and tested for 12 mutations.

Kaiser has developed an informational videotape that is being tested on control and experimental groups to determine its adequacy for educational use. In addition, several psychosocial survey instruments are being used to assess patients' understanding of the progression and genetics of CF both before and after screening.

The pilot program will end after 5,000 samples have been analyzed. At that time, Kaiser Permanente will make a decision as to how to proceed with general screening for members of its health plan. As of March 1992, 78 percent of women offered CF mutation analysis elected it (82).

# SUMMARY AND CONCLUSIONS

The prospect of a highly sensitive, inexpensive assay for CF carrier status is not far into the future. As the sensitivity approximates 90 percent[1] for the general population, demand for carrier screening is likely to increase as the medical profession concomitantly recognizes its increasing duty to inform patients about the availability of the information. The ambiguous nature of the information, however, requires that the consequences of screening be fully understood.

Public education can go a long way toward preparing individuals for the decision of whether and when to be screened. However, public education campaigns related to family planning issues, such as CF carrier screening, are unlikely to be sponsored by the Federal Government. Thus, the clinical genetics community will have to work with allied health professionals and educators in designing and delivering information regarding CF carrier screening and, for that matter, other genetic tests to come.

In addition, the clinical genetics community will need to train other health care providers to help bear the educational burden as CF and other genetic tests become widely used. This expansion must maintain the nondirective philosophy of traditional genetic counseling. The Federal Government, through more support for training and genetic services, could facilitate this effort.

Because the current genetics infrastructure is built around the concept of entry into genetic services during the prenatal period or following the birth of an affected child, adults involved in a pregnancy will likely be the first population to undergo routine CF carrier screening—this despite recognition that preconceptional screening is considered by most to be the optimal situation. Although adolescent screening programs appear to be successful in Canada, they are as yet unproven in the United States. At this time, widespread newborn CF carrier screening is unlikely.

Privately funded pilot studies have contributed groundbreaking and timely data about test sensitivity, target population, participant education, and patient response. As well, the Federal Government, through its clinical assessments of CF carrier screening, is playing an important role in examining the factors important to widespread carrier screening. The lessons of the past, from Tay-Sachs, sickle cell, and MSAFP screening programs, are instructive for the future.

# CHAPTER 6 REFERENCES


1. Al-Jader, L.N., Goodchild, M.C., Ryley, H.C., et al., "Attitudes of Parents of Cystic Fibrosis Children Towards Neonatal Screening and Antenatal Diagnosis," *Clinical Genetics* 38:460-465, 1990.
2. Angastiniotis, M., Cyprus Thalassaemia Centre, Archbishop Makarios III Hospital, Nicosia, Cyprus, "Development of Genetics Services From Disease Oriented National Genetics Programs," presentation at the 8th International Congress of Human Genetics, Washington, DC, October 1991.
3. Barathur, R., Roche Biomedical Laboratories, Research Triangle Park, NC, personal communication, September 1991.
4. Benkendorf, J.L., and Bodurtha, J.M., "Human Genetics Education Videoshare: Round One," *American Journal of Human Genetics* 44:611-615, 1989.
5. Biesecker, L., Bowles-Biesecker, B., Collins, F., et al., "General Population Screening for Cystic Fibrosis Is Premature," *American Journal of Human Genetics* 50:439, 1992.
6. Botkin, J.R., "Prenatal Screening: Professional Standards and the Limits of Parental Choice," *Obstetrics and Gynecology* 75:875-880, 1990.
7. Bowling, F., Cleghorn, G., Chester, A., et al., "Neonatal Screening for Cystic Fibrosis," *Archives of Disease in Childhood* 63:196-198, 1988.
8. Bowman, J.E. "Invited Editorial: Prenatal Screening for Hemoglobinopathies," *American Journal of Human Genetics* 48:433-438, 1991.


[1] As noted previously, dF508+6-12 detects 85 to 90 percent of CF carriers in the general population, but 95 percent in Ashkenazic Jews.

9. Brock, D.J., "Cystic Fibrosis," *Antenatal and Neonatal Screening*, N.J. Wald (ed.) (New York, NY: Oxford University Press, 1984).
10. Cao, A., "A Disease Oriented National Health Service," *American Journal of Human Genetics* 49(Supp.):31, 1991.
11. Caskey, C.T., Kaback, M.M., and Beaudet, A.L., "The American Society of Human Genetics Statement on Cystic Fibrosis," *American Journal of Human Genetics* 46:393, 1990.
12. Chase, G.A., Faden, R.R., Holtzman, N.A., et al., "Assessment of Risk by Pregnant Women: Implications for Genetic Counseling and Education," *Social Biology* 33:57-64, 1986.
13. Clarke, A., "Is Non-Directive Genetic Counselling Possible?," *Lancet* 338:998-1001, 1991.
14. Collins, D.L., University of Kansas Medical Center, Kansas City, KS, personal communication, September 1991.
15. Cox, T.K., and Chakravarti, A., "Detection of Cystic Fibrosis Gene Carriers: Comparison of Two Screening Strategies by Simulations," *American Journal of Human Genetics* 49(Supp.):327, 1991.
16. Crisis Counseling and HIV Antibody Testing, Report from the First Interdisciplinary Conference on Human Immunodeficiency Virus Antibody Testing and Counseling, sponsored by Sarah Lawrence College and Memorial Sloan-Kettering Cancer Center, Nov. 17, 1987.
17. Cunningham, G.C., and Kizer, K.W., "Maternal Serum Alpha-Fetoprotein Screening Activities of State Health Agencies: A Survey," *American Journal of Human Genetics* 47:899-903, 1990.
18. Davis, J.G., "Invited Editorial: State-Sponsored Maternal Serum Alpha-Fetoprotein Activities: Current Issues in Genetics and Public Health," *American Journal of Human Genetics* 47:896-898, 1990.
19. Davis, J.G., New York Hospital, Cornell University Medical College, personal communication, March 1992.
20. Earley, K.J., Blanco, J.D., Prien, S., et al. "Patient Attitudes Toward Testing for Maternal Serum Alpha-Fetoprotein Values When Results Are False-Positive or True-Negative," *Southern Medical Journal* 84: 439-442, 1991.
21. Edwards, J.H., and Miciak, A., "The Slash Sheet: A Simple Procedure for Risk Analysis in Cystic Fibrosis," *American Journal of Human Genetics* 47:1024-1028, 1990.
22. Elias, S.E., Annas, G.J., and Simpson, J.L., "Carrier Screening for Cystic Fibrosis: Implications for Obstetrics and Gynecologic Practice," *American Journal of Obstetrics and Gynecology* 164:1077-1083, 1991.
23. Evers-Kiebooms, G., and van den Berghe, H., "Impact of Genetic Counseling: A Review of Published Follow-Up Studies," *Clinical Genetics* 15:465-474, 1987.
24. Fernbach, S.D., Baylor College of Medicine, Houston, TX, personal communication, December 1991.
25. Fibison, W.J., "The Nursing Role in the Delivery of Genetic Services," *Issues in Health Care of Women* 4:1-15, 1983.
26. Fine, B.A., Northwestern University Medical School, Chicago, IL, personal communication, September 1991.
27. Forsman, I., "Education of Nurses in Genetics," *American Journal of Human Genetics* 43:552-558, 1988.
28. Friedman, J.M., and Blitzer, M., "ASHG/NSGC Activities Related to Education: Workshop on Human Genetics Education," *American Journal of Human Genetics* 49:1127-1128, 1991.
29. Harper, P.S., and Clarke, A., "Viewpoint: Should We Test Children For 'Adult' Genetic Diseases?," *Lancet* 335:1205-1206, 1990.
30. Hodgkinson, K.A., Kerzin-Storrar, L., Watters, E.A., et al., "Adult Polycystic Kidney Disease: Knowledge, Experience, and Attitudes to Prenatal Diagnosis," *Journal of Medical Genetics* 27:552-558, 1990.
31. Holtzman, N.A., *Proceed With Caution: Predicting Genetic Risks in the Recombinant DNA Era* (Baltimore, MD: The Johns Hopkins University Press, 1989).
32. Jones, S.L., "Decision Making in Clinical Genetics: Ethical Implications for Perinatal Practice," *Journal of Perinatal and Neonatal Nursing* 111-23, 1988.
33. Jones, S.L., Genetics & IVF Institute, Fairfax, VA, personal communication, December 1991.
34. Kahneman, D., and Tversky, A., "The Psychology of Preference," *Scientific American* 246:160-171, 1982.
35. Kaplan, F., Clow, C., and Scriver, C.R., "Cystic Fibrosis Carrier Screening by DNA Analysis: A Pilot Study of Attitudes Among Participants," *American Journal of Human Genetics* 49:240-243, 1991.
36. Keenan, K.L., Basso, D., Goldkrand, J., et al., "Low Level of Maternal Serum Alpha-Fetoprotein: Its Associated Anxiety and the Effects of Genetic Counseling," *American Journal of Obstetrics and Gynecology* 164:54-56, 1991.
37. Kessler, S. (ed.), *Genetic Counseling: Psychological Dimensions* (New York, NY: Academic Press, Inc., 1979).
38. Lin-Fu, J.S., Genetic Services Branch, Maternal and Child Health Bureau, U.S. Department of Health and Human Services, Rockville, MD, personnal communication, April 1992.
39. Lipkin, M., Fisher, L., Rowley, P.T., et al., "Genetic Counseling of Asymptomatic Carriers in a Primary Care Setting," *Annals of Internal Medicine* 105:115-123, 1986.

40. Lippman-Hand, A., and Fraser, F.C., "Genetic Counseling—The Postcounseling Period: Parents' Perceptions of Uncertainty," *American Journal of Medical Genetics* 4:51-71, 1979.
41. Loader, S., Sutera, C.J., Walden, M., et al., "Prenatal Screening for Hemoglobinopathies. II. Evaluation of Counseling," *American Journal of Human Genetics* 48:447-451, 1991.
42. Marteau, T.M.., "The Impact of Prenatal Screening and Diagnostic Testing Upon the Cognitions, Emotions, and Behaviour of Pregnant Women," *Journal of Psychosomatic Research* 33:7-16, 1989.
43. Marteau, T.M., "Reducing the Psychological Costs," *British Medical Journal* 301:26-28, 1990.
44. Meaney, F.J., "CORN Report on Funding of State Genetic Services Programs in the United States, 1990," contract document prepared for the U.S. Congress, Office of Technology Assessment, April 1992.
45. McNeil, B.J., Pauker, S.G., Sox, H.C., et al., "On the Elicitation of Preferences for Alternative Therapies," *New England Journal of Medicine* 306:1259-1262, 1982.
46. Miller, S.R., and Schwartz, R.H., "Attitudes Toward Genetic Testing of Amish, Mennonite, and Hutterite Families With Cystic Fibrosis," *American Journal of Public Health* 82:236-242, 1992.
47. Mischler, E., et al., "Progress Report: Neonatal Screening for Cystic Fibrosis in Wisconsin," *Wisconsin Medical Journal* 88:14-18, 1989.
48. Modell, B., "Cystic Fibrosis Screening and Community Genetics," *Journal of Medical Genetics* 27:475-479, 1990.
49. Modell, B., Ward, R.H., and Fairweather, D.V., "Effect of Introducing Antenatal Diagnosis on Reproductive Behavior of Families at Risk for Thalassemia Major," *British Medical Journal* 280:1347-1350, 1980.
50. Nance, W.E., "Statement of the American Society of Human Genetics on Cystic Fibrosis Carrier Screening," in press, 1992.
51. Nance, W.E., Rose, S.P., Conneally, P.M., et al., "Opportunities for Genetic Counseling Through Institutional Ascertainment of Affected Probands," *Genetic Counseling*, F. de la Cruz and H.A. Lubs (eds.) (New York, NY: Raven Press, 1977).
52. National Center for Human Genome Research, press release, "NIH Collaboration Launches Research on Education and Counseling Related to Genetic Tests," Oct. 8, 1991, Bethesda, MD.
53. National Institutes of Health, Workshop on Population Screening for the Cystic Fibrosis Gene, "Statement From the National Institutes of Health Workshop on Population Screening for the Cystic Fibrosis Gene," *New England Journal of Medicine* 323:70-71, 1990.
54. National Society of Genetic Counselors, Wallingford, PA, personal communication, June, 1991.
55. National Society of Genetic Counselors Cystic Fibrosis Ad Hoc Committee, *Genetic Testing for Cystic Fibrosis: A Handbook for Professionals* (Wallingford, PA: National Society of Genetic Counselors, 1990).
56. *Perspectives in Genetic Counseling* 12:6, spring 1990.
57. Riccardi, V.M., and Schmickel, R.D., "Human Genetics as a Component of Medical School Curricula: A Report to the American Society of Human Genetics," *American Journal of Human Genetics* 42:639-643, 1988.
58. Roberts, L., "Cystic Fibrosis Pilot Projects Go Begging," *Science* 250:1076-1077, 1990.
59. Rowley, P.T., Loader, S., Sutera, C.J., "Do Pregnant Women Benefit From Hemoglobinopathy Carrier Detection?," *Annals of the New York Academy of Sciences* 565:152-160, 1989.
60. Rowley, P.T., Loader, S., Sutera, C.J., et al., "Prenatal Screening for Hemoglobinopathies. I. A Prospective Regional Trial," *American Journal of Human Genetics* 48:439-446, 1991.
61. Schild, S., "Psychological Issues in Genetic Counseling of Phenylketonuria," *Genetic Counseling: Psychological Dimensions*, S. Kessler (ed.) (New York, NY: Academic Press, 1979).
62. Schull, W.J., and Hanis, C.L., "Genetics and Public Health in the 1990s," *Annual Review of Public Health* 11:105-125, 1990.
63. Schulman, J.D., Genetics & IVF Institute, Fairfax, VA, personal communication, March 1992.
64. Shaw, M.W., "Conditional Prospective Rights of the Fetus," *Journal of Legal Medicine* 5:63-115, 1984.
65. Shiloh, S., and Sagi, M., "Effect of Framing on the Perception of Genetic Recurrence Risks," *American Journal of Medical Genetics* 33:130-135, 1989.
66. Singer, E., "Public Attitudes Toward Genetic Testing," in *Population Research and Policy Review* 10:235-255, 1991.
67. Sorenson, J.R., Levy, H.L., Mangione, T.W., et al., "Parental Response to Repeat Testing of Infants With 'False-Positive' Results in a Newborn Screening Program," *Pediatrics* 73:183-187, 1984.
68. Sujansky, E., Kreutzer, S.B., Johnson, A.M., et al., "Attitudes of At Risk and Affected Individuals Regarding Presymptomatic Testing for Autosomal Dominant Polycystic Kidney Disease," *American Journal of Medical Genetics* 35:510-515, 1990.
69. Tabor, A., Norgaard-Pedersen, B., and Jacobsen, J.C., "Low Maternal Serum AFP and Down Syndrome," *Lancet* 2(8395):161-162, 1984.
70. Thompson, L., "Cell Test Before Implant Helps Ensure Healthy 'Test-Tube' Baby," *Washington Post*, Apr. 27, 1992.

71. U.S. Congress, Office of Technology Assessment, *Cystic Fibrosis and Genetic Screening: Policies, Practices and Attitudes of Genetic Counselors—Results of a Survey*, OTA-BP-BA-97 (Washington, DC: U.S. Government Printing Office, forthcoming 1992).
72. U.S. Department of Education, *1990 Science Report Card, National Assessment of Educational Progress* (Washington, DC: U.S. Government Printing Office, 1992).
73. Walker, A.P., Scott, J.A., Biesecker, B.B., et al., "Report of the 1989 Asilomar Meeting on Education in Genetic Counseling," *American Journal of Human Genetics* 46:1223-1230, 1990.
74. Waples, C.M., Buswell, B.E., Martz, J.C., et al., "Resources for Genetic Disorders," *Genetics Applications: A Health Perspective* (Lawrence, KS: Learner Managed Designs, 1988).
75. Watson, J.D., testimony before the House Appropriations Committee, Subcommittee on Labor—Health and Human Services—Education, Mar. 25, 1992.
76. Wertz, D.C., and Fletcher, J.C., "Attitudes of Genetic Counselors: A Multinational Survey," *American Journal of Human Genetics* 42:592-600, 1988.
77. Wertz, D.C., Janes, S.R., Rosenfeld, J.M., et al., "Attitudes Toward the Prenatal Diagnosis of Cystic Fibrosis: Factors in Decision Making Among Affected Families," *American Journal of Human Genetics* 50:1077-1085, 1992.
78. Wertz, D.C., Rosenfeld, J.M., Janes, S.R., et al., "Attitudes Toward Abortion Among Parents of Children With Cystic Fibrosis," *American Journal of Public Health* 81:992-996, 1991.
79. Wertz, D.C., Sorenson, J.R., and Heeren, T.C., "Clients' Interpretation of Risks Provided in Genetic Counseling," *American Journal of Human Genetics* 39:253-264, 1986.
80. Wertz, D.C., Sorenson, J.R., and Heeren, T.C., "Communication in Health Professional-Lay Encounters: How Often Does Each Party Know What the Other Wants to Discuss?," *Information and Behavior*, vol. 2, B.D. Ruben (ed.) (New Brunswick, NJ: Transaction Books, 1988).
81. Wilfond, B.S., and Fost, N., "The Cystic Fibrosis Gene: Medical and Social Implications for Heterozygote Detection," *Journal of the American Medical Association* 263:2777-2783, 1990.
82. Witt, D., Kaiser Permanente Medical Group, San Jose, CA, personal communication, March 1992.
83. Zeesman, S., Clow, C.L., Cartier, L., et al., "A Private View of Heterozygosity: Eight Year Follow-up Study on Carriers of the Tay-Sachs Gene Detected by High School Screening in Montreal," *American Journal of Medical Genetics* 18:769-778, 1984.

# Chapter 7

# Financing

# Contents



### *Boxes*



### *Tables*

# Chapter 7
# Financing

Health care financing in the United States is not monolithic: There are several forms of private financing, as well as public financing. This chapter provides a general review of health care financing in the United States. It briefly discusses how each entity determines eligibility for coverage and describes how each is regulated. OTA has examined the U.S. health insurance industry in greater detail elsewhere (27).

This chapter also describes a 1991 OTA survey of U.S. commercial health insurers, Blue Cross and Blue Shield (BC/BS) plans, and the largest health maintenance organizations (HMOs). This chapter focuses on survey results of private sector insurers' general attitudes towards genetic tests and their reimbursement practices for genetic tests and genetic services. Chapter 8 reports on results from this survey that pertain to the potential impact of genetic information or genetic tests on access to health care coverage.

## OVERVIEW OF U.S. HEALTH CARE FINANCING

Health care financing in the United States totaled more than $800 billion in 1991 (15). Public funding includes Medicare and Medicaid programs, as well as the Civilian Health and Medical Program of the Uniformed Services (CHAMPUS), which insures military personnel and their dependents. Private funding mechanisms include self-funded plans (which generally are plans administered by large employers), commercial health insurance plans, BC/BS plans, and HMOs. Finally, membership in a State high-risk pool—in the 25 States that have them—is an option that is increasingly available to individuals who cannot obtain private health insurance (table 7-1).

### *Public Financing*

Most public spending for health services covers six populations: low income individuals and others eligible for Medicaid, those over age 65, military personnel and their dependents, veterans, Federal civilian employees, and Native Americans. As with most workers in the United States, Federal civilian employees receive benefits through their employer, the Federal Government, through plans similar to private sector plans. Native Americans are covered under the Indian Health Service.

#### Medicaid

Medicaid is a joint State and federally funded program for low income citizens and people with disabilities. Administered by the States, it provides medical assistance to an estimated 6 percent of the U.S. population (25). Operating within Federal guidelines, each State designs and administers its own Medicaid program. Thus, Medicaid eligibility requirements, services offered, and methods and levels of payment to providers vary widely among States, although a minimum Federal standard of services must be covered. The adequacy of Medicaid in ensuring access to health care in general, and genetic services specifically, depends on these State-specific features.

#### Medicare

For people over the age of 65 and some disability recipients under age 65, Medicare is the primary source of health insurance, covering about 12.6 percent of the U.S. population (32). People below age 65 who are totally and permanently disabled can become eligible for Medicare coverage after a minimum waiting period. In this way, some adults with cystic fibrosis (CF) (who have worked and contributed to the Social Security system for a period) can receive medical coverage through Medicare under the program's disability provisions.

**Table 7-1—Health Care Coverage in the United States**

| U.S. population | 245 million |
|---|---|
| Persons with health care coverage | 214 million |
| Persons with private coverage[a] | 189 million |
| Persons with public coverage[b] | 45 million |
| Number of uninsured persons | 31 million |

[a]Persons with private coverage could be covered under commercial insurance plans, BC/BS plans, HMOs, or self-funded plans that offer these options. Of those covered by commercial plans, people can be covered under group plans, medically underwritten group plans, or individual plans.
[b]Some persons with public coverage also have private coverage.

SOURCE: Office of Technology Assessment, 1992, based on Health Insurance Association of America, *Source Book of Health Insurance 1991* (Washington, DC: Health Insurance Association of America, 1991); and U.S. Department of Commerce, Bureau of the Census, Current Population Reports Series P-70, No. 17, 1990.

### CHAMPUS

Medical treatment is available for all active and retired military personnel and their dependents at Department of Defense (DOD) medical sites through the Military Health Services System (MHSS). CHAMPUS, a component of MHSS, provides health care for certain dependents of active duty personnel, military retirees, and their dependents. In 1989, expenditures for medical care comprised $2.8 billion of the $13 billion DOD budget (14).

### State Pools

In response to citizens' difficulties in obtaining health care coverage, several States have established health insurance pools for underinsured and uninsurable persons. As of December 1990, State legislation creating State high-risk pools for such individuals had been created in 25 States, but not all are operating (14). Several additional States are considering legislation.

State insurance pools provide an opportunity for many to purchase health insurance regardless of circumstance or physical condition, although generally at a rate considerably higher than most other individual plans. Although eligibility for the plans varies from State to State, the basic criterion for participation is denial of coverage by other insurers. To qualify for a State high-risk pool, an individual typically must have been rejected for health care coverage at least three times for reasons related to medical risk factors (17,20).

State pools vary greatly in the type and amount of coverage they provide. Premiums are paid by enrollees, but are capped at a certain level. Enrollee premiums help fund pools, with the balance of costs financed by State revenues and insurers. Insurance companies contribute funds to pools proportional to their market share in the State. Self-funded plans (described in a following section) are the largest payers of health care in the United States, but are not assessed premiums for State high-risk pools.

Because State pools insure individuals with the highest risks for medical needs and do not have broad-based financing, they have not been without problems. The high-risk insurance pool in Florida, for example, covers 7,600 people, but was closed for new enrollments in April 1991 because of budget problems (12). State high-risk pools often have large deductibles, high premiums, and maximum lifetime benefits.

## *Private Financing*

For the majority of Americans, access to health care—and the health insurance that makes such access possible—is provided through the private sector. Privately financed health insurance for medical expenses covers more than 189 million persons through self-funded companies, commercial insurance companies, BC/BS plans, and managed care programs (e.g., HMOs and preferred provider organizations (PPOs)) (14). Although the term health insurance broadly includes various types of insurance—e.g., disability income or accident—this chapter focuses on health insurance for medical expenses (also known as major medical expense policies).

Private health insurance exists in a variety of forms. The majority of Americans obtain health insurance coverage through employment—either directly as employees or as family members of the employed. The employer, in turn, contracts with a commercial insurer, a BC/BS plan, or an HMO. Such groups are both large, with no diagnostic tests or physical examinations required for entry (i.e., no medical underwriting) or small (i.e., require some diagnostic tests or physical examinations, on which the insurance contract's coverage and costs are based). An employer can also be self-funded, meaning it does not pay premiums to an outside insurer but instead pays its employees' medical claims out of its own resources—although self-funded companies can buy claims processing services from outside insurers (box 7-A). Finally, persons without group coverage can seek individual health insurance from commercial insurers, BC/BS plans, or HMOs.

BC/BS plans provide both individual and group coverage to more than 80 million Americans (16). Nationwide, 73 BC/BS plans operate on a regional basis—many enjoying significant shares of their local health care coverage market—and all offer some form of individual health coverage. Market share and regional focus can play a pivotal role in how a BC/BS plan underwrites its policies. That is, unlike commercial insurers, BC/BS plans are regional and do not sell coverage outside a particular State, metropolitan area, or region. The market share of many BC/BS plans—though decreasing in recent years—has historically overshadowed that of any individual commercial carrier, so that in some States as much as half the population are BC/BS subscribers. A secure market position can shape underwrit-

### Box 7-A—Self-Funded Employee Health Benefit Plans

Since enactment of the Employee Retirement Income Security Act of 1974 (ERISA) (29 U.S.C. 1131 et seq.), many companies have found it beneficial to self-fund their employee health insurance benefits. Under ERISA, a company is viewed as an employer providing benefits, not an insuring entity, and so escapes State insurance regulation. Self-funded plans, for example, need not comply with State laws that mandate health insurance contracts to include specified benefits (e.g., minimum maternity coverage or alcohol and drug addiction treatment), nor comply with certain antidiscrimination standards applicable to insured plans. Self-funded plans are also exempt from State insurance premium taxes and need not participate in insurance pools for high-risk individuals. Self-funding is particularly attractive to multi-State employers that do not want to tailor their benefit plans to each set of State laws. Today, the majority of the large group market is self-funded, leaving most of the group benefits marketplace virtually unregulated by the States (33).

Self-funding means benefits are provided by an employer, which directly assumes most or all of the financial risk for its employees' health care expenses. Self-funded employers can use and retain earnings on a pretax basis on money they must otherwise set aside in claims reserves. The actual value of these reserves varies from company to company, but can represent a sizable portion of the annual premium. Many employers prefer to have the use of their capital instead of holding it in reserve.

Although some self-funded companies administer their own plans, most use independent third-party administrators—often other commercial insurers, BC/BS plans, or independent claims processors. In addition to administrative services, some commercial insurers and BC/BS plans also provide stop-loss insurance, which allows employers to self-insure their plan up to a certain dollar amount. Should a company have employee health care claims exceeding this amount, the stop-loss plan becomes effective and the policy pays additional claims—i.e., stop-loss insurance protects an employer from a catastrophic claim (7).

SOURCE: Office of Technology Assessment, 1992.

ing policies allowing a plan, for example, to enroll high-risk applicants because the plan can spread risks over a broader base.

BC/BS plans often operate under considerably different regulatory conditions from commercial carriers. Currently, BC/BS plans in 12 States have an open enrollment period, during which all individuals who apply for coverage are accepted regardless of their health status (16) (box 7-B), although most contracts have waiting periods for preexisting conditions. Some open enrollment plans are continuous (accept all applicants throughout the year), whereas others limit open enrollment to a designated number of weeks.

One of the fastest growing areas of health insurance in the last decade has been managed care groups such as PPOs[1] and HMOs. HMOs are health care organizations that provide comprehensive services to enrolled members for a fixed, prepaid amount that is independent of the number of services actually used. The market share for these plans has increased at the expense of conventional health insurance plans (14), although the net number of HMOs declined by 22 in 1990, reflecting industry consolidation. As of December 1990, there were 569 HMOs in the United States, with enrollment exceeding 36.5 million members (11).

As with other health insurers, HMOs are paid a fixed premium for each member. Unlike other insurers, however, an HMO is financially responsible for its members' medical costs only if the HMO's affiliated providers are used (except for emergencies) (24). By assuming not only the insurance risk but also the responsibility for providing their members' health care, HMOs operate under significantly different conditions from either BC/BS plans or commercial carriers. Another important distinction is that while commercial insurers and BC/BS plans are governed solely by State regulations, many HMOs voluntarily also adhere to Federal qualification standards. In order to become federally qualified, HMOs must meet certain financial, underwriting, and rate-setting standards and provide specified medically necessary health services (10). More than

[1] PPOs are similar to HMOs, but provide more flexibility in physician selection. They involve contractual arrangements with specific physicians or provider organizations.

### *Box 7-B—Open Enrollment and Blue Cross and Blue Shield Associations*

When BC/BS plans were first offered in the 1930s, all applicants were accepted for coverage regardless of their health status—i.e., open enrollment. Today, plans in 12 States have an open enrollment period, although most contracts have waiting periods for preexisting conditions. The implications of such plans for the underwriting process are significant. Because no individual standards of insurability are applied to open enrollment applicants, adverse selection exists. Adverse selection occurs when applicants seek coverage because they are aware of medical problems (and hence medical expenses) that are not yet evident to the underwriter.

Most plans attempt to hold down premium rates for open enrollment subscribers by providing less comprehensive benefits than are offered in other plans. Other BC/BS plans require open enrollment subscribers to pay higher premiums than underwritten applicants for identical coverage. Finally, open enrollment coverage of high-risk applicants usually entails a waiting period before initial benefits are paid, and they often impose limitations on coverage of preexisting conditions.

Some applicants to BC/BS open enrollment plans must furnish evidence of their health status, even though the plans never deny an application. Individuals seeking health care coverage through an open enrollment program often have the option of undergoing medical underwriting, and even a physical exam, to determine whether they qualify for a more comprehensive benefit package at a lower rate. Additionally, health information may be required by the underwriter to develop benefit limits, exclusion riders, waiting periods for preexisting conditions, or premium rates.

SOURCE: Office of Technology Assessment, 1992.

half the Nation's HMOs are federally qualified, and 74 percent of HMO enrollment is in federally qualified plans (11). Federal qualification can be important to consumers: If an HMO accepts non-Medicare individual members, they must be either accepted at a community rate or rejected altogether. Exclusion riders and rated premiums are prohibited. Waiting periods as well as preexisting condition waivers are not allowed. However, medical screening of individual applicants is permitted.

Photo credit: Carlton Agee

A health maintenance organization in Washington, DC.

### Individual Health Insurance

Despite the fact that most people in the United States obtain health care through group plans, many have no access to an employer-sponsored plan because they are unemployed, self-employed, or employed by companies that do not provide health benefits. Thirty-six percent of companies with fewer than 25 employees offered their workers health insurance in 1990, compared to 87 to 99 percent of larger employers (13). An individual who is unable to obtain health care coverage through his or her employer must generally seek individual health insurance.

Persons who obtain health care through individual health insurance policies—from 10 to 15 percent of all persons with health insurance—usually have their health status evaluated by the insurer to determine whether they are insurable, and if so, at what price (a process called rating). This evaluation of the applicant's risk is commonly referred to as medical underwriting, and relies at a minimum on a medical history questionnaire, and less frequently, on other sources of information such as an attending physician's statement or medical tests. Applicants in groups of 10 or fewer employees are often individually medically underwritten as well (8,21). Medium to large groups with 10 to 25 or more members are seldom, if ever, medically underwritten. Risk classification is also generally not used in employer-sponsored/

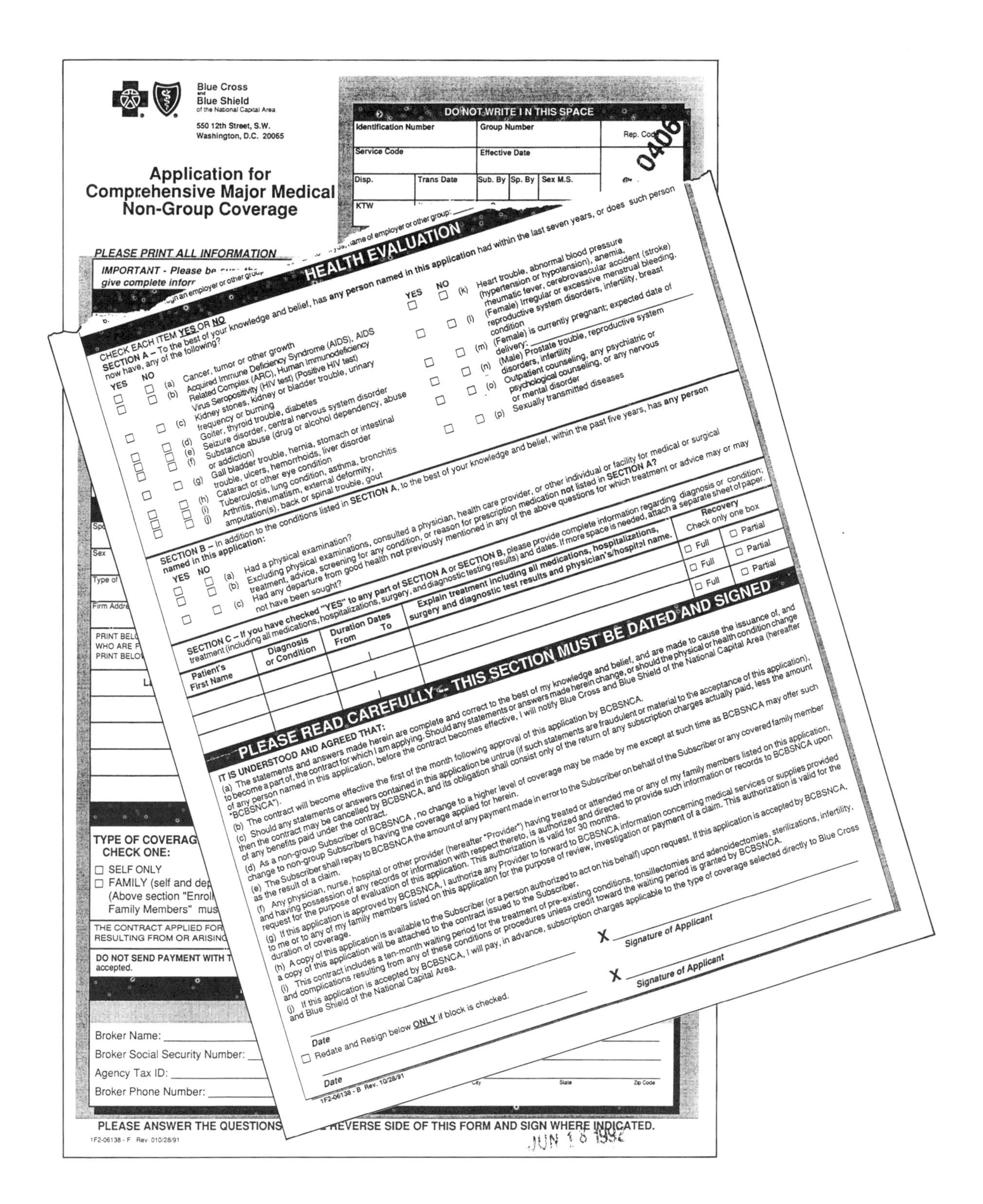

Blue Cross and Blue Shield of the National Capital Area
550 12th Street, S.W.
Washington, D.C. 20065

**Application for Comprehensive Major Medical Non-Group Coverage**

DO NOT WRITE IN THIS SPACE

| Identification Number | | Group Number | | | Rep. Code |
|---|---|---|---|---|---|
| Service Code | | Effective Date | | | |
| Disp. | Trans Date | Sub. By | Sp. By | Sex M.S. | |
| KTW | | | | | |

PLEASE PRINT ALL INFORMATION

IMPORTANT - Please be ... give complete infor...

TYPE OF COVERAGE CHECK ONE:
☐ SELF ONLY
☐ FAMILY (self and dep... (Above section "Enroll... Family Members" mus...

THE CONTRACT APPLIED FOR ... RESULTING FROM OR ARISING ...

DO NOT SEND PAYMENT WITH T... accepted.

Broker Name: ____
Broker Social Security Number: ____
Agency Tax ID: ____
Broker Phone Number: ____

PLEASE ANSWER THE QUESTIONS ... REVERSE SIDE OF THIS FORM AND SIGN WHERE INDICATED.

1F2-06138 - F Rev. 010/28/91

JUN 18 1992

**HEALTH EVALUATION**

CHECK EACH ITEM YES OR NO

SECTION A – To the best of your knowledge and belief, has any person named in this application had within the last seven years, or does such person now have, any of the following?

| YES | NO | |
|---|---|---|
| ☐ | ☐ | (a) Cancer, tumor or other growth |
| ☐ | ☐ | (b) Acquired Immune Deficiency Syndrome (AIDS), AIDS Related Complex (ARC), Human Immunodeficiency Virus Seropositivity (HIV test) (Positive HIV test) |
| ☐ | ☐ | (c) Kidney stones, kidney or bladder trouble, urinary frequency or burning |
| ☐ | ☐ | (d) Goiter, thyroid trouble, diabetes |
| ☐ | ☐ | (e) Seizure disorder, central nervous system disorder |
| ☐ | ☐ | (f) Substance abuse (drug or alcohol dependency, abuse or addiction) |
| ☐ | ☐ | (g) Gall bladder trouble, hernia, stomach or intestinal trouble, ulcers, hemorrhoids, liver disorder |
| ☐ | ☐ | (h) Cataract or other eye condition |
| ☐ | ☐ | (i) Tuberculosis, lung condition, asthma, bronchitis |
| ☐ | ☐ | (j) Arthritis, rheumatism, external deformity, amputation(s), back or spinal trouble, gout |
| ☐ | ☐ | (k) Heart trouble, abnormal blood pressure (hypertension or hypotension), anemia, rheumatic fever, cerebrovascular accident (stroke) |
| ☐ | ☐ | (l) (Female) Irregular or excessive menstrual bleeding, reproductive system disorders, infertility, breast condition |
| ☐ | ☐ | (m) (Female) is currently pregnant; expected date of delivery: ____ |
| ☐ | ☐ | (n) (Male) Prostate trouble, reproductive system disorders, infertility |
| ☐ | ☐ | (o) Outpatient counseling, any psychiatric or psychological counseling, or any nervous or mental disorder |
| ☐ | ☐ | (p) Sexually transmitted diseases |

SECTION B – In addition to the conditions listed in SECTION A, to the best of your knowledge and belief, within the past five years, has any person named in this application:

| YES | NO | |
|---|---|---|
| ☐ | ☐ | (a) Had a physical examination? |
| ☐ | ☐ | (b) Excluding physical examinations, consulted a physician, health care provider, or other individual or facility for medical or surgical treatment, advice, screening for any condition, or reason for prescription medication not listed in SECTION A? |
| ☐ | ☐ | (c) Had any departure from good health not previously mentioned in any of the above questions for which treatment or advice may or may not have been sought? |

SECTION C – If you have checked "YES" to any part of SECTION A or SECTION B, please provide complete information regarding diagnosis or condition; treatment (including all medications, hospitalizations, surgery, and diagnostic testing results) and dates. If more space is needed, attach a separate sheet of paper.

| Patient's First Name | Diagnosis or Condition | Duration Dates From | To | Explain treatment including all medications, hospitalizations, surgery and diagnostic test results and physician's/hospital name. | Recovery Check only one box |
|---|---|---|---|---|---|
| | | | | | ☐ Full ☐ Partial |
| | | | | | ☐ Full ☐ Partial |
| | | | | | ☐ Full ☐ Partial |

**PLEASE READ CAREFULLY - THIS SECTION MUST BE DATED AND SIGNED**

IT IS UNDERSTOOD AND AGREED THAT:

(a) The statements and answers made herein are complete and correct to the best of my knowledge and belief, and are made to cause the issuance of, and to become a part of, the contract for which I am applying. Should any statements or answers made herein change, or should the physical or health condition change of any person named in this application, before the contract becomes effective, I will notify Blue Cross and Blue Shield of the National Capital Area (hereafter "BCBSNCA").

(b) The contract will become effective the first of the month following approval of this application by BCBSNCA.

(c) Should any statements or answers contained in this application be untrue (if such statements are fraudulent or material to the acceptance of this application), then the contract may be cancelled by BCBSNCA, and its obligation shall consist only of the return of any subscription charges actually paid, less the amount of any benefits paid under the contract.

(d) As a non-group Subscriber of BCBSNCA, no change to a higher level of coverage may be made by me except at such time as BCBSNCA may offer such change to non-group Subscribers having the coverage applied for herein.

(e) The Subscriber shall repay to BCBSNCA the amount of any payment made in error to the Subscriber on behalf of the Subscriber or any covered family member as the result of a claim.

(f) Any physician, nurse, hospital or other provider (hereafter "Provider") having treated or attended me or any of my family members listed on this application, and having possession of any records or information with respect thereto, is authorized and directed to provide such information or records to BCBSNCA upon request for the purpose of evaluation of this application. This authorization is valid for 30 months.

(g) If this application is approved by BCBSNCA, I authorize any Provider to forward to BCBSNCA information concerning medical services or supplies provided to me or to any of my family members listed on this application for the purpose of review, investigation or payment of a claim. This authorization is valid for the duration of coverage.

(h) A copy of this application is available to the Subscriber (or a person authorized to act on his behalf) upon request. If this application is accepted by BCBSNCA, a copy of this application will be attached to the contract issued to the Subscriber.

(i) This contract includes a ten-month waiting period for the treatment of pre-existing conditions, tonsillectomies and adenoidectomies, sterilizations, infertility, and complications resulting from any of these conditions or procedures unless credit toward the waiting period is granted by BCBSNCA.

(j) If this application is accepted by BCBSNCA, I will pay, in advance, subscription charges applicable to the type of coverage selected directly to Blue Cross and Blue Shield of the National Capital Area.

X ____ Signature of Applicant

Date ____

☐ Redate and Resign below ONLY if block is checked.

X ____ Signature of Applicant

Date ____

1F2-06138 - B Rev. 10/28/91

Photo credit: Office of Technology Assessment

A Blue Cross and Blue Shield Association application for individual health insurance.

group plans. Large group plans accept all eligible employees regardless of characteristics such as age, sex, or health. However, most individual, medically underwritten groups and large group plans have restrictions on preexisting conditions.

Individual insurance operates on the principle that the cost of insurance should be proportional to the risks involved. Most applicants for individual insurance receive standard rates (22). Individuals applying for insurance whose potential losses might be large, however, can expect to pay higher premiums—often called rated premiums—than those whose potential losses are expected to be less. Individuals might also be accepted at standard rates, but with certain waivers excluding health care coverage related to specific conditions. Some applicants might be accepted with both an exclusion waiver and at a rated premium. Applicants with significant disease can be denied insurance altogether.

Insurers are particularly concerned about applicants for individual insurance who seek to purchase coverage because they are aware of a medical problem that is not yet evident to the underwriter. Such applicants pose a financial threat to the insurer, and the situation is referred to as "adverse selection" or "antiselection" (of an insurer by high-risk applicants). It refers to the situation where, in the absence of any controls, persons who seek to obtain insurance will tend to be those who will use it most—that is, those with a greater than average probability of loss to the insurer. Insurers' ability to accept applicants and their rating structure are influenced by the threat of adverse selection (16). The potential for adverse selection is most relevant for small group and individual insurance, because large groups tend to have an even distribution of low- and high-risk individuals.

### The Medical Information Bureau and Individual Health Insurance

In 1902, a group of 15 life insurance companies established the Medical Information Bureau (MIB). Now located in Westwood, MA, MIB strives to discourage fraud when companies are called on to write insurance for applicants with conditions significant to longevity or insurability. MIB acts as a medical information clearinghouse where member companies can determine if an applicant for health or life insurance has previously been denied coverage for medical reasons. About 750 U.S. and Canadian life insurance companies at 1,150 locations belong to MIB (18). BC/BS companies are not members of MIB, and therefore do not use its data (4).

Although MIB was set up by and for life insurance companies, a member can also access MIB for health or disability insurance purposes if the member sells those products. Perhaps more importantly, information about persons applying for individual health insurance through an MIB member company can be entered into MIB.

Applications for individual insurance—health, life, or disability—carry an explanation about MIB. If the insurance company finds something in an applicant's history that could affect longevity, the member company must file a report with MIB about the applicant's insurability. The potential insurer also may request an MIB check to see if past reports about the applicant have been filed by other companies; MIB makes about 22 million such checks each year. MIB's reports alert a potential insurer to omissions or misrepresentation of facts by an applicant. In principle, an applicant can refuse to allow his or her information to be communicated to MIB. The penalty to the applicant, however, typically means refusal by the insurance company to process the application.

MIB enters approximately 3 million coded records a year and has information on about 15 million persons in the United States (6). Information about applicants is encoded into a broad-based set of 210 medical categories and 5 nonmedical codes (e.g., aviation, hazardous sport) at the time an individual applies for medically underwritten life, health, or disability insurance from a member company. Not all information entered into MIB is negative information about an applicant, as normal results of tests are also submitted to MIB. For example, if an applicant has a previous record for high blood pressure, an entry might be made at a later date reflecting a normal blood pressure reading. Insurance claims made by individuals are not a source of records and codes for MIB.

MIB codes include a few for classifying genetic diseases. A specific category exists for CF and a broad category also exists for family histories of hereditary disease. Currently, MIB has no plans for expanding genetic disease categories to classify information that could become available from new genetic tests (6). Coded information on an applicant is released to authorized personnel at the member company requesting the information through computer terminals dedicated exclusively to MIB activities.

Any individual can inquire whether MIB retains a record on him or her. Individuals can inspect and seek correction of their own records. On average, 19,000 people request disclosure annually, and about 400 make corrections to their records. MIB retains records on an individual for 7 years; if no additional applications or inquiries come to MIB on a person after that period of time, the record is purged.

MIB emphasizes that its reports are not used as the basis for a decision to reject an application or to increase the cost of insurance premiums (18). Actual underwriting decisions are based on information from the applicant and from medical professionals, hospital records, and laboratory results. In 12 States it is illegal under the National Association of Insurance Commissioners Insurance Information and Privacy Protection Model Act to make underwriting decisions solely on the content of an MIB record; the act also is adhered to by some insurers in States that have not enacted it. Another deterrent to using MIB codes to deny coverage: Insurers must disclose the basis for an adverse underwriting decision under the Federal Fair Credit Reporting Act (Public Law 101-50).

## REGULATION AND UNDERWRITING

Regulation of insurance in the United States is largely State-based, although some Federal laws apply. Within State laws, private insurers have some discretion in determining insurance coverage and how the costs will be distributed. For individual and medically underwritten groups, how much a policy costs, what type of coverage is available, and even whether insurance can be bought at all are determined, in large part, by how a group or individual is classified by insurers—i.e., rating, coverage, and underwriting decisions.

### *Regulation of Insuring Entities*

The McCarran-Ferguson Act of 1945 (Public Law 79-15) accords States the principal regulatory responsibilities with regard to the business of insurance, although some Federal laws (e.g., the Employee Retirement Income Security Act, or ERISA) affect health benefit plans, particularly group plans. Besides ERISA, the Federal tax code, through such things as the exclusion of employer contributions for health benefits from the taxable income of workers, has an important impact on health insurance.

All 50 States and the District of Columbia have insurance laws that require insurers to meet a variety of financial and other requirements in order to obtain a license to do business. These laws do not apply to self-funded plans. The exact requirements vary widely from State to State, but ordinarily stipulate certain amounts of financial resources needed to establish solvency as an insurer. BC/BS plans are treated somewhat differently. Although they do not have to adhere to State commercial insurance law, they are subject to a rate-making process that does not generally apply to commercial insurers.

### *Regulation of Insurance Contracts*

All States require that individual health insurance policy forms be filed with the appropriate regulatory authority before being used. Some States require filing and approval of premium rates for new business as well as for renewal rate changes for individual coverage. Most States also require similar filings of group insurance contracts. Insurance laws generally authorize a State insurance commissioner (or comparable authority) to disapprove policies if they contain unjust, unfair, inequitable, misleading, or deceptive provisions. Many States also permit their regulators to disapprove contracts on the grounds that the benefits provided are unreasonable in relation to the premium charged. Similar to the policies directed at commercial insurers, many BC/BS plans are required to obtain prior approval of individual subscriber rate schedules.

Certain types of practices in issuing, continuing, or canceling insurance polices are also prohibited and monitored by States. Certain factors can be barred from use in making underwriting decisions for individual coverages. Even though rating classification schemes must be submitted to State insurance authorities for review, insurers are not required (as they are with rating) to submit the criteria used in underwriting to regulatory authorities (35), and little empirical work on what State regulators actually do in reviewing rating classifications is available (26).

### *Underwriting*

Underwriting is the process by which an insurer determines whether and on what basis it will accept an application for insurance. Evaluating whether an insurance applicant will be covered on a standard or substandard basis—or not at all—is called risk classification. Because of potentially large differences in the health status and potential risks presented of individual applicants, insurers evaluate individuals using criteria different from those for groups. Individuals generally are placed in classes with about the same expectation of loss. Those with higher than average risks might be accepted, but under special conditions. Seventy-three percent of applicants for individual policies from commercial insurers are classified as standard (26). Those with the highest expectation of loss are declined and deemed uninsurable, except in some States where BC/BS is required to accept all applicants (i.e., open enrollment periods are required).

## OTA SURVEY OF HEALTH INSURERS

Although genetic tests and information are important to companies that offer disability and life insurance (box 7-C), the 1991 OTA survey focused on health insurance. Specifically, OTA conducted a survey in 1991 of the commercial insurers, BC/BS plans, and largest HMOs that write individual health insurance policies or medically underwritten groups to assess their practices and attitudes toward genetic tests. Complete data from the survey, as well as details about its methodology, are presented elsewhere (31). This section summarizes those findings most pertinent to carrier screening for CF.

### *Demographics of the Survey Population*

OTA's survey population was derived from three sources: a Health Insurance Association of America (HIAA) database of member companies that offer individual health coverage, the BC/BS Association directory (3), and the Group Health Association of America 1991 National Directory of HMOs (11). Again, third-party payors' policies and attitudes for two particular populations were examined:

**Individuals**—those who seek insurance independently and without any association with an employer or membership group of any kind.

**Medically underwritten groups**—those groups whose members must be medically underwritten.

Members of these populations are required by insurers to undergo diagnostic tests or physical examination before a policy will be issued. Some large group policies might require tests or physical examinations for cases of late applicants (i.e., employees who are eligible for group health insurance but choose not to sign up until after the normal enrollment period when they know they will soon

### *Box 7-C—Life Insurance and Cystic Fibrosis*

Life insurance does not provide access to health care, and the interaction of genetic tests with the life insurance industry could differ considerably from that with the health insurance industry. In the United States, about 156 million people are covered by some type of life insurance. For those covered under group plans as part of employee benefit packages, there is typically no medical testing or screening. For those who obtain individual life insurance, however, some medical screening invariably accompanies an application, and its thoroughness usually reflects the amount of coverage being sought. Companies reject about 3 percent of all applications for life insurance.

Life insurance generally is unavailable for individuals with cystic fibrosis (CF)—hence the current availability of DNA-based assays has no impact on life insurance for people with CF. Similarly, since CF carrier status does not affect life expectancy, being identified as a carrier should have no bearing on access to life insurance. Nevertheless, life insurance considerations in the context of CF carrier screening arise because of concerns about the effect other DNA tests could have, generally. Thus, while identifying CF carrier status represents a case with no risk of premature mortality related to that status (and having CF represents a case where an obvious risk of premature mortality would preclude life insurance), other genetic screening that could occur in the future—e.g., for breast or colon cancer—could influence the type and cost of life insurance coverage available to an affected individual.

SOURCE: Office of Technology Assessment, 1992, based on American Council of Life Insurance, *1990 Life Insurance Fact Book* (Washington, DC: American Council of Life Insurance, 1990); and R. Bier, American Council of Life Insurance, "Questions and Answers: Genetic Information and Insurance," June 1990.

Table 7-2—Number of People Insured by OTA Survey Respondents

| | | Commercial insurers | BC/BS plans | HMOs |
|---|---|---|---|---|
| Individual policies | *Total:* | 2.0 million | 1.7 million | 306,861 |
| | *Individual respondent:* | Range: 171 to 240,000 | Range: 1,500 to 690,559 | Range: 350 to 258,945 |
| Medically underwritten group policies | *Total:* | 2.3 million | 2.4 million | 4.0 million |
| | *Individual respondent:* | Range: 1,000 to 382,000 | Range: 1,039 to 1,592,000 | Range: 1,501 to 2 million |

SOURCE: Office of Technology Assessment, 1992.

have expenses), but this survey does not encompass such cases. Thus, results from OTA's survey apply to the 12.7 million people who have individual or medically underwritten group coverage provided through survey respondents (table 7-2).[2]

## Commercial Health Insurers

In the United States, approximately 1,250 for-profit companies are in the business of writing health insurance policies (15). Increasingly, however, few commercial health insurers write policies for individuals or medically underwritten groups—the focus of OTA's survey. The OTA survey was sent to 225 health insurers that had recently offered individual coverage, as identified by the HIAA, and OTA received responses from 132 commercial health insurers (59 percent response rate). The list OTA obtained was 4 years old, and in that time period well over half of those companies had stopped offering individual coverage (22), which is confirmed by the 81 commercial insurance companies responding that they no longer wrote individual or medically underwritten group policies. Fifty-one commercial insurers responding to the OTA survey said they write individual or medically underwritten group contracts. Of these respondents, 29 companies offer individual coverage and 37 companies offer medically underwritten group policies. Fifteen companies offer both types of policies.

## Blue Cross/Blue Shield Plans

Both the chief underwriters and the chief medical directors for 72 of the 73 BC/BS plans were surveyed. (Puerto Rico's plan was excluded.) Twenty-nine chief underwriters completed a survey (40 percent response rate), as did 18 chief medical directors (25 percent response rate). Of the 29 BC/BS plans represented by the chief underwriters, 25 write individual policies; 21 of 29 write medically underwritten group contracts. To represent a larger pool of plans and because a number of underwriters specified that their survey was a joint underwriter/medical director response, only data from the chief underwriters' survey are used in this chapter to describe BC/BS responses. Complete data are presented elsewhere (31).

## Health Maintenance Organizations

OTA sent surveys to the 50 largest local and national HMOs, as well as a sample of 28 plans that were the largest HMO within a State or the largest by HMO model type. (Four HMO model types exist: the staff model plan, group model plan, network model plan, and independent practice association model plan.) Forty-three surveys were returned (55 percent response rate); 20 of the responding HMOs offered policies neither to individuals nor medically underwritten groups. Of the 23 HMOs responding that do offer such coverage, 11 HMOs accept individuals and 20 medically underwrite groups.

### *General Attitudes Towards Genetic Tests*

OTA's survey findings indicate that insurers generally believe that it is fair for them to use genetic tests to identify those at increased risk of disease, and that they should decide how to use that information in risk classification (table 7-3). A majority of medical directors from commercial insurers (34 respondents, 67 percent) said they "agree strongly" or "agree somewhat" with the statement that "it's fair for insurers to use genetic tests to identify individuals with increased risk of disease"; 15 disagree to some extent (30 percent). Similar responses were obtained from survey respondents from BC/BS plans and HMOs (table 7-3).

Survey respondents were also asked whether "an insurer should have the option of determining how to use genetic information in determining risks." Thirty-eight commercial respondents (74 percent)

[2] Throughout the discussion in this chapter, "no response" is not reported in the text, but the percentages presented account for them, as indicated in the tables.

**Table 7-3—General Attitudes of Insurers Toward Genetic Tests**

| Statement | Respondent | Agree strongly | Agree somewhat | Disagree somewhat | Disagree strongly | No response[a] |
|---|---|---|---|---|---|---|
| It's fair for insurers to use genetic tests to identify individuals with increased risk of genetic disease. | *Commercials* | 11 (22%) | 23 (45%) | 11 (22%) | 4 ( 8%) | 2 ( 4%) |
| | *HMOs* | 3 (13%) | 14 (61%) | 2 ( 9%) | 2 ( 9%) | 2 ( 9%) |
| | *BC/BS plans* | 4 (14%) | 17 (59%) | 4 (14%) | 2 ( 7%) | 2 ( 7%) |
| An insurer should have the option of determining how to use genetic information in determining risks. | *Commercials* | 19 (37%) | 19 (37%) | 9 (22%) | 3 ( 6%) | 1 ( 2%) |
| | *HMOs* | 2 ( 9%) | 15 (65%) | 4 (17%) | 0 ( 0%) | 2 ( 9%) |
| | *BC/BS plans* | 9 (31%) | 15 (52%) | 4 (14%) | 0 ( 0%) | 1 ( 3%) |

[a]Percentages may not add to 100 due to rounding.

SOURCE: Office of Technology Assessment, 1992.

agreed strongly or somewhat with this statement; 12 respondents (28 percent) disagreed to some extent. Responses from HMOs indicated similar sentiments: 17 medical directors (74 percent) agreed strongly or somewhat compared to 4 (17 percent) who disagreed somewhat. For BC/BS plans, 24 respondents (83 percent) agreed with the statement to some extent, against 4 respondents (14 percent) who disagreed somewhat (table 7-3).

As genetic tests become widely available, under what conditions do insurers believe a negative financial impact would occur for their company? The majority of commercial insurers (30 respondents; 59 percent) said a negative financial impact would not occur if genetic tests were, in general, widely available to the medical/provider community. In contrast, 34 respondents from commercial insurers (67 percent) thought a negative financial impact would occur under such circumstances if constraints were placed on insurers' access to the results. Forty-seven respondents (92 percent) thought a negative impact would occur if there were adverse claims or underwriting results due to adverse selection (table 7-4). Similar results were obtained from the BC/BS and HMO survey respondents (31).

OTA found that no commercial insurer had conducted an economic analysis of the costs and benefits of carrier tests as part of applicant screening or genetic tests as part of applicant screening. One commercial company reported it had done an analysis of the costs and benefits of carrier tests as part of prenatal coverage, but 48 companies had not. Similar data were found for both the BC/BS plans and for the HMOs (table 7-5). It is clear that few companies have considered genetic tests or services in terms of the costs and benefits of coverage. It is a particularly important finding that companies had not done such an analysis of the costs and benefits of using genetic tests for the purpose of applicant screening. For reimbursement purposes, it is also important to note that most insurers have not looked

**Table 7-4—Impact of Genetic Tests on Insurers**

| Question | Respondent | Yes | No | No response[a] |
|---|---|---|---|---|
| **Under what conditions would a negative financial impact be likely to occur for your company (check all that apply):** | | | | |
| Widespread availability of genetic tests to the medical provider community. | *Commercials* | 19 (37%) | 30 (59%) | 2 ( 4%) |
| | *HMOs* | 10 (44%) | 10 (44%) | 3 (13%) |
| | *BC/BS plans* | 7 (24%) | 20 (69%) | 2 ( 7%) |
| Widespread availability of genetic tests with constraints on insurers' access to results. | *Commercials* | 34 (67%) | 15 (29%) | 2 ( 4%) |
| | *HMOs* | 16 (70%) | 4 (17%) | 3 (13%) |
| | *BC/BS plans* | 17 (59%) | 10 (35%) | 2 ( 7%) |
| Adverse claims or underwriting results from antiselection. | *Commercials* | 47 (92%) | 2 ( 4%) | 2 ( 4%) |
| | *HMOs* | 18 (78%) | 2 ( 9%) | 3 (13%) |
| | *BC/BS plans* | 27 (93%) | 0 ( 0%) | 2 ( 7%) |

[a]Percentages may not add to 100 due to rounding.

SOURCE: Office of Technology Assessment, 1992.

Table 7-5—Economic Analyses of Genetic Tests and Genetic Counseling by Insurers

| Question | Respondent | Yes | No | No response[a] |
|---|---|---|---|---|
| **Has your company ever conducted an economic analysis of:** | | | | |
| Carrier testing as part of applicant screening? | *Commercials* | 0 ( 0%) | 50 (98%) | 1 ( 2%) |
| | *HMOs* | 0 ( 0%) | 20 (87%) | 3 (13%) |
| | *BC/BS plans* | 0 ( 0%) | 28 (94%) | 1 ( 3%) |
| Carrier testing as part of prenatal coverage? | *Commercials* | 1 ( 2%) | 48 (94%) | 2 ( 4%) |
| | *HMOs* | 0 (10%) | 20 (87%) | 3 (13%) |
| | *BC/BS plans* | 1 (13%) | 27 (94%) | 1 (13%) |
| Genetic testing as part of applicant screening? | *Commercials* | 0 ( 0%) | 49 (96%) | 2 ( 4%) |
| | *HMOs* | 0 ( 0%) | 20 (87%) | 3 (13%) |
| | *BC/BS plans* | 0 ( 0%) | 28 (97%) | 1 ( 3%) |
| Genetic counseling of carriers who are covered? | *Commercials* | 0 ( 0%) | 49 (96%) | 2 ( 4%) |
| | *HMOs* | 1 ( 4%) | 19 (83%) | 3 (13%) |
| | *BC/BS plans* | 1 ( 3%) | 27 (94%) | 1 ( 3%) |

[a]Percentages may not add to 100 due to rounding.

SOURCE: Office of Technology Assessment, 1992.

into the costs and benefits of providing carrier screening or genetic counseling as part of a benefits package.

# REIMBURSEMENT FOR GENETIC SERVICES—OTA SURVEY RESULTS

Will insurers pay for voluntary screening and followup counseling? And will insurance companies authorize payment for prenatal screening or testing of newborn children? Answers to these questions carry significant cost implications. They also will likely affect the degree to which carrier screening for CF becomes commonplace, since many people will be unwilling to pay out-of-pocket costs for the assays.

Insurance industry representatives assert that companies will not pay for most genetic tests unless they are "medically indicated." Thus, many health insurance companies do not pay for what they consider to be "screening" tests (28). Currently, the trend is toward closer evaluation of tests' medical necessity before insurance companies agree to pay for them. For example, a BC/BS task force evaluates 30 or 40 different procedures and devices each year and shares the results with the 73 independent BC/BS plans, each of which makes its own decisions about reimbursement (4).

More broadly, an increasing number of health insurance plans require that patients receive approval for procedures, including diagnostic tests, before the company will reimburse the cost. As more people become aware of carrier screening for CF, insurance companies are likely to receive more requests for reimbursement. In addition to uncertainty about reimbursement for the test, uncertainty also exists as to who will pay for the genetic counseling that must accompany CF carrier screening. Third-party insurers often have a policy of not reimbursing for counseling unless performed by physicians, which means the costs are reimbursed as general medical consultation fees or absorbed as part of costs on research grants (28).

From the perspective of the commercial laboratory that provides genetic tests to medical providers and patients, the issue of reimbursement is crucial to the level of their potential business—current and future. Few efforts have been made to assess the degree that CF carrier screening is being reimbursed by insurers and self-funded companies, but some individuals have been successful in obtaining reimbursement even in the absence of family history.

One private genetic service provider surveyed 66 patients about this issue in February 1991, and 27 responded (40 percent). After CF carrier screening, each patient had been given a letter explaining the CF carrier assay to submit with their claim. Third-party payors covered all costs of CF carrier screening for 11 of the 27 patients who responded; costs for 5 patients were covered in part and 11 received no reimbursement. Three of the eleven patients who received no reimbursement did not submit the letter to their insurer (9). Two individuals who were originally denied coverage appealed the decision and received full coverage. All patients who were partially covered had 80 percent coverage or had not yet met their deductible, which is compatible with CF carrier screening being treated as a compensable procedure. While the survey data represent a small sample size at one clinic, the information collected shows that some patients have obtained reimbursement when CF mutation analysis is done for screening purposes.

On balance, however, it appears that, for now, if no medical indication for the test exists, a third-party payor generally will not pay for the assay. However, an appeal can usually be made and is sometimes successful for CF carrier screening when the specifics of mutation frequencies are documented (2). Nevertheless, lack of reimbursement is likely to influence the number of individuals who opt to be screened. Thus, the concept of medical necessity is particularly important to CF carrier screening and revolves around the issue of standards of care (ch. 5); insurers are likely to continue refraining from reimbursement for tests not judged to be customary physician practice. If CF carrier screening becomes commonplace, especially in the context of obstetric/prenatal care, the current situation of third-party payment for CF mutation assay could change.

To analyze the extent to which genetic tests and services were being, or might be, reimbursed by third-party payors, OTA collected data from three populations: genetic counselors and nurses in genetics, health insurers, and State Medicaid directors.

### *Experiences of Genetic Counselors and Nurses*

In June 1991, members of the National Society of Genetic Counselors and the International Society of Nurses in Genetics who said they were currently in clinical practice were asked about the health care coverage of their patients (30). Approximately half of the respondents (198 respondents, 51 percent) reported that their patients have health care coverage very often or always (defined as between 75 and 100 percent of their patients). However, 43 respondents (11 percent) said that their patients sometimes or seldom if ever had coverage (between 0 and 50 percent of their patients).

Survey respondents were asked to recount their experience with reimbursement for various genetic services they performed. For general genetic counseling services, 22 (5 percent) responded they seldom if ever were covered, 56 (13 percent) said they sometimes were covered, 53 (12 percent) said they often were covered, 42 (10 percent) said they very often were covered, and 67 (16 percent) said they almost always were covered.

Where there was a positive family history for CF, genetic counseling was reported to be seldom if ever covered by 17 respondents (4 percent), sometimes covered and often covered by 86 (20 percent), very often covered by 26 (6 percent), and almost always covered by 65 (15 percent) respondents. Where there was no family history for CF, genetic counseling was reported to be seldom if ever covered by 35 respondents (8 percent), sometimes or often covered by 69 respondents (16 percent), very often covered by 10 respondents (3 percent), and almost always covered by 16 respondents (4 percent).

When asked if they knew of a patient's insurance claims for DNA analysis being rejected, 96 respondents (27 percent) said that they knew of such denials. One respondent to OTA's 1991 survey of genetic counselors and nurse geneticists gave this reason for the denial of a client's claim:

> In one family, the husband had an affected first cousin. This insurance would not pay for his screening because it is only a risk if the woman is a carrier and that the father's carrier status did not affect the pregnancy.

It is clear in this case that the insurance company falsely assumed that the father's carrier status was not relevant to the condition. At least two other surveys were conducted recently that also dealt with the issue of reimbursement for genetic screening services (1,19). One of these found a majority of respondents obtained full or partial reimbursement for CF mutation analysis. Reimbursement was more likely if a pregnancy was involved or when there was a family history of CF (1).

### *Health Insurers' Approaches*

OTA's survey of health insurers inquired whether certain genetic tests or services—again, for individual and medically underwritten groups—are covered "at patient request" (no family history, i.e., screening), "only if medically indicated" (family history), or "not covered." No commercial company reimburses for CF carrier tests for screening purposes. The survey also found that carrier tests for CF—as well as Tay-Sachs and sickle cell (31)—are not covered for any reason by 12 of 29 commercial insurers that offer individual coverage. Twelve respondents (41 percent) cover CF carrier assays if medically indicated. With respect to prenatal tests for CF, about 41 percent (12 respondents) that write individual policies reimburse for such tests when medically indicated (table 7-6).

For the 37 commercial companies offering medically underwritten group policies, carrier tests for CF (and, again, sickle cell or Tay-Sachs (31)) are not covered by any company when done solely at patient request. CF mutation analysis is covered by 24 of 37 companies if medically indicated. Ten companies offering medically underwritten group coverage do not cover any of the carrier or prenatal tests in the OTA survey. Sixty-two percent of companies (23 respondents) that offer medically underwritten group policies cover prenatal tests for CF (table 7-6).

Two of 25 BC/BS plans offering individual coverage would reimburse CF carrier screening at patient request. Sixteen of these BC/BS plans (64 percent) cover them if they are medically indicated and seven do not cover them. For prenatal tests for CF, 3 of these companies cover them at a patient's request, 19 if medically indicated, and 3 not at all. Of 21 BC/BS plans offering coverage to medically underwritten groups, CF carrier screening is covered at patient request by 2 companies (10 percent), only if medically indicated by 11 companies (52 percent), and not at all by 8 companies (38 percent) (table 7-6). Data for reimbursement for prenatal CF tests by BC/BS companies that medically underwrite groups are also presented in table 7-6.

Of the 11 HMOs that offer health insurance under individual policies, 1 respondent (9 percent) covers CF carrier tests at patients' requests and 7 HMOs (64 percent) reimburse for them if medically indicated. For the 20 HMOs that offer medically underwritten group contracts, 1 HMO (5 percent) covers CF carrier tests at patients' request, 13 respondents (45 percent) reimburse for them if medically indicated, and 2 (10 percent) do not cover them at all. Table 7-6 presents these results as well as how HMOs cover prenatal tests for CF.

OTA's survey results reveal that carrier and prenatal tests often are not covered under individual and medically underwritten group policies unless they are medically necessary—i.e., a family history exists. Such lack of reimbursement could have a

**Table 7-6—Reimbursement for Cystic Fibrosis Carrier Tests and Genetic Counseling**

Question: Do your standard individual policies and medically underwritten policies provide coverage for:

| | Respondent | At patient request | Medically indicated only | Not covered | No response[a] |
|---|---|---|---|---|---|
| **Individual policies** | | | | | |
| Carrier tests for CF? | *Commercials* | 0 ( 0%) | 12 (41%) | 12 (41%) | 5 (17%) |
| | *HMOs* | 2 (18%) | 7 (64%) | 0 ( 0%) | 2 (18%) |
| | *BC/BS plans*[b] | 2 ( 8%) | 16 (64%) | 7 (28%) | 0 ( 0%) |
| Prenatal tests for CF? | *Commercials* | 0 ( 0%) | 12 (41%) | 14 (48%) | 3 (10%) |
| | *HMOs* | 1 ( 9%) | 7 (64%) | 1 ( 9%) | 2 (18%) |
| | *BC/BS plans* | 3 (12%) | 19 (76%) | 3 (12%) | 0 ( 0%) |
| Genetic counseling? | *Commercials* | 2 ( 7%) | 6 (21%) | 18 (62%) | 3 (10%) |
| | *HMOs* | 1 ( 9%) | 6 (55%) | 1 ( 9%) | 3 (27%) |
| | *BC/BS plans* | 1 ( 4%) | 9 (36%) | 13 (52%) | 2 ( 8%) |
| **Medically underwritten policies** | | | | | |
| Carrier tests for CF? | *Commercials* | 0 ( 0%) | 24 (65%) | 10 (27%) | 3 ( 8%) |
| | *HMOs* | 1 ( 5%) | 13 (65%) | 2 (10%) | 4 (20%) |
| | *BC/BS plans* | 2 (10%) | 11 (52%) | 8 (38%) | 0 ( 0%) |
| Prenatal tests for CF? | *Commercials* | 1 ( 3%) | 23 (62%) | 10 (27%) | 3 ( 8%) |
| | *HMOs* | 2 (10%) | 14 (70%) | 0 ( 0%) | 4 (20%) |
| | *BC/BS plans* | 3 (14%) | 14 (67%) | 4 (19%) | 0 ( 0%) |
| Genetic counseling? | *Commercials* | 2 ( 5%) | 16 (43%) | 17 (46%) | 2 ( 5%) |
| | *HMOs* | 2 (10%) | 12 (60%) | 1 ( 5%) | 5 (25%) |
| | *BC/BS plans* | 1 ( 5%) | 7 (33%) | 12 (57%) | 1 ( 5%) |

[a]Percentages may not add to 100 due to rounding.
[b]OTA also inquired about reimbursement practices for BC/BS open enrollment nongroup polices and reports these data elsewhere (31).

SOURCE: Office of Technology Assessment, 1992.

significant impact on the ultimate utilization of CF mutation analysis.

OTA found that genetic counseling was not covered by 18 of 29 commercial companies offering individual coverage and 17 of 37 offering medically underwritten group coverage. Six insurance companies offering individual policies and 16 that medically underwrite groups cover genetic counseling only if it is medically indicated. Two companies offering each type of coverage will reimburse for genetic counseling at the patient's request (table 7-6). Similar results for BC/BS plans and HMOs are also presented in table 7-6.

Finally, respondents were asked to indicate whether they agreed or disagreed with the following scenario:

> Through prior genetic testing, the husband is known to be a carrier for CF. Before having children, the wife seeks genetic testing for CF. The insurance company declines to pay for the testing, since there is no history of CF in her family.

For commercial insurers, 21 medical directors (41 percent) agreed strongly or somewhat. Twenty-nine respondents (47 percent) disagreed somewhat or strongly with this scenario. For respondents from BC/BS plans, 12 agreed strongly or somewhat (41 percent) and 15 disagreed strongly or somewhat (52 percent). Four respondents from HMOs (17 percent) agree somewhat compared to 17 who disagreed somewhat or strongly (74 percent). These results indicate that insurers are split in their attitudes (or in their understanding of genetics) towards financing CF carrier screening as a part of reproductive decisionmaking.

### *Medicaid Reimbursement*

For some low income citizens, Medicaid provides access to genetic tests and genetic counseling. Medicaid reimbursement for genetics and pregnancy-related services has been reported to vary from State to State (34). To examine the current state of such reimbursement, OTA surveyed directors of State Medicaid programs in June 1991 to assess which of seven services—amniocentesis, ultrasound, chorionic villus sampling (CVS), maternal serum alpha-fetoprotein (MSAFP) tests, DNA analysis, chromosomal analysis, and genetic counseling—were covered. OTA also asked for information about reimbursement amounts for each service.

Respondents were asked to indicate if their State guidelines stipulated whether a procedure was "covered," "not covered," "coverage based on individual consideration," or "unknown." There was no attempt to determine how completely these guidelines were followed by each State, and there have been reports that people have experienced difficulties in getting any Medicaid reimbursement for the types of services OTA inquired about (29). In total, 47 States and the District of Columbia responded (94 percent response rate). Two States responded to OTA's survey, but are not included in this analysis. Arizona's program differs from all other States, and OTA could not obtain comparable data for it. Connecticut returned a survey, but said budget restraints precluded it from completing the survey.

State coverage of genetic procedures clearly varies (tables 7-7, 7-8). Of the 46 States[3] in the analysis, 45 cover amniocentesis, with an average reimbursement of $59.32. Fetal ultrasound is covered in 44 of 46 States, with 2 States covering it only by individual consideration. The average reimbursement for fetal ultrasound is $83.13. CVS is covered by 31 States (67 percent) and not covered in 10 States (22 percent), with 1 State reporting unknown coverage and 4 States reporting individual consideration only. The average reimbursement for CVS is $145.90. MSAFP testing is covered in 44 States and by individual consideration in 2 States. Average reimbursement for MSAFP is $21.76.

DNA analysis is covered by 26 States (57 percent) and not covered in 6 States (13 percent), with unknown coverage in 8 States (17 percent) and 6 States (13 percent) covering it based on individual consideration. Average reported reimbursement for DNA analysis is $33.39. Chromosome analysis, from amniotic fluid or chorionic villus, is covered by 41 States (89 percent), not covered by 1 State, with 4 States (9 percent) reporting individual consideration only. Average reimbursement for chromosome analysis is $235.68.

Whether the State covered genetic counseling clearly posed the most difficult question for Medicaid program directors. A substantial percentage indicated that if the service were coded as an office

[3] Hereinafter, "States" refers to the 45 States and the District of Columbia that completed a questionnaire used in OTA's analysis.

## Table 7-7—Medicaid Reimbursement for Genetic Procedures By State

| State | Amniocentesis | Ultrasound | Chorionic villus sampling | Maternal serum alpha-fetoprotein | DNA analysis | Chromosome analysis | Genetic counseling |
|---|---|---|---|---|---|---|---|
| Alabama | $ 45.00 | $ 58.50 | Not covered | $19.60 | $ 24.34 | $199.99 | Not covered |
| Alaska | 220.00 | 126.00 | $100.00 | 24.00 | 31.00 | 270.00 | Covered if part of office visit |
| Arizona | Managed care plans offer different coverages. See text for explanation. | | | | | | |
| Arkansas | 49.16 | 54.62 | Covered[a] | 34.00 | 29.50 | 275.21 | Not covered |
| California | 46.94 | 80.98 | Not covered | 12.03 | Not covered | 273.18 | $200.56 complete 133.40 interim 100.28 followup |
| Colorado | 167.00 | 103.00 | 112.00 | 24.28 | 29.25 | 275.12 | Not covered |
| Connecticut | Questionnaire not completed due to budgetary constraints. | | | | | | |
| Delaware | Did not respond. | | | | | | |
| District of Columbia | 41.00 | 100.71 | 44.65 | 15.87 | ? | 80.00 | Not covered |
| Florida | 23.00 | 137.00 | 23.00 | 24.50 | 14.50 | 243.50 | Not covered |
| Georgia | 107.00 | 80.00 | Covered[a] | 6.85 | Not covered | 28.69 | Not covered |
| Hawaii | 75.60 | 81.25 | Covered[a] | 23.52 | ? | 164.50[b] | Covered if part of office visit |
| Idaho | 41.90 | 88.90 | Covered[a] | 26.55 | Covered[a] | 281.83 | Not covered |
| Illinois | 59.95 | 70.65 | 105.00[b] | 24.41 | 14.40 | 87.10 | Not covered |
| Indiana | [b] | [b] | [b] | [b] | [b] | [b] | Not covered |
| Iowa | 56.58 | 84.17 | 71.84 | 22.11 | 30.57 | 278.69 | 52.15/15 minutes |
| Kansas | 100.00 | 120.00 | Not covered | 20.25 | Not covered | Not covered | Not covered |
| Kentucky | 75.00-100.00 | 97.50-130.00 | 375.00-500.00 | 24.41[b] | 29.50[b] | 268.94[b] | 100.00-300.00[b] |
| Louisiana | 39.48 | 80.00 | Not covered | 15.70 | ? | 275.21 | ? |
| Maine | 23.00 | 25.30-59.40 | 101.80 | 15.00 | ? | 251.00 | Covered as part of office visit |
| Maryland | 31.00 | 56.00 | 31.00 | 27.26 | 54.00 | 215.25 | 13.00-40.50 |
| Massachusetts | 49.43 | 92.00 | 481.07 | 16.73 | 24.76 | 225.73 | Covered if part of office visit |
| Michigan | 36.80 | 66.12 | 358.17 | 20.60 | ? | 167.31 | Covered as part of office visit; 11.00-54.00 |
| Minnesota | 55.00 | 70.00 | 153.00 | 25.28 | 30.57 | 278.71 | 75% of office visit rates |
| Mississippi | 41.90 | 69.30 | Not covered | 6.15 | Not covered | 260.56 | Not covered |
| Missouri | 25.00 | 65.00 | Not covered | 24.41 | 16.43 | 150.00 | Not covered |
| Montana | 51.91 | 68.68 | 65.2% of charges | 42.30 | ? | 309.79 | Covered if part of office visit |
| Nebraska | Did not respond. | | | | | | |
| Nevada | 69.70 | 152.36 | Covered[a] | 42.89-55.76 | [b] | 400-520 | 47.46-156.66 |
| New Hampshire | 25.00 | 64.00 | [b] | 14.00 | 29.00 | 14.00 | [b] |
| New Jersey | 37.00 | 55.00 | Not covered | 10.20 | [b] | 230.00 | Covered[a] |
| New Mexico | 59.60 | 52.87 | ? | 23.41 | 29.50[b] | 268.50[b] | Not covered |
| New York | 20.00 | 55.00 | Covered[a] | 6.50 | 31.39 | 90.00 | Covered[a] |
| North Carolina | 119.20 | 73.44[b] | Not covered | 20.80 | Not covered | 297.65 | Not covered |
| North Dakota | 39.28 | 109.93 | 52.20 | 21.73 | 110.49 | 239.42 | Not covered |
| Ohio | 75.00-98.00 | 95.77-102.65 | 250.00-402.00 | 24.41 | [b] | 268.94 | 16.88-20.00 |
| Oklahoma | 59.50 | 92.70 | Not covered | 24.41 | Covered[a] | 268.93 | Not covered |
| Oregon | 44.48 | 74.82 | 38.05 | 22.94 | Covered[a] | 268.93 | Covered if part of office visit |
| Pennsylvania | 50.00 | 97.50 | 59.00 | 20.00 | 14.50-30.80 | 275.20 | Covered if part of office visit; 30.00-49.00 |
| Rhode Island | Did not respond. | | | | | | |
| South Carolina | 31.80 | 66.00 | 75.00 | 6.20 | 28.50 | 300.00 | Not covered |
| South Dakota | 63.00 | 100.00 | 50% of usual and customary charges | 24.41 | 29.50 | 275.21 | ? |
| Tennessee | 57.00-60.00 | 51.00-88.00 | 178.75 | 24.98 | 109.68 | 275.21 | Covered[a] |
| Texas | 81.22 | 116.41 | 94.82 | 23.77 | ? | 200.00 | 100.00 initial 25.00 followup |
| Utah | 46.45 | 47.29 | 111.60 | 20.87 | 11.86 | Covered[a] | Covered if part of office visit |
| Vermont | 22.00 | 75.00 | Not covered | 25.00 | Not covered | 400.00 | Not covered |
| Virginia | 110.00 | 90.00 | 66.00 | 25.00 | 10.50 | 135.00 | Covered if part of office visit |
| Washington | 31.54 | 61.10 | Covered[a] | 24.38 | Covered[a] | 251.91 | Not covered |
| West Virginia | 43.00 | 36.00 | [b] | 24.98 | 20.97-31.39 | 275.20 | Covered if part of office visit; 10.00 |
| Wisconsin | 47.64 | 115.68 | 189.40 | 24.13 | Covered[a] | 281.47 | Covered[a] |
| Wyoming | 50.00 | 127.95 | Covered[a] | 22.00 | ? | 198.00 | ? |

[a]No dollar amount reported to OTA.
[b]Individual consideration.
?Respondent did not indicate whether explicitly covered or not.

SOURCE: Office of Technology Assessment, 1992.

**Table 7-8—Average Medicaid Reimbursement for Genetic Procedures**

| | Amniocentesis | Ultrasound | Chorionic villus sampling | Maternal serum alpha-fetoprotein | DNA analysis | Chromosome analysis | Genetic counseling |
|---|---|---|---|---|---|---|---|
| Number of States reporting dollar amounts of reimbursement. . . . . | 45 | 45 | 22 | 45 | 2 | 43 | 10 |
| Average amount reimbursed. . . . | $59.32 | $83.13 | $145.90 | $21.76 | $33.39 | $235.68 | $68.87 |

SOURCE: Office of Technology Assessment, 1992.

visit or consultation, it might be covered; in such cases, however, the service of genetic counseling is hidden in a general visit code. Eleven States (24 percent) reported covering genetic counseling; 11 (24 percent) reported covering it only if part of an office visit or consultation; 19 States (41 percent) do not cover genetic counseling, 2 States cover it by individual consideration, and 3 States (7 percent) reported unknown coverage. The average reimbursement amount, in large measure, reflects the range of reimbursements for different levels of office visits. As such, the average amount given ($68.87) cannot be viewed as accurate for genetic counseling services only. It should also be noted that "family DNA testing" is covered in some States (e.g., New York).

In addition to finding that some States do not cover certain services, the survey indicates the amounts reimbursed by States that do pay fall well short of charges for the procedures (5,23) (ch. 9). Hence, genetic service providers that accept Medicaid patients must subsidize the costs.

## SUMMARY AND CONCLUSIONS

Because the U.S. insurance industry is not homogeneous in its composition and policies, interest in new technologies (e.g., CF carrier screening) will vary according to both the type of insuring entity and the specific company or plan involved. The majority of the insured U.S. population obtains health insurance through the workplace under group policies. Such policies do not require diagnostic tests or physical examinations. Some Americans, however, obtain health insurance through medically underwritten group policies or obtain it on an individual basis. These individuals typically undergo risk classification and might pay higher rates. Yet little data exist on how commercial insurers, Blue Cross and Blue Shield plans, and health maintenance organizations factor genetic tests in the risk classification process. Chapter 8 reports OTA survey data related to this issue.

How insurers view genetic tests, generically, might affect their utilization. OTA's 1991 survey of commercial insurers, BC/BS plans, and HMOs that offer individual policies or medically underwrite groups sheds some light on how these populations view genetic tests, generally, and CF carrier tests, specifically. Clearly, they want the option of determining how to use genetic tests in determining risks. OTA's survey also found that insurers generally agree that it is fair for them to use genetic tests to identify persons with increased risk of disease.

Finally, the issue of who pays for CF carrier tests, prenatal tests for CF, and genetic counseling is important to the frequency at which people will opt for CF carrier screening. OTA survey results indicate that the costs of carrier tests or prenatal tests for CF (as well as sickle cell anemia and Tay-Sachs) are rarely covered by an insurer when carried out at the patient's request. Insurers either covered those costs when medically indicated (family history) or not at all. With respect to public financing for genetic tests, OTA surveyed State Medicaid directors to determine which services were covered and at what levels. Medicaid reimbursement for genetic services varies widely from State to State and does not approach full reimbursement of the actual amount charged for the service.

## CHAPTER 7 REFERENCES

1. Bernhardt, B.A., and Eierman, L.A., "Reimbursement for Cystic Fibrosis (CF) DNA Testing," submitted abstract, American Society of Human Genetics, 42nd Annual Meeting, June 1992.
2. Bernsten, C., Vivigen, Inc., Santa Fe, NM, personal communication, April 1991.
3. *Blue Cross and Blue Shield Association Directory* (Chicago, IL: Blue Cross and Blue Shield Association, 1990).
4. Conway, L., Blue Cross and Blue Shield Association, Washington, DC, personal communications, February 1991, December 1991.
5. Davis, J.G., New York Hospital, New York, NY, personal communication, March 1992.

6. Day, N., Medical Information Bureau, Inc., Westwood, MA, personal communications, September 1991, December 1991.
7. DiCarlo, S., and Gabel, J., "Conventional Health Insurance: A Decade Later," *Health Care Financing Review* 10(3):77-90, 1989.
8. Gallagher, T., National Association of Insurance Commissioners, "Health Insurance Rating," testimony before the Subcommittee on Commerce, Consumer Protection and Competitiveness, Committee on Energy and Commerce, House of Representatives, U.S. Congress, Apr. 30, 1991.
9. Genetics & IVF Institute, Fairfax, VA, "Survey of Insurance Coverage of Cystic Fibrosis Testing," personal communication, May 1991.
10. Group Health Association of America, *Employers, HMOs, and Dual Choice* (Washington, DC: Group Health Association of America, 1984).
11. Group Health Association of America, *1991 National Directory* (Washington, DC: Group Health Association of America, 1991).
12. Halifax, J., "Bill Would Close Health Insurance Pool to Newcomers," *Associated Press*, Apr. 16, 1991.
13. Hall, M., "The Purpose of Reforms in the Small-Group Market," *New England Journal of Medicine* 326:565-570, 1992.
14. Health Insurance Association of America, *Source Book of Health Insurance Data 1991* (Washington, DC: Health Insurance Association of America, 1991).
15. Iglehart, J.K., "The American Health Care System—Private Insurance," *New England Journal of Medicine* 326:1715-1720, 1992.
16. Jost, D.C., Blue Cross and Blue Shield Association, "Health Insurance Rating Practices," hearing before the Subcommittee on Commerce, Consumer Protection and Competitiveness, Committee on Energy and Commerce, House of Representatives, U.S. Congress, Apr. 30, 1991.
17. Kass, N.E., "The Ethical, Legal and Social Issues Concerning the Use of Genetic Tests by Insurers: Toward the Development of Appropriate Public Policy," manuscript prepared for the NIH/DOE Ethical, Legal, and Social Issues Working Group, November 1990.
18. Medical Information Bureau, Inc., *A Consumer's Guide* (Westwood, MA: Medical Information Bureau, Inc., 1991).
19. O'Connor, K., letter to the editor, *Perspectives in Genetic Counseling* 13(2):10, 1991.
20. Payne, J., Health Insurance Association of America, Washington, DC, personal communication, December 1991.
21. Pokorski, R., North American Reassurance Comp., Westport, CT, personal communication, February 1991.
22. Raymond, H., Health Insurance Association of America, Washington, DC, personal communication, December 1991.
23. Schulman, J.D., Genetics & IVF Institute, Fairfax, VA, personal communication, March 1992.
24. U.S. Congress, Congressional Research Service, *Health Care Financing and Health Insurance: A Glossary of Terms* (Washington, DC: Congressional Research Service, August 1988).
25. U.S. Congress, Congressional Research Service, *Health Insurance*, IB900005 (Washington, DC: Congressional Research Service, May 1991).
26. U.S. Congress, General Accounting Office, *Issues and Needed Improvements in State Regulation of the Insurance Business*, GAO/PAD-79-72 (Gaithersburg, MD: U.S. General Accounting Office, October 1979).
27. U.S. Congress, Office of Technology Assessment, *Medical Testing and Health Insurance*, OTA-H-384 (Washington, DC: U.S. Government Printing Office, August 1988).
28. U.S. Congress, Office of Technology Assessment, "Cystic Fibrosis, Genetic Screening, and Insurance," transcript of workshop proceedings, Feb. 1, 1991.
29. U.S. Congress, Office of Technology Assessment, "Cystic Fibrosis, Genetic Screening, and ERISA," transcript of workshop proceedings, Aug. 6, 1991.
30. U.S. Congress, Office of Technology Assessment, *Cystic Fibrosis and Genetic Screening: Policies, Practices, and Attitudes of Genetic Counselors-Results of a Survey*, OTA-BP-BA-97 (Washington, DC: U.S. Government Printing Office, forthcoming 1992).
31. U.S. Congress, Office of Technology Assessment, *Cystic Fibrosis and Genetic Screening: Policies, Practices, and Attitudes of Health Insurers—Results of a Survey*, OTA-BP-BA-98 (Washington, DC: U.S. Government Printing Office, forthcoming 1992).
32. U.S. Department of Health and Human Services, Public Health Service, National Center for Health Statistics, *Characteristics of Persons With and Without Health Care Coverage: United States, 1989*, No. 201, June 1991.
33. Vranka, L.A., "Defining the Contours of ERISA Preemption of State Insurance Regulation: Making Employee Benefit Plan Regulation an Exclusively Federal Concern," *Vanderbilt Law Review* 42(579):607-638, 1989.
34. Weiner, J., and Bernhardt, B.A., "A Survey of State Medicaid Policies for Coverage of Abortion and Prenatal Diagnostic Procedures," *American Journal of Public Health* 90:717-720, 1990.
35. Wortham, L., "Insurance Classification: Too Important To Be Left to the Actuaries," *University of Michigan Journal of Law Reform* 19:349-423, 1986.

Chapter 8

# Discrimination Issues

# Contents



### *Boxes*



### *Figure*



### *Tables*

# Chapter 8
# Discrimination Issues

**dis•crim•i•na•tion** /dis-krim-ə-nā-shən/ *n* differential treatment or favor with a prejudiced outlook or action.

**stig•ma•ti•za•tion** /stig-mə-tə-zā-shən/ *n* branding, marking, or discrediting because of a particular characteristic.

Stigmatization of, or discrimination against, persons with certain diseases is not unique to genetic conditions. Persons with certain infectious diseases (e.g., leprosy, tuberculosis, or AIDS) have often borne the brunt of social ostracism, as have people with conditions such as cancer or schizophrenia (for which genetic components are now known to exist). As technologies for predicting genetic disorders expand, so do concerns about behavior toward people who have such conditions, or who are carriers for them.

The primary effect of any screening is to provide information (18), but how will the information be used? What is "genetic discrimination," and will it increase (27,33,48)? Will the new knowledge elucidated through the Human Genome Project positively or negatively affect how Americans obtain or retain health care coverage?

This chapter examines aspects of discrimination from several perspectives: societal stigmatization, access to health care coverage, insurers' views toward genetic information, and genetics and new Federal antidiscrimination law (i.e., the Americans With Disabilities Act of 1990 (ADA); Public Law 101-336; 42 U.S.C. 12101 et seq.).[1] For some areas, the discussion is limited to carrier status. In others, the analysis encompasses the broader issue of the role genetic information and tests—whether to reveal carrier status or diagnose illness—play in discrimination issues.

## STIGMATIZATION AND CARRIER STATUS

Increased knowledge about human genetics challenges existing public perceptions of "genetic normalcy." In his or her genome, each person harbors stretches of DNA that silently code for recessive, lethal, or debilitating genetic disorders or that predispose—with or without certain environmental factors—future illness. As the Human Genome Project progresses, the capacity to reveal these silent genes will increase. What will the social and psychological effect of knowing such information be to an individual? Will carriers be viewed as flawed—by themselves or others—or as blameworthy for having children despite identified genetic risks? Because genetic diseases sometimes cluster in ethnic or racial groups, will the potential for discrimination and stigmatization be compounded (27,38)? Public misinformation, for example, can lead to a "courtesy stigma" applied to those affiliated only by common ancestry to the stigmatized individuals (25,38).

Some express concern that routine carrier screening for cystic fibrosis (CF) (or other disorders) might be viewed as a tacit acknowledgment that the birth of children with genetic conditions should be avoided. They express concern that if emphasis is placed on preventing the births, less effort will be made, or fewer funds allocated, to create a climate of greater tolerance and social inclusion of people with disabilities (3,63). Similarly, concern is raised that a focus on prenatal CF carrier screening raises questions about further stigmatization of pregnant women (41); one 1992 survey found 70 percent of respondents viewed all pregnant women as "people who are acutely sick" (44,52).

While some relationship exists between a characteristic's visibility and the amount of stigma it arouses (29,57), nonvisible characteristics (e.g., carrier status) are also stigmatized (25). Individuals who reveal hidden differences often encounter hostility, aversion, or discomfort (25). People with epilepsy, for example, have been ostracized—even in the absence of a visible seizure—when others have found out about their condition (60, 1983). Thus, because CF carrier status is not observable, the condition could be less stigmatized than some

[1] The use of genetic monitoring and screening assays specifically in the workplace, and the broad array of legal implications arising from such use, were discussed in an earlier OTA report (68). This chapter expands on developments since publication of that report. It focuses on the ADA and its implications for employability of carriers or persons with genetic disease.

Photo credit: American Philosophical Society

Eugenics Building, Kansas Free Fair, 1929.

attributes, but some negative reactions might well result from carrier identification.

In fact, stigmatization of carriers is likely to focus on beliefs that it is irresponsible and immoral for people who could transmit disability to their children to reproduce (box 8-A) (23,54). Embodied in this notion is the view expressed by one philosopher that: "If reproductive partners are informed they both carry a dread disease such as Tay-Sachs or CF, and even so conceive with the intention of bringing every conceptus to birth, their supposed right to reproduce becomes ethically invalid" (23).

While this sentiment represents one pole in the gradient of views on reproductive decisionmaking and genetic information, it is not inconsistent with the views of many Americans. A 1990 general population survey found 39 percent said "every woman who is pregnant *should* be tested to determine if the baby has any serious genetic defects." Twenty-two percent responded that regardless of what they would want for themselves, "a woman *should* have an abortion if the baby has a serious genetic defect," with nearly 10 percent believing a woman should be required by law to have an abortion rather than have the government help pay for the child's care if the parents are poor (64).

How CF is viewed by the American populace obviously will affect perceptions and potential reproductive stigma associated with CF carriers. Increased public awareness and education as screening becomes more common could reduce problems of stigmatization of carriers, generally, with CF

### *Box 8-A—Bree Walker Lampley and Preventing Versus Allowing Genetic Disability*

In July 1991, Los Angeles radio talk show host Jane Norris launched a firestorm of controversy when she solicited listener comments on Los Angeles television anchorwoman Bree Walker Lampley's pregnancy. Making her disapproval clear, Norris said:

> We're going to talk about a woman in the news and I mean that literally. She's a very beautiful, very pregnant news anchor, and Bree Walker also has a very disfiguring disease. It's called syndactyly [sic] and the disease is very possibly going to be passed along to the child that she's about to have. And our discussion this evening will be, is that a fair thing to do? Is it fair to pass along a genetically disfiguring disease to your child?

Bree Walker Lampley has ectrodactyly, a genetic condition manifest as the absence of one or more fingers or toes. It is an autosomal dominant disorder; hence her potential offspring have a 50-50 chance of inheriting ectrodactyly. Norris' show highlighted the public tension that exists over attitudes toward preventing genetic disability, illness, and disease.

Some listeners agreed with Norris' opinion against knowingly conceiving a child who would be at 1 in 2 risk of "this deformity—webbed hands. . . ." One caller stated she would "rather not be alive than have a disease like that when it's a 50-50 chance." Other callers compared her comments to racism and eugenic genocide: ". . .this tone of yours that just kind of smacks of eugenics and selective breeding. . . . Are you going to talk in the next hour about whether poor women should have kids?"

The opinions offered illustrate the concern over the potential for discrimination or stigmatization as personal knowledge of one's genetic makeup increases. Shortly after the program aired, one disability rights activist pointed out that the radio show reminded her of her discomfort with the Human Genome Project.

On August 28, 1991, Bree Walker Lampley delivered a healthy baby boy, who has ectrodactyly. In October 1991, arguing that a biased presentation with erroneous information was broadcast, Walker Lampley was joined by her husband, several groups, and other individuals in filing a complaint with the Federal Communications Commission (FCC). Norris and the radio station stand by their right to raise the issue and "have no regrets." The FCC rejected Walker Lampley's complaint in February 1992, and no appeal is planned.

SOURCE: Office of Technology Assessment, 1992, based on *Associated Press*, "FCC Rejects Anchorwoman's Complaint Over Call-In Radio Show," Feb. 14, 1992; J. Mathews, "The Debate Over Her Baby: Bree Walker Lampley Has a Deformity. Some People Think She Shouldn't Have Kids," *Washington Post*, Oct. 20, 1991; and J. Seligmann, "Whose Baby Is It, Anyway?," *Newsweek*, Oct. 28, 1991.

carrier screening serving as a model. Such awareness and education might avert a case such as one described by a respondent to OTA's 1991 survey of genetic counselors and nurses in genetics:

> Carrier screening can be a loaded gun; just this week one of our patients learned he was a carrier of the ΔF508 mutation and his fiancee broke off their engagement. Now not only has he been dealt the bad news of being a carrier, his personal life is in a shambles and we have spent a great deal of time addressing his feelings of guilt, anger, and betrayal.

## *Empirical Studies*

A few empirical studies addressing stigmatization and carrier status have been conducted in the United States, most in conjunction with Tay-Sachs screening during the 1970s (12-15,35; app. B). Data indicate that the majority of carriers felt they were not stigmatized, but one program found that 10 percent of noncarriers reported they would not marry a carrier (14,15). A small percentage of noncarriers expressed attitudes of superiority (11). One survey, conducted about 2 years after carriers were identified through Tay-Sachs screening, reported they and their spouses were initially "upset" when they learned the results, but that only a small minority considered themselves adversely affected (11).

Little current data on stigmatization and genetics exist; few are specific to CF. Research funded by the National Center for Human Genome Research, National Institutes of Health (NIH), however, is under way (47). One pilot study on attitudes toward CF carriers was recently conducted among high school students in Montreal, Canada who had been screened for ΔF508. In general, carriers expressed positive views about their new awareness of carrier status. Most (68 percent) would want their partner tested, and 60 percent said if the partner were a carrier, it would not affect the relationship. Sixty-three percent of persons negative for ΔF508 believed there would be no harm to their self-image should subsequent screening reveal they were actually carriers of a non-ΔF508 CF mutation. The study is only preliminary, however, with a small sample size. Further, some speculate that social values and the structure of Canada's health care system might render these data nontransferable to attitudes in the United States (36), although others believe the Canadian experience can be applied here (62).

### *Screening Without Increasing Stigma*

Concern about stigmatization arising from widespread CF carrier screening does not necessarily translate to unequivocal opposition by advocates for individuals with disabilities. A coherent effort that includes successful education and counseling could offer CF carrier screening without stigmatizing people with CF mutations as being disabled. Achieving this requires a commitment by health professionals and government that only those wanting to be screened will be screened; that all who want screening can have access to it; that results will remain confidential; and that individuals will not be coerced—overtly or covertly—into making any particular reproductive decision following screening (3,39).

Reducing perceived biases—so individuals can autonomously access the data needed to make informed choices about bearing children with CF—is of paramount concern if widespread CF carrier screening is to be viewed acceptable to those concerned about disability rights and the potential for stigmatization (3,13,55). Despite the commitment to nondirective genetic counseling, biases can sometimes emerge in the choice of words used to describe conditions, the questions asked, and the information provided (3,19,55,73).

Finally, as mentioned earlier, public education appears to resolve some potential stigmatization associated with carrier screening. Experience with massive public education efforts for β-thalassemia, for example, demonstrates that such outreach can reduce stigmatization (2,10). On the other hand, when public education is insufficient (e.g., targeted only to the screened population and not to all individuals), stigmatization can be exacerbated, as witnessed by sickle cell carrier screening (38).

## HEALTH CARE COVERAGE

Many view good health care—and access to it—as a moral right, not a privilege (box 8-B) (46). Perhaps the most widely raised social question stemming from the Human Genome Project is what effect genetic tests have had (and will have) on health care access in the United States. Because for most citizens health care access involves private health insurance, concern focuses on this market.

Consumers fear exclusion from health care coverage due to genetic or other factors. Such fears are not unfounded. Health insurance in the United States is largely employment-based: 147 million Americans secure health insurance as part of a benefits package from their or another family member's employer (31). A nationwide survey revealed 3 in 10 Americans say they or someone in their household have stayed in a job they wanted to leave mainly to preserve health care coverage (17). This so-called "job lock" freezes an employee with medical problems (or one with a dependent with medical problems) in place, because a change in employment (and health insurance) would likely result in preexisting medical conditions being excluded from health care coverage—totally or for some period of time. Job lock also occurs when an individual cannot secure a new job because of a potential employer's fear of increased health care costs. This is particularly true in small businesses, where a single employee with costly health care needs can result in cancellation of the company's policy or premium increases that become unmanageable for the remaining employees.

A 1989 OTA survey of Fortune 500 companies and a random sample of 1,000 businesses with at least 1,000 employees found 11 percent of respondents assess the health insurance risk of job applicants on a routine basis; another 25 percent assess health risks sometimes. Of these, 9 percent of employers surveyed also take into account dependents' potential expenses when considering an individual's employment application. Forty-two percent of respondents said the health insurance risk of a job applicant reduced the likelihood of an otherwise healthy, able job applicant being hired. Whether the company was self-funded, used a private carrier, or some combination of both was not predictive of response (69). Consternation about restricted health care for a nongenetic factor has already been voiced in court (box 8-C).

### Box 8-B—Ethics, Genetics, and Health Insurance

As with many issues involving public policy, discussions about the use of genetic information or tests and health insurance do not center solely on legal considerations. While ethical and legal analyses can share common ground, the overlap between law and ethics is limited. The law does not reflect all moral values held by members of society, nor can it necessarily be used to resolve ethical dilemmas. Ethical arguments about health care access and health insurance, for example, often address obligations, rights, or values not explicitly covered by law, and are used to express incumbency the law does not acknowledge.

In 1983, the President's Commission for the Study of Ethical Problems in Medicine and Biomedical and Behavioral Research concluded that health care is a need, and that:

> . . .society has an ethical obligation to ensure equitable access to health care for all. This obligation rests on the special importance of health care, which derives from its role in relieving suffering, preventing premature death, restoring functioning, increasing opportunity, providing information about an individual's condition, and giving evidence of mutual empathy and compassion (53).

The Commission also pointed out that determining that health care access is a need does not determine the mechanism for distributing it, only that the system or combination of systems (i.e., public and private) should meet the need.

In philosophy, justice concerns the distribution of social goods (e.g., health care access) and ills—basically, that similar cases should be treated alike and unlike cases should be treated differently. If a particular case of just or unjust treatment arises, the philosophical question becomes, "What makes these cases like or unlike in a morally relevant way?" Failing to state a morally relevant reason for treating the case (or people) differently lays way to the charge that action is arbitrary, capricious, or unjust.

Human genetics can be viewed as a science of inequality—a study of human particularity and difference. Genetic factors can be used as answers to the question: What makes these individuals alike or different? For some cases, the genetic difference provides a morally persuasive answer. Height, for example, is largely determined by genetics and an important factor in some jobs (e.g., playing professional basketball) but not in other jobs (e.g., computer programming). It would not be unjust to use a particular genetic difference—height—in selecting basketball players, but it would be unjust to use it in hiring computer programmers.

Thus, genetic differences are sometimes good moral reasons for treating different people differently, but in some cases they are not. Are genetic differences in propensity toward disease, for example, good moral reasons for treating people differently with respect to their access to health care? To the extent that health insurance is a mechanism for obtaining access to health care, then arguments about justice and access to health care hold equally for justice and access to health insurance. If the President's Commission was correct in concluding that there is a social obligation to provide equitable access to health care for all, then an obligation exists to ensure that people who need health care can obtain health insurance, or get access by some other means. If genetic characteristics like cystic fibrosis make it difficult for people to get the health care they need, then using genetic characteristics to disqualify individuals for health care coverage would be morally unjust.

SOURCE: Office of Technology Assessment, 1992, based on T.H. Murray, "Genetics, Ethics, and Health Insurance," contract document prepared for the U.S. Congress, Office of Technology Assessment, July 1991.

## Insurers' Attitudes Toward Genetic Risk Factors: OTA Survey Results

Risk classification and the world of insurance underwriting are arcane to most people. Persons without insurance, especially those who recently lost coverage, are puzzled (and, indeed often panicked) by interactions with the insurance industry. Any obstacles—real or perceived—encountered as they attempt to obtain individual coverage can lead to a situation of misunderstanding and mistrust. In many respects, citizens' generic concerns about health care access (49) increase concern about health insurers' use of genetic information.

As detailed in chapter 7, organizations offering health care coverage medically underwrite some policies—i.e., they classify risks of an individual or group based on actuarial data. Currently, about 10 to 15 percent of individuals with health care coverage are medically underwritten. This selection process—i.e., differentiation based on medical characteristics—is an integral part of the insurance mechanism. Risk classification is the foundation, in fact, for the concept of private insurance.

### *Box 8-C—McGann* v. *H & H Music Co.*

In August 1988, H & H Music Co. in Houston, TX, faced with rising health insurance premiums, decided it would switch from purchasing coverage from a commercial plan and become self-funded. At the same time, the company eliminated drug and alcohol treatment benefits and lowered the benefits cap for AIDS to $5,000, compared to a $1 million cap that was available for other catastrophic problems.

John McGann, who had worked for H & H Music for 5 years, learned he had AIDS in December 1987. In a suit filed in U.S. District Court, McGann held that the change in coverage was differential treatment aimed at him. He contended that dropping the coverage was discriminatory because the plan capped the AIDS benefit after he was stricken and after he informed his employer he had AIDS—the latter claim disputed by H & H Music.

Although Texas insurance law prohibits the denial of health insurance coverage for AIDS and AIDS-related illnesses, self-funded plans (like that of H & H Music) are exempt from State law because their regulation falls under Federal jurisdiction defined by the Employee Retirement Income Security Act of 1974 (ERISA; 29 U.S.C. S 1131 et seq.). Thus, McGann could not use State law to support his case. ERISA does provide that a company cannot discriminate against people for the purposes of keeping them from attaining their benefits. McGann argued that H & H Music's cap on AIDS-related benefits violated ERISA because it constituted discrimination motivated by his prior filing of AIDS-related claims, or was discrimination designed to prevent him from using health benefits to which he would have been entitled.

The district court ruled against McGann, finding that it was permissible for a self-funded plan to cover any disease it wanted, and to deny benefits for diseases for which it did not want to offer benefits; McGann died in June 1991. In November 1991, the U.S. Court of Appeals for the 5th Circuit upheld the district court opinion. The decision has clear implications for cystic fibrosis or other genetic conditions. Under current law, any self-funded company can cap, modify, or eliminate employees' health care benefits for a particular condition at any time, as long as the company complies with the notice requirements in the plan agreement. The Americans With Disabilities Act of 1990 does not address this issue. In March 1992, the U.S. Supreme Court agreed to hear the case.

SOURCE: Office of Technology Assessment, 1992, based on *Associated Press*, "AIDS Victim's Case May Define Health Plan Caps on Expensive Illnesses," Houston, TX, June 24, 1991; M.A. Bobinski, University of Houston, Health Law and Policy Institute, Houston, TX, personal communication, August 1991.

Over the last decade, health insurers have exhibited a tendency to avoid risks, rather than to find ways to spread risks over a broader base (40). Some commentators speculate that, overall, genetic analyses will mean fewer people will have access to health insurance because tests identify or refine risks. They argue genetic tests will provide the best reason yet for a nationalized health care system (4,32,65). Others contend, however, that genetic assays could detect noncarriers or rule out an individual's risk for a disorder and hence increase access to health care coverage (51). That is, making use of genetic information allows insurers to better assess risks, such that individuals at elevated risk will pay more (or be denied access), but people with low risk will pay less (30). Still others point out that as the number of genes identified increases, so will the number of potentially adverse conditions that apply to the general population, which could spread risk (45). All positions depend on the practices and attitudes insurance carriers actually have toward tests for genetic disorders, as well as the morbidity and mortality of a particular condition.

OTA found no data on how third-party payors view genetic information, generally, and the use of genetic assays for testing and screening specifically. To fill this void, OTA undertook a survey[2] in 1991 to determine how third-party payors might use genetic information in risk classification, how they would view presymptomatic, carrier, and prenatal testing, and what impact insurers project genetic tests could have on their future practices. This section uses results from the OTA survey to report how medical directors at commercial insurance companies, Blue Cross/Blue Shield (BC/BS) plans,[3]

[2] Complete data from OTA's survey of commercial insurers, HMOs, and BC/BS plans are published separately (71).

[3] OTA surveyed both chief underwriters and medical directors of BC/BS plans to see whether responses would differ. Eighteen medical directors responded and 29 chief underwriters responded. To represent a larger pool of plans, only data from the chief underwriters' survey are used in this chapter to describe BC/BS responses. Small sample size and a poor response rate from the BC/BS medical directors makes analyzing statistically significant differences in responses between the two BC/BS populations impossible. A separate report addresses this issue and presents relevant medical director responses (71).

Table 8-1—Genetic Information as Medical Information or Preexisting Conditions

| Question | Respondent | Agree strongly | Agree somewhat | Disagree somewhat | Disagree strongly | No response[a] |
|---|---|---|---|---|---|---|
| Genetic information is no different than other types of medical information | *Commercials* | 17 (33%) | 10 (20%) | 12 (23%) | 10 (20%) | 2 ( 4%) |
| | *HMOs* | 7 (30%) | 6 (26%) | 5 (22%) | 3 (13%) | 2 ( 9%) |
| | *BC/BS plans* | 6 (21%) | 14 (48%) | 6 (21%) | 1 ( 3%) | 2 ( 7%) |
| Genetic conditions such as cystic fibrosis or Huntington disease are preexisting conditions | *Commercials* | 14 (28%) | 9 (18%) | 17 (33%) | 8 (16%) | 3 ( 6%) |
| | *HMOs* | 12 (52%) | 8 (35%) | 1 ( 4%) | 0 ( 0%) | 2 ( 9%) |
| | *BC/BS plans* | 8 (28%) | 7 (24%) | 8 (28%) | 5 (17%) | 1 ( 3%) |
| Carrier status for genetic conditions such as cystic fibrosis or Tay-Sachs are preexisting conditions | *Commercials* | 8 (16%) | 12 (24%) | 16 (31%) | 13 (25%) | 2 ( 4%) |
| | *HMOs* | 5 (22%) | 12 (52%) | 0 ( 0%) | 4 (17%) | 2 ( 9%) |
| | *BC/BS plans* | 4 (14%) | 6 (21%) | 7 (24%) | 9 (31%) | 3 (10%) |

[a] Percentages may not add to 100 due to rounding.

SOURCE: Office of Technology Assessment, 1992.

and health maintenance organizations (HMOs) view genetic risk factors. As with survey results in chapter 7, these results represent attitudes and practices for insurers that write individual and medically underwritten group policies only.

The information presented in the following sections should not be construed to represent either numbers or percentages of commercial entities, BC/BS plans, or HMOs that have dealt with the issues presented. Respondents were asked how they *would* treat certain conditions or scenarios presented (currently or in the future, depending on the question), not whether they, in fact, *had* made such decisions.

## Medical Information Versus Genetic Information

Do insurers view genetic information as just another type of medical information? At first glance it would appear they do. In OTA's survey, 27 medical directors (53 percent) from commercial insurers said they "agree strongly" or "agree somewhat" with the statement that "genetic information is no different than other types of medical information"; 22 (43 percent) disagreed to some extent. For medical directors of HMOs, 13 respondents (57 percent) generally agreed, compared to 8 (35 percent) who generally disagreed. Chief underwriters for BC/BS plans responded similarly: OTA found 20 respondents (69 percent) who agreed against 7 (24 percent) who did not. Similarly, OTA found that, collectively, the majority of respondents "agree strongly" or "somewhat" that genetic conditions (e.g., CF or Huntington disease) are preexisting conditions, but carrier status (e.g., for CF or Tay-Sachs) is not (table 8-1).[4]

Yet these general views are not wholly consistent with what factors insurers view as important to insurability (not rating). Personal and family medical histories were the most important factors in determining insurability whether the respondent was from a commercial insurer, HMO, or BC/BS plan, and personal history appears to outweigh family medical history. All 29 commercial vendors (100 percent) offering individual policies in OTA's survey said personal medical history of significant conditions was "very important," and 36 (97 percent) who sell medically underwritten group policies answered similarly. The large majority of HMOs and BC/BS plans also take personal medical history into account (table 8-2).

In contrast, OTA found medical directors and underwriters felt less strongly about "genetic predisposition to significant conditions" as a facet of insurability than they did about medical history. Genetic predisposition was a "very important" criterion to 4 medical directors (14 percent) from commercial insurers of individual policies, "important" to 6 (21 percent), unimportant to 3 (10 percent), and never used by 16 (55 percent). No commercial-based respondent whose company offers coverage to medically underwritten groups considered genetic predisposition to significant conditions an important factor for insurability—18 (49 percent) never used it, 6 (16 percent) considered

[4] Throughout this chapter, survey results might not add to 100 percent because of rounding and because 'no response' is not included in the text (but is included in the tables).

**Table 8-2—Medical History, Genetic Factors, and Insurability**

Question: For each category of coverage, please indicate the importance of each of the following factors in determining insurability (not in rating):

| | Respondent | Very important | Important | Unimportant | Never used | No response[a] |
|---|---|---|---|---|---|---|
| **Individual policies** | | | | | | |
| Personal medical history of significant conditions | *Commercials* | 29(100%) | 0 ( 0%) | 0 ( 0%) | 0 ( 0%) | 0 ( 0%) |
| | *HMOs* | 9( 82%) | 0 ( 0%) | 0 ( 0%) | 1 ( 9%) | 1 ( 9%) |
| | *BC/BS plans* | 22( 88%) | 1 ( 4%) | 0 ( 0%) | 1 ( 4%) | 1 ( 4%) |
| Family medical history of significant conditions | *Commercials* | 5 (17%) | 11 (38%) | 9 (31%) | 4 (14%) | 0 ( 0%) |
| | *HMOs* | 1 ( 9%) | 0 ( 0%) | 2 (18%) | 7 (64%) | 1 ( 9%) |
| | *BC/BS plans* | 0 ( 0%) | 6 (24%) | 4 (16%) | 14 (56%) | 1 ( 4%) |
| Genetic predisposition to significant conditions | *Commercials* | 4 (14%) | 6 (21%) | 3 (10%) | 16 (55%) | 0 ( 0%) |
| | *HMOs* | 0 ( 0%) | 3 (27%) | 1 (18%) | 6 (55%) | 1 ( 9%) |
| | *BC/BS plans* | 1 ( 4%) | 2 ( 8%) | 5 (20%) | 16 (64%) | 1 ( 4%) |
| Carrier risk for genetic disease | *Commercials* | 2 ( 7%) | 5 (17%) | 6 (21%) | 16 (55%) | 0 ( 0%) |
| | *HMOs* | 0 ( 0%) | 2 (18%) | 1 (18%) | 7 (64%) | 1 ( 9%) |
| | *BC/BS plans* | 0 ( 0%) | 2 ( 8%) | 5 (20%) | 17 (68%) | 1 ( 4%) |
| **Medically underwritten groups** | | | | | | |
| Personal medical history of significant conditions | *Commercials* | 36 (97%) | 1 ( 3%) | 0 ( 0%) | 0 ( 0%) | 0 ( 0%) |
| | *HMOs* | 15 (75%) | 1 ( 5%) | 0 ( 0%) | 3 (15%) | 1 ( 5%) |
| | *BC/BS plans* | 18 (86%) | 1 ( 5%) | 0 ( 0%) | 2 (10%) | 0 ( 0%) |
| Family medical history of significant conditions | *Commercials* | 3 ( 8%) | 14 (38%) | 10 (27%) | 9 (24%) | 1 ( 3%) |
| | *HMOs* | 4 (20%) | 3 (15%) | 2 (10%) | 10 (50%) | 1 ( 5%) |
| | *BC/BS plans* | 1 ( 5%) | 3 (14%) | 4 (19%) | 13 (62%) | 0 ( 0%) |
| Genetic predisposition to significant conditions | *Commercials* | 0 ( 0%) | 12 (32%) | 6 (16%) | 18 (49%) | 1 ( 3%) |
| | *HMOs* | 0 ( 0%) | 3 (15%) | 2 (10%) | 13 (65%) | 2 (10%) |
| | *BC/BS plans* | 1 ( 5%) | 1 ( 5%) | 4 (19%) | 15 (71%) | 0 ( 0%) |
| Carrier risk for genetic disease | *Commercials* | 1 ( 3%) | 9 (24%) | 9 (24%) | 17 (46%) | 1 ( 3%) |
| | *HMOs* | 0 ( 0%) | 3 (15%) | 2 (10%) | 13 (65%) | 2 (10%) |
| | *BC/BS plans* | 1 ( 5%) | 0 ( 0%) | 5 (24%) | 15 (71%) | 0 ( 0%) |

[a] Percentages may not add to 100 due to rounding.

SOURCE: Office of Technology Assessment, 1992.

it unimportant, and 12 (32 percent) considered it important.

With respect to CF carrier screening, OTA found that "carrier risk for genetic disease"—where the individual has no symptoms of the disease—was "very important" (2 respondents; 7 percent) or "important" (5 respondents; 17 percent) in individual policy insurability by commercial insurers. For medically underwritten groups, carrier risk was viewed as "very important" or "important" by 10 commercial respondents (27 percent). Response for HMOs and BC/BS plans are also presented in table 8-2.

On the other hand, when specifically asked how an individual's application would be treated if the applicant were asymptomatic but had a family history of CF, 27 medical directors (93 percent) from commercial insurers would accept the person with standard rates; 1 respondent would accept the applicant at standard rates, but with an exclusion waiver; and 1 would decline coverage. All 11 HMOs (100 percent) offering individual coverage would accept CF carriers at standard rates. But for BC/BS plans, 16 chief underwriters (55 percent) would accept at standard rates, while 6 (21 percent) would accept at the standard rate with a waiting period, and 2 (7 percent) would decline to cover the carrier (71).[5] Thus, the mere fact that a medical director or underwriter considers carrier status important to insurability does not appear to translate into difficulties in obtaining health care coverage (rating). For

[5] The fact that the underwriters' responses are used here to report BC/BS data versus the medical directors' responses is not relevant. For the 18 BC/BS medical directors who responded, 9 (50 percent) would accept at standard rates, 3 (17 percent) would accept at standard rates but require a waiting period, and 2 (11 percent) would decline to cover.

those who responded they would accept with a waiting period or decline to cover, reluctance to offer standard insurance might stem from not wanting to pay for possible children or from a misunderstanding of the meaning of carrier status.

Given these results, do commercial insurers, HMOs, and BC/BS plans view genetic information differently than medical information? In response to the direct question comparing the two, apparently not. On the other hand, "genetic predisposition to significant conditions" is clearly part of "personal/family medical history of significant conditions." So if genetic information is viewed as a subset of personal and family medical history, why was it accorded less weight than medical history in deciding insurability? A few explanations seem plausible.

Medical directors of insurance companies, HMOs, and BC/BS might have accounted for the probabilistic nature of genetics, and therefore viewed genetics as "important," but not "very important." They also might have weighted the importance of genetic information to determining insurability as less important than personal and family medical history, although OTA did not ask them to do so. It also might be that no single risk—e.g., genetic risk—is as important as general medical risk and so entire family history was weighted more heavily. In any case, OTA's survey reveals, not surprisingly, that genetic history is used in assessing risk for individual policies and medically underwritten groups. In making decisions on insurability and rating based on genetic history, what seems important is the particular condition and its health care costs—e.g., CF, diabetes, sickle cell anemia (ch. 7), not that the consequence is genetically based.

### Genetics and Coverage Decisions: One Scenario

Because information derived from CF carrier screening is primarily useful for reproductive decisionmaking, OTA sought the reactions of commercial insurers, HMOs, and BC/BS plans to a hypothetical situation based on a real-life case (described in a following section). One alteration from the actual incident was made in this hypothetical case: Rather than refusing to pay for all health care of the child, the scenario was constructed so the insurer refused to pay for CF-related costs of the child. This change was made because, as described later, OTA was aware that insurers in all States and the District of Columbia must cover (or offer the option to include, with or without conditions) a newborn child if a valid insurance contract for the parent exists, and felt it unethical to ask respondents about breaking the law, even unknowingly. Specifically, respondents were asked to indicate whether they "agree strongly," "agree somewhat," "disagree somewhat," or "disagree strongly," with:

> Prenatal diagnosis indicates the fetus is affected with cystic fibrosis; the couple decides to continue the pregnancy. The health insurance carrier, which paid for the tests, informs the couple they will have no financial responsibility for the CF-related costs for the child.

For commercial vendors, three medical directors (6 percent) who responded to the OTA question agreed strongly or somewhat. Thirteen individuals (25 percent) in this population disagreed somewhat and 34 (67 percent) disagreed strongly. Among medical directors at HMOs, 3 respondents (13 percent) agreed to some extent with the decision in the hypothetical case, but 18 medical directors (78 percent) disagreed, 15 (65 percent) of them strongly. For chief underwriters of BC/BS plans, 6 respondents agreed (21 percent), either strongly or somewhat, with the decision in the scenario. OTA's survey revealed 8 chief underwriters (28 percent) indicated they disagreed somewhat, and 14 (48 percent) disagreed strongly.

### Perspectives on the Future of Genetic Tests

Third-party payors already use genetic information in making decisions about individual policies or medically underwritten groups. Applicants for such coverage reveal genetic information when responding to the battery of questions in personal and family histories. OTA is unaware of any insurer who underwrites individual or medically underwritten groups and requires carrier or presymptomatic tests (e.g., for Huntington or adult polycystic kidney diseases). Preliminary data from a 1991 Health Insurance Association of America survey also reveal no health insurer requires genetic tests in underwriting (56), and ordering tests to review an application appears remote at this time (1). What will be the practice in the next 5 or 10 years?

Even a decade from now, OTA's survey found that the majority of respondents do not expect to require genetic tests of applicants who have a family history of serious genetic conditions, nor do they anticipate requiring carrier assays (table 8-3). OTA finds that a minority of commercial insurers who

**Table 8-3—Projected Use of Genetic Tests by Insurers in 5 and 10 Years**

| | Respondent | Very likely | Somewhat likely | Somewhat unlikely | Very unlikely | No response[a] |
|---|---|---|---|---|---|---|
| **How likely do you think it is that your company/HMO will in the next 5 years:** | | | | | | |
| Require genetic testing for applicants with family histories of serious conditions? | *Commercials* | 1 ( 2%) | 3 ( 6%) | 16 (31%) | 31 (61%) | 0 ( 0%) |
| | *HMOs* | 1 ( 4%) | 4 (17%) | 7 (39%) | 9 (39%) | 2 ( 9%) |
| | *BC/BS plans* | 0 ( 0%) | 1 ( 3%) | 11 (38%) | 15 (52%) | 2 ( 7%) |
| Require carrier tests for applicants at risk of transmitting serious genetic disease to offspring? | *Commercials* | 2 ( 4%) | 13 (25%) | 35 (69%) | 1 ( 2%) | 0 ( 0%) |
| | *HMOs* | 2 ( 9%) | 3 (13%) | 5 (22%) | 11 (48%) | 2 ( 9%) |
| | *BC/BS plans* | 0 ( 0%) | 1 ( 3%) | 12 (41%) | 14 (48%) | 2 ( 7%) |
| Require genetic testing for applicants with no known risk of genetic disease? | *Commercials* | 0 ( 0%) | 0 ( 0%) | 4 ( 8%) | 47 (92%) | 0 ( 0%) |
| | *HMOs* | 1 ( 4%) | 0 ( 0%) | 2 ( 9%) | 18 (78%) | 2 ( 9%) |
| | *BC/BS plans* | 0 ( 0%) | 1 ( 3%) | 6 (21%) | 20 (69%) | 2 ( 7%) |
| Offer optional genetic testing and carrier testing? | *Commercials* | 0 ( 0%) | 3 ( 6%) | 18 (35%) | 30 (59%) | 0 ( 0%) |
| | *HMOs* | 4 (17%) | 6 (26%) | 6 (26%) | 5 (22%) | 2 ( 9%) |
| | *BC/BS plans* | 1 ( 3%) | 5 (17%) | 9 (31%) | 12 (41%) | 2 ( 9%) |
| **How likely do you think it is that your company/HMO will in the next 10 years:** | | | | | | |
| Require genetic testing for applicants with family histories of serious conditions? | *Commercials* | 2 ( 4%) | 17 (33%) | 14 (28%) | 18 (35%) | 0 ( 0%) |
| | *HMOs* | 3 (13%) | 5 (22%) | 9 (39%) | 3 (13%) | 3 (13%) |
| | *BC/BS plans* | 0 ( 0%) | 10 (34%) | 8 (28%) | 9 (31%) | 2 ( 7%) |
| Require carrier tests for applicants at risk of transmitting serious genetic disease to offspring? | *Commercials* | 1 ( 2%) | 13 (25%) | 16 (31%) | 21 (41%) | 0 ( 0%) |
| | *HMOs* | 3 (13%) | 4 (17%) | 9 (39%) | 4 (17%) | 3 (13%) |
| | *BC/BS plans* | 0 ( 0%) | 9 (31%) | 9 (31%) | 9 (31%) | 2 ( 7%) |
| Require genetic testing for applicants with no known risk of genetic disease? | *Commercials* | 0 ( 0%) | 4 ( 8%) | 8 (16%) | 39 (76%) | 0 ( 0%) |
| | *HMOs* | 1 ( 4%) | 0 ( 0%) | 6 (26%) | 13 (57%) | 3 (13%) |
| | *BC/BS plans* | 0 ( 0%) | 3 (10%) | 9 (31%) | 15 (52%) | 2 ( 7%) |
| Offer optional genetic testing and carrier testing? | *Commercials* | 0 ( 0%) | 12 (24%) | 17 (33%) | 22 (43%) | 0 ( 0%) |
| | *HMOs* | 5 (22%) | 7 (30%) | 6 (26%) | 2 ( 9%) | 3 (13%) |
| | *BC/BS plans* | 3 (10%) | 10 (34%) | 5 (17%) | 9 (31%) | 2 ( 7%) |

[a] Percentages may not add to 100 due to rounding.

SOURCE: Office of Technology Assessment, 1992.

responded believe it will be "very likely" (2 respondents; 4 percent) or "somewhat likely" (17 respondents; 33 percent) that they will, in 10 years, require genetic testing for applicants who have a family history of serious conditions. Over the next decade, no BC/BS chief underwriter considered it "very likely" that his or her company would require genetic testing for applicants who had family histories of serious disorders; 10 (34 percent) replied they viewed it as "somewhat likely." Of medical directors at HMOs, 3 (13 percent) thought their HMO would require applicants to have a genetic test if a family history of a serious disorder existed, and 5 (22 percent) said they considered it "somewhat likely" tests would be required in this manner—again, in the next 10 years. A similar distribution of responses was revealed when respondents were queried about requiring carrier tests for applicants at risk of passing on serious genetic conditions to their offspring (table 8-3). Requiring carrier screening as a condition of consideration for insurance appears even more remote than using genetic assays on those who have family histories of serious disorders (table 8-3).

Few respondents believe their companies will require genetic tests in either 5 or 10 years, but what about optional testing? Commercial health insurers and BC/BS plans do not anticipate that optional testing or screening will be part of their company's policy in 5 or 10 years. It is interesting to note that a majority of HMO-based medical directors who responded to OTA's survey said they considered it "very likely" or "somewhat" likely that their

**Table 8-4—Projected Use of Genetic Information by Insurers in 5 and 10 Years**

| | Respondent | Very likely | Somewhat likely | Somewhat unlikely | Very unlikely | No response[a] |
|---|---|---|---|---|---|---|
| **How likely do you think it is that your company/HMO will in the next 5 years:** | | | | | | |
| Use information derived from genetic tests for underwriting? | *Commercials* | 7 (14%) | 12 (24%) | 16 (31%) | 16 (31%) | 0 ( 0%) |
| | *HMOs* | 1 ( 4%) | 5 (22%) | 9 (26%) | 6 (26%) | 2 ( 9%) |
| | *BC/BS plans* | 3 (10%) | 8 (28%) | 10 (34%) | 6 (21%) | 2 ( 7%) |
| **In the next 10 years:** | | | | | | |
| Use information derived from genetic tests for underwriting? | *Commercials* | 12 (24%) | 20 (39%) | 11 (22%) | 7 (14%) | 1 ( 2%) |
| | *HMOs* | 3 (13%) | 6 (26%) | 8 (35%) | 3 (13%) | 3 (13%) |
| | *BC/BS plans* | 5 (17%) | 13 (45%) | 3 (10%) | 6 (21%) | 2 ( 7%) |

[a] Percentages may not add to 100 due to rounding.

SOURCE: Office of Technology Assessment, 1992.

HMO would offer optional genetic testing and carrier testing in 10 years (12 respondents; 52 percent) (table 8-3). The difference in response between the HMO population versus the commercial insurers and BC/BS plans could reflect HMOs' longer history with and emphasis on managed and preventive care.

Health insurers do not need genetic tests to find out genetic information. It is less expensive to ask a question or request medical records. Thus, whether or not genetic information is available to health insurers hinges on whether individuals who seek personal policies, or are part of medically underwritten groups, become aware of their genetic status because of general family history, because they have sought a genetic test because of family history, or because they have been screened in some other context. Even then, a majority of respondents to OTA's survey reported they thought it "somewhat unlikely" or "very unlikely" that they would be using information derived from genetic tests for underwriting (table 8-4).

### *Access to Health Insurance After Genetic Tests: OTA Survey Results*

Existing information about how genetic test results affect individuals' health care coverage is largely anecdotal (7). Quantifying such information has proved difficult, and verifying it, impossible. Reported insurance rejections on genetic bases sometimes fail to distinguish among insurance product lines—e.g., health, life, or disability. Some cases reflect insurers' longstanding risk classification practices to decline coverage (or reduce coverage or offer it at increased rates) to individuals in ill health, regardless of whether it has a genetic basis. One case from the Baylor College of Medicine, Houston, TX, however, illustrates why concern is continually expressed about health-insurer uses of genetic tests.

> A couple in their 30s has a 6-year-old son with CF. Prenatal diagnostic studies of the current pregnancy indicate the fetus is affected. The couple decides to continue the pregnancy. The HMO indicated it should have no financial responsibility for the prenatal testing and that the family could be dropped from coverage if the mother did not terminate the pregnancy. The HMO felt this to be appropriate since the parents had requested and utilized prenatal diagnosis ostensibly to avoid a second affected child. After a social worker for the family spoke with the local director of the HMO, the company rapidly reversed its position (22).

Industry representatives acknowledge that an individual company could exercise poor judgment, but contend the problem is not widespread: If the problem were prevalent, ample court cases could be cited because patients and their attorneys would not be passive recipients of ill-based judgments such as occurred in the case just described (56). Clients and patient advocates argue to the contrary (6,43) and maintain that cases like those just mentioned represent the tip of an iceberg.

Do individuals who avail themselves of genetic tests subsequently have difficulty obtaining or retaining health insurance? To explore this issue, OTA decided not to survey either party with a direct stake in the answer, but chose instead to ask third parties—genetic counselors and nurses in genetics—for their firsthand experiences (70). In contrast to the survey of health insurers, which asked respondents to speculate about accepting applicants with certain conditions, this survey attempted to measure actual occurrence. Specifically, in June 1991, OTA surveyed 794 members of the National Society of Genetic Counselors and the International Society of Nurses in Genetics and asked:

> Have any of your patients experienced difficulties in obtaining or retaining health insurance coverage as a result of genetic testing? If yes, please provide details.

Four-fifths (347) of the 431 respondents to OTA's inquiry currently perform genetic counseling. Fifty respondents (14 percent) reported they had clients who had experienced difficulties obtaining or retaining health care coverage as a result of genetic testing (table 8-5). Because some respondents described more than one case, the number of affirmative answers understates the actual number of cases. Examination of the qualitative responses, some of which are presented in table 8-6, reveals affirmative responses represent, at minimum, 68 individual cases. (Where the term "patients" was used with specifics not described, a single event was recorded.)

Test results for some conditions where positive results led to reported difficulties—such as for Huntington disease, adult polycystic kidney disease, and Marfan syndrome—were cited by more than one respondent. Since genetic tests for conditions such as Huntington disease or Marfan syndrome are available at a limited number of sites, OTA attempted to ascertain whether surveys reporting patient insurance difficulties were geographically consistent with known test sites. With few exceptions, the respondent resided in a State where the test was available. In exceptions, the counselor resided in a neighboring State.

In addition to affirmative answers, several respondents reported that although they had no direct experience with a patient's difficulty in obtaining or retaining health care coverage, they had clients who

**Table 8-5—Difficulties in Obtaining or Retaining Health Insurance After Genetic Tests**

Question[a]: Have any of your patients experienced difficulties in obtaining or retaining health insurance coverage as a result of genetic testing?

| | Number | Percent |
|---|---|---|
| No | 281 | 81.0 |
| Yes | 50 | 14.4 |
| No answer | 16 | 4.6 |

[a]1991 OTA survey of genetic counselors and nurses in genetics. Sample base of 347 represents individuals currently in clinical practice.

SOURCE: Office of Technology Assessment, 1992.

**Table 8-6—Case Descriptions of Genetic Testing and Health Insurance Problems[a]**

Positive test for adult polycystic kidney disease resulted in canceled policy or increased rate for company of newly diagnosed individual.

Positive test for Huntington disease resulted in canceled policy or being denied coverage through a health maintenance organization.

Positive test for neurofibromatosis resulted in canceled policy.

Positive test for Marfan syndrome resulted in canceled policy.

Positive test for Down syndrome resulted in canceled policy or increased rate.

Positive test for alpha-1-antitrypsin defined as preexisting condition; therapy related to condition not covered.

Positive test for Fabry disease resulted in canceled policy.

Woman with balanced translocation excluded from future maternity coverage.

Positive Fragile X carrier status and subsequent job change resulted in no coverage.

After prenatal diagnosis of hemophilia-affected fetus, coverage denied due to preexisting condition clause.

Denied coverage or encountered difficulty retaining coverage after birth of infant with phenylketonuria.

Woman diagnosed with Turner's syndrome denied coverage for cardiac status based on karyotype. Normal electrocardiogram failed to satisfy company.

Family with previous Meckel-Gruber fetus denied coverage in subsequent applications despite using prenatal diagnosis and therapeutic abortion.

Mother tested positive as carrier for severe hemophilia A. Prenatal diagnosis revealed affected boy; not covered as preexisting condition when pregnancy carried to term.

After a test revealed that a woman was a balanced translocation carrier, she was initially denied coverage under spouse's insurance because of risk of unbalanced conception. Subsequently overturned.

Woman without prior knowledge that she was an obligate carrier for X-linked adrenoleukodystrophy found out she was a carrier. She had two sons, both of whom were healthy, but each at 50 percent risk. Testing was done so they could be put on an experimental diet to prevent problems that can arise from mid- to late childhood or early adulthood. One boy tested positive. The family's private pay policy (Blue Cross/Blue Shield) is attempting to disqualify the family for failing to report the family history under preexisting conditions.

After birth of child with CF, unable to insure unaffected siblings or themselves.

[a]1991 OTA survey of genetic counselors and nurses in genetics. Not all cases, or multiple cases involving same disorder, listed.

SOURCE: Office of Technology Assessment, 1992.

feared their coverage would be dropped if they requested payment for tests from insurers. One respondent commented that greater than 80 percent of her clients who have tests for Huntington disease self-pay. Similarly, others with no direct experience said they often advise patients not to request reimbursement for a test so that an insurer would not learn that testing had occurred. One counselor offered the information that a patient had refused testing for adult polycystic kidney disease because of concern over health insurance. Another respondent reported that a patient with a CF-affected child had been dropped by one insurance company and would not consider prenatal testing in the future for fear her current insurer would not cover the child should she decide to continue the pregnancy.

Such fears persist despite the fact that most contracts for individual health insurance coverage preclude blanket nonrenewal (37,56). Similarly, an insurer cannot raise rates for an individual who has been continuously covered if the person develops a new condition (37). On the other hand, it is legal for an insurer not to renew a group contract based on the results of one individual's genetic or other medical test. Group policies are rarely guaranteed renewable (37). In lieu of not offering to renew, an insurer might opt to levy a steep premium increase at renewal time.

OTA's survey reports—conservatively—consumer difficulties in obtaining or retaining health care coverage after genetic tests. OTA has no basis for evaluating whether the nonrespondents would be more or less familiar with patients' insurance difficulties and potentially know of additional cases. The data permit neither extrapolation about the total number of cases that have occurred in the United States, nor speculation about any trends.

OTA did not attempt to ascertain whether or not patients had challenged—or were challenging—insurers' rulings. Thus, OTA cannot determine whether some of the disputes reported in table 8-6 were resolved fully in favor of the consumer because the initial judgment was deemed improper or illegal. Some cases, for example, reported a fetus or newborn had tested positive and coverage had been denied. In all 50 States and the District of Columbia, insurers must cover (or offer the option to include) a newborn child if a valid insurance contract for the parent exists. However, whether the insurance company can deny certain benefits for the newborn by evoking the preexisting condition clause generally contained in all insurance contracts is unclear.

In presenting table 8-6, OTA does not judge the validity—positively or negatively—of the claim. Some cases might have been settled in favor of the individual. Others might have been cases where an applicant attempted to select against an insurer by misrepresenting his or her health history, which would have been resolved against the individual.

In 1991, at least 50 genetic counselors or nurses in clinical practice knew of at least 68 actual incidents where their own patients reported difficulties with health insurance due to genetic tests. OTA estimates, based on the average number of patients directly counseled, that genetic counselors and nurses responding to the survey collectively saw about 110,600 individuals in 1990. However, OTA did not advise respondents to limit descriptions of clients' insurance difficulty to 1990. Thus, it is unlikely that all reported cases occurred in 1990; assuming all reported cases occurred in 1990, the 68 cases represent 0.06 percent of patients seen by survey respondents.

Critics, however, question whether the data—especially the qualitative descriptions—merely represent more anecdotal stories that unfairly present one side of the story and for which no response can be developed (56). Skeptics point out that some of the cases might fall into the gray area of whether exclusion or increased rates resulted because an adverse medical condition was revealed through a diagnostic test that just happened to be genetic. The border between what conditions are genetic or not is blurred, however, and will become increasingly diffuse. Because genetic-based predictive tests promise to have a profound impact on clinical medicine (28)—and because access to medical care is inextricably linked to private health insurance in this country—these cases underscore certain policy dilemmas arising from the increased availability of genetic assays. For genetic testing or screening to detect genetic illness (or the potential for illness), the possibilities for problems are not remote, but real indeed.

Finally, it is important to note that most of the cases revealed through the OTA survey do not involve recessive disorders and carrier screening for conditions like CF. And while one assumption might have been that health care coverage for CF carriers would not be an issue because the individuals are asymptomatic, OTA's survey of health insurers reveals that a few insurers would require a waiting period or deny coverage for these individuals.

# GENETICS, DISCRIMINATION, AND U.S. LAW

Federal, State, and local laws provide only incomplete protection against invidious genetic discrimination (48). Overall, explicit safeguards have not been enacted in most jurisdictions to protect against discrimination, or to allow favoritism, specifically on the basis of genetic characteristics. This section examines how State statutes and the Americans With Disabilities Act of 1990 (Public Law 101-336; 42 U.S.C. 12101 et seq.) apply—or could apply—to discrimination with a genetic basis.

Is genetic discrimination a form of racial or gender discrimination? Or is it best classified as a form of disability differentiation? Some argue that CF carrier screening, specifically, is unlikely to raise issues under Title VII of the Civil Rights Act (42 U.S.C. 2000e), or its 1991 amendments (Public Law 102-166). That legislation focuses primarily on protecting groups historically the subject of discrimination. Thus, some argue that only if it can be shown that CF carrier screening has a disparate impact on women (67) would the Civil Rights Act and other protections be pertinent, and that since CF mutation analysis is done on both men and women, it does not have such an impact. Others disagree, asserting that if discriminatory consequences arise from CF carrier screening, a case could be made that the Civil Rights Act applies.

## *State Statutes and Hereditary Conditions*

Several States—Arizona, California, Florida, Illinois, Iowa, Kentucky, Louisiana, Maryland, Missouri, New Jersey, New York, Oregon, Virginia, and Wisconsin—have statutes that specifically mention testing, counseling, or employment of persons with hereditary conditions.[6] The California (box 8-D), Maryland, and New Jersey laws recognize that disease-specific language could prove too rigid and instead have broad application to "any hereditary disorder." California also draws the distinction between carriers and those who experience manifestations of the disease. A handful of statutes narrowly target specific conditions or traits, such as sickle cell (Florida, Louisiana, New York), hemophilia (Florida, Missouri), CF (Missouri), Tay-Sachs (New York), or Cooley's anemia (β-thalassemia; New York).

Statutes in at least seven States—Florida, Oregon, Louisiana, New Jersey, New York, North Carolina, and Wisconsin—prohibit employment discrimination against persons with any atypical hereditary or blood trait or with named genetic conditions or traits. New York law, for example, says that persons with sickle cell trait and carriers of Tay-Sachs disease or Cooley's anemia may not be denied opportunities for employment unless their disorder would prevent them from performing the job. Some State statutes address issues beyond employment discrimination. Wisconsin and Arizona, for example, prohibit certain forms of insurance discrimination. Genetic-related State laws also prohibit certain types of screening (Florida), provide funding for research or treatment (Florida, Iowa), or require information on or testing of genetic disorders be given to marriage applicants (California, Illinois, Kentucky, New Jersey, Virginia). Other State laws are concerned with genetic counseling and confidentiality (Missouri).

## *The Americans With Disabilities Act of 1990*

In 1990, Congress enacted the ADA, a comprehensive civil rights bill to prohibit discrimination on the basis of disability.[7] Unlike the Rehabilitation Act of 1973, which is still in force, the ADA extends antidiscrimination protection of persons with disabilities to private sector employment, public services, public accommodations, and telecommunications.

State and local antidiscrimination legislation supplement Federal law and are not preempted by the ADA. All 50 States have disability statutes, 48 of

[6] Arizona Rev. Stat. Sec. 20-448 (1991); California Health and Safety Code, Sec. 150, 151, 155, 309, 341 (West 1990); Florida Statute, Sec. 385.206 (1989); Illinois 1990 Public Act 86-1028; Iowa Code Sec. 136A.2 (1989); Kentucky Rev. Stat. Ann. Sec. 402.320 (Banks-Baldwin 1991); Louisiana Rev. Stat. Ann. Sec. 46.2254 (West 1982); Maryland Health-Gen. Code Ann. Sec. 13-101; Missouri Rev. Stat. Sec. 191 (1989); New Jersey Rev. Stat. Sec. 26:5B-3 (1987); New York Laws 900 (1990); Virginia Code Ann. Sec. 32.1-68. (1990); 1991 Wisc. Act 177 (signed Mar. 5, 1992).

[7] Two Federal disability laws other than the ADA have potential application to broad considerations of genetic discrimination but are not discussed in this report—the Education for All Handicapped Children Act of 1975 (Public Law 94-142, 89 Stat. 773; renamed the Individuals With Disabilities Education Act of 1990, Public Law 101-476, 104 Stat. 1142) giving all school-aged children with disabilities the right to a free public education in the least restrictive environment appropriate to their needs and the Fair Housing Amendments Act of 1988 (Public Law 100-430; 102 Stat. 1619).

### *Box 8-D—The California Hereditary Disorders Act of 1990*

California law is the most comprehensive of State statutes specifically addressing discrimination and hereditary conditions. It touches on access to health care services, professional and public education about genetic disorders, confidentiality of genetic information, voluntariness of genetic screening, and continued reproductive freedom for those at risk of passing on a disabling genetic trait. The 1990 Hereditary Disorders Act finds that "In order to minimize the possibility for the reoccurrence of abuse of genetic intervention in hereditary disorders programs . . . [t]he Legislature finds it necessary to establish a uniform statewide policy for screening for hereditary disorders."

The statute mandates that:

- The public . . . should be consulted before any rules, regulations, and standards are adopted by the State Department of Health Services.
- Clinical testing procedures established for use in programs, facilities, and projects be accurate and provide maximum information, and that the testing procedures produce results that are subject to minimum misinterpretation.
- No test(s) shall be performed on any minor over the objection of the minor's parents or guardian, nor may any tests be performed unless such parent or guardian is fully informed of the purposes of testing and is given reasonable opportunity to object to such testing.
- No testing, except initial screening for phenylketonuria and other diseases that may be added to the newborn screening program, shall require mandatory participation. No testing programs shall require restriction of childbearing, and participation in a testing program shall not be a prerequisite to eligibility for, or receipt of, any other service or assistance from, or to participate in, any other program, except where necessary to determine eligibility for further programs of diagnoses of or therapy for hereditary conditions.
- Counseling services for hereditary disorders shall be available through the program or a referral source for all persons determined to be or who believe themselves to be at risk for a hereditary disorder as a result of screening programs. Such counseling shall be nondirective, emphasize informing the client, and not require restriction of childbearing.
- All participants in programs on hereditary disorders shall be protected from undue physical and mental harm. Those determined to be affected shall be informed of the nature, and where possible, the cost of available therapies or maintenance programs, and be informed of the possible benefits and risks associated with such therapies and programs.
- All testing results and personal information generated from hereditary disorders programs shall be made available to an individual over 18 years of age, or to the individual's parent or guardian.
- All testing results and personal information from hereditary disorders programs obtained from any individual, or from specimens from any individual, shall be held confidential. An individual whose confidentiality has been breached may recover compensatory damages. In addition, he or she may recover civil damages not to exceed $10,000, reasonable attorney's fees, and the costs of litigation.

California's law, any violation of which is a misdemeanor, authorizes the State Department of Health Services to administer a statewide prenatal screening program for genetic disorders. The Department shall also develop an education program designed to educate physicians and the public concerning the uses of prenatal testing, as well as to set quality control standards for clinics offering prenatal screening. Funding is to be assured for screening services for low income women. To emphasize the noneugenic purposes of the program, special attention is paid to voluntary participation. In addition, repeated mention is made that screening and counseling should emphasize information delivery, and should not be directed toward persuading or coercing individuals to forego childbearing or conception.

Although Maryland has a law with similar goals, the California legislation has not been widely adopted, nor have its broad goals been fully realized due to funding constraints. It could serve as a model for State efforts to ensure that genetic tests take their place beside other, voluntary health services, while simultaneously discouraging third parties from discriminating against those who suffer from or carry traits for genetic disorders. Conversely, it can also be viewed as a poor model that represents unnecessary government intervention and control of medical and laboratory genetics.

SOURCE: Office of Technology Assessment, 1992, based on California Health and Safety Code, Sec. 150, 151, 155, 309, 341 (West 1990); 1990 California Senate Bill 1008 (ch. 26).

them prohibiting discrimination in the private, as well as public, sector; Alabama's and Mississippi's laws extend to the public sector only (58). State and local disability laws are enforced by human rights organizations that are often successful at both public education and at alternative dispute resolution, resulting in settlement of over four-fifths of their cases (27). Thus, such laws might remain important protections against genetically based discrimination, although only a few State cases directly involve genetic discrimination (58).

The following sections examine how genetic illness, predisposition to genetic conditions, or carrier status appear to be treated under provisions of the ADA and Equal Employment Opportunity Commission (EEOC) regulations that define disability and impairment.

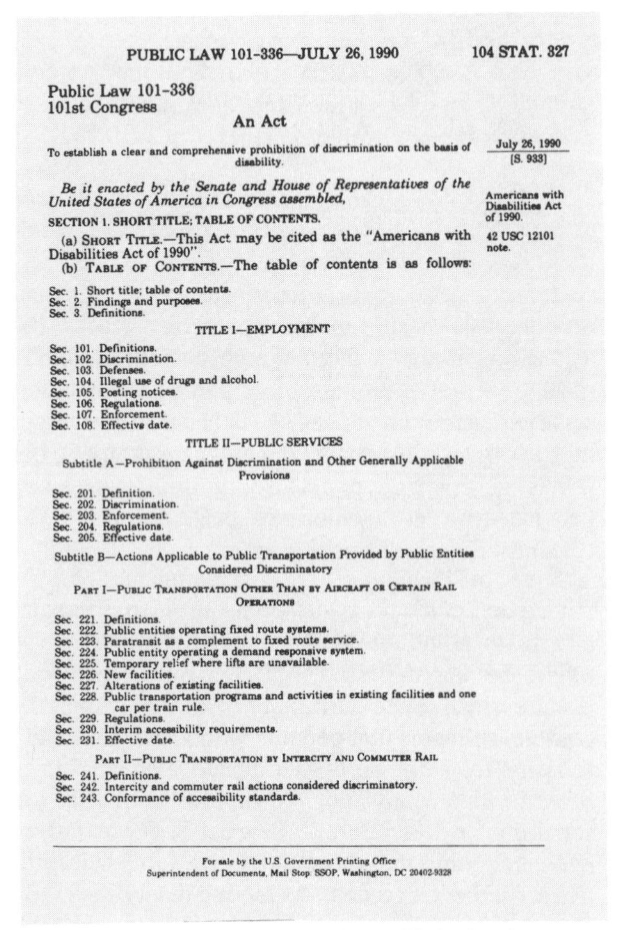

PUBLIC LAW 101-336—JULY 26, 1990 104 STAT. 327

Public Law 101-336
101st Congress

An Act

To establish a clear and comprehensive prohibition of discrimination on the basis of disability.

July 26, 1990
[S. 933]

*Be it enacted by the Senate and House of Representatives of the United States of America in Congress assembled,*

Americans with Disabilities Act of 1990.

SECTION 1. SHORT TITLE; TABLE OF CONTENTS.

(a) SHORT TITLE.—This Act may be cited as the "Americans with Disabilities Act of 1990".

42 USC 12101 note.

(b) TABLE OF CONTENTS.—The table of contents is as follows:



For sale by the U.S. Government Printing Office
Superintendent of Documents, Mail Stop: SSOP, Washington, DC 20402-9328

*Photo credit: Office of Technology Assessment*

The Americans With Disabilities Act (Public Law 101-336).

## What Is Disability?

Disability is defined broadly in the ADA to mean:

> (A) a physical or mental impairment that substantially limits one or more of the major life activities. . . . (B) a record of such impairment, or (C) being regarded as having such an impairment (42 U.S.C.A. Sec. 12102(2)).

This definition is based on the term "handicap" in the Rehabilitation Act of 1973 (29 U.S.C. Sec. 706(7)(B) (1988)) and Fair Housing Amendments Act of 1988 (45 U.S.C. Sec. 3602(h) (Supp. 1990)). Congress intended that regulations implementing the Rehabilitation Act and the Fair Housing Amendments Act apply in interpreting the term "disability" in the ADA (21).

In spelling out the meaning of subsection (A), the Senate Report states:

> "Physical or mental impairment" includes the following: any physiological disorder or condition, disfigurement, or anatomical loss affecting any of the major bodily systems, or any mental or psychological disorder such as mental retardation, mental illness or dementia . . . . The term physical or mental impairment does not include simple physical characteristics, such as blue eyes or black hair . . . [nor does it include] environmental, cultural, and economic disadvantages [in and of themselves] (72).

A person with a disability includes someone who has a "record" of or is "regarded" as having a disability, even if there is no actual incapacity. Further, a "record" of disability means that the person has a history of impairment, or has been misclassified as having an impairment (72). This provision protects those who have recovered from a disability that previously impaired their life activities. In this manner, Congress recognized that people who have recovered from diseases such as cancer—or have diseases under control such as diabetes—can face discrimination based on misunderstanding, prejudice, or irrational fear (21,42,50), and so still merit protection.

Additionally, individuals regarded as having disabilities include those who have an impairment, but do not have limitations in their major life functions and yet are treated as if they did have such limitations. Thus, the ADA encompasses people who are discriminated against based on a false belief that they have disabilities (61). This provision is particularly important for individuals who are per-

**Figure 8-1—Genetics and the Americans With Disabilities Act of 1990**

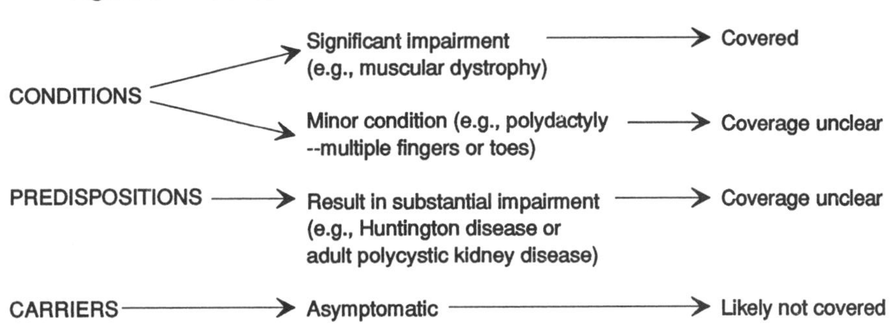


SOURCE: Office of Technology Assessment, 1992.

ceived to have stigmatic conditions. That is, society's reaction, rather than the disability itself, deprives the person of equal enjoyment of rights and services (27).

## The ADA and Genetics

Given the dearth of State legislation specific to genetic characteristics, if the ADA addressed this issue it would create a nationwide standard of protection. The ADA is silent with respect to genetics, per se, however, as are the EEOC's regulations implementing it (29 CFR part 1630; 56 FR 35726) (58,59). Furthermore, the legislative history of the ADA indicates that little attention was given to the role of genetics in discrimination. During debate on the ADA, the Congressional Biomedical Ethics Advisory Committee was informed that genetic discrimination was "not raised or discussed," and so could not be addressed by the conference committee—although several Representatives supported the argument that the ADA will also benefit individuals who are identified through genetic tests as being carriers of a disease-associated gene (27). The following sections examine the apparent coverage—or lack of coverage—for different genetic statuses under the ADA (figure 8-1).

*Genetic Conditions.* Disability under the ADA is defined only according to the degree of impairment, with no distinction between disabilities with genetic origins and those without. Congress and the courts have long recognized disabilities of primarily or partial genetic origin, including Down syndrome, CF, muscular dystrophy, multiple sclerosis, heart disease, schizophrenia, epilepsy, diabetes, and arthritis (8,24,34,72). In fact, the legislative history of the ADA cites muscular dystrophy as an example of a condition covered by the ADA.

That the condition is genetic, then, is not the defining event. At issue is how severely the disability interferes with life activities, not its origins. In defining disability, the courts require a "substantial" limitation of one or more major life activities. A genetic condition that does not cause substantial impairment might not constitute a disability, unless others treat the person as disabled. Thus, significant cosmetic disfigurements from burns or neurofibromatosis could be classified as disabilities if public prejudices act to limit the life opportunities of those with the cosmetic problem (72).

*Genetic Predisposition.* The ADA expressly protects not only individuals who actually have disabilities, but also those who are "regarded" or perceived as having them. It judges disability not by an objective measure of inability to perform tasks, but also subjectively by the degree to which the public makes the condition disabling through misunderstanding or prejudice (21,50). This definition might then apply to individuals who are asymptomatic but predicted to develop disease in the future—i.e., persons who are sometimes referred to as the "healthy ill" or "at risk" (16,27).

One commentator argues that the ADA's legislative history indicates that genetic predisposition might be encompassed (27). For example, one Congressman stated during the 1990 debate over the conference report that persons who are theoretically at risk "may not be discriminated against simply because they may not be qualified for a job sometime in the future." Several Representatives agreed, arguing that those at risk for future disabilities are to be "regarded" as having disabilities (136 Congressional Record H4614, H4623, H4624, H4626). On the other hand, no further substantive discussion on the issue occurred (58), and as described later, the EEOC rejects the premise that genetic predisposition is covered.

***Carrier Status.*** The ADA's prohibition of discrimination and case law generally hold that employment decisions must be based on reasonable medical judgments showing that the disability prevents the individual from meeting legitimate performance criteria (9,26,61,66). For asymptomatic carriers of recessive genetic conditions such as CF, sickle cell anemia, the thalassemias, and Tay-Sachs, there is no disability per se. Carriers appear not to be covered by the ADA. Such individuals are, however, at risk of having an affected child if their partners also carry the trait, and are often themselves misunderstood to be affected by the disease. Discrimination against asymptomatic carriers, therefore, arguably can constitute discrimination based on a perception of disability.

### The Equal Employment Opportunity Commission Regulations

Although the ADA does not provide explicit guidance about how genetic information should be viewed, the lack of this type of specificity in the literal reading of a law is not unusual. Instead, the executive branch (the EEOC in this case), relying on the bill's legislative history and congressional intent, interprets the legislation and issues regulations for executing the law. Thus, speculation about any perceived vagueness of the law could be addressed through public comment on EEOC's proposed rules.

In February 1991, EEOC proposed regulations for implementing the ADA (56 FR 8578). EEOC's proposed regulations did not specifically prohibit discrimination against carriers or persons who are identified presymptomatically for a late-onset genetic condition (e.g., adult polycystic kidney disease or Huntington disease). This perceived void led the Joint Working Group on Ethical, Legal, and Social Issues (ELSI) of the NIH and Department of Energy (DOE) to urge that the EEOC revise its proposed rule to explicitly include such individuals (74). Similarly, the members of the NIH/DOE Joint Subcommittee on the Human Genome endorsed the ELSI Working Group's action, and recommended that EEOC make explicit the protection of carriers or those diagnosed presymptomatically (i.e., that no individual shall be discriminated against on the basis of genetic makeup) (5).

In July 1991, EEOC published the final rule addressing the definitions of disability and impairment under the ADA (56 FR 35726). The final rules did not reflect the suggestions of either the ELSI Working Group or the Joint Subcommittee on the Human Genome. In fact, EEOC specifically amended its interpretive guidance "to note that the definition of the term 'impairment' does not include characteristic predisposition to illness or disease" (56 FR 35727).

Additionally, in correspondence to the Joint Subcommittee on the Human Genome, EEOC stated that "the ADA does not protect individuals, who are not otherwise impaired, from discrimination based on genotype alone" (20). Thus, from EEOC's perspective, asymptomatic carriers are not encompassed by the ADA's provisions. With respect to individuals diagnosed presymptomatically, EEOC has concluded that "such individuals are protected, either when they develop a genetic disease that substantially limits one or more of their major life activities, or when an employer regards them as having a genetic disease that substantially limits one or more of their major life activities" (20). Again, as with carriers, EEOC's interpretation is that individuals who are identified as at risk for a late-onset adult disorder are not protected by the ADA until the condition is manifest. Some argue that the ADA might need amending if carriers or presymptomatic individuals come to be widely perceived as having a disability, thus invoking the law's broader definition.

### The ADA and Health Insurance

An employer's fear of future disability in an applicant's family that would affect the individual's usage of health insurance and leave time would also appear to be a prohibited basis for discrimination under the ADA. Nevertheless, the ADA does not speak to this point directly, and so leaves open for future interpretation whether employers may discriminate against carriers who are perceived as more likely to incur extra costs due to illnesses likely to occur in their future children. The ADA specifically states that it does not restrict insurers, health care providers, or other benefit plan administrators from carrying out existing underwriting practices based on risk classification (27,59). Nor does the ADA make it clear whether such employers may question individuals about their marital or reproductive plans prior to offering employment or enrollment in an insurance plan. (As discussed earlier and described in box 8-C, however, after a person is hired, self-funded insurance plans can alter benefits to exclude or limit coverage for specific conditions.)

## SUMMARY AND CONCLUSIONS

Among the many issues raised by prospects of both routine carrier screening for CF and increased availability of DNA-based diagnostic or predictive tests are stigmatization and discrimination. For CF carrier screening, stigmatization might focus on the notion that it is irresponsible for people who are at risk of having offspring who might have a genetic condition to have affected children. With respect to discrimination, CF carrier screening raises questions about access to health care coverage and Federal discrimination law.

Few empirical studies have examined stigmatization of CF carriers directly, but several projects are underway. Existing research on stigmatization and carriers for Tay-Sachs or sickle cell anemia have a bearing on carrier screening for CF, but only in a limited manner. These studies can guide efforts to help clients avoid feelings of guilt or shame that could be associated with being identified as a CF carrier, but provide less concrete approaches that must be taken to educate the public. Historical perspectives can assist health care professionals in counseling clients identified as carriers, but the greatest barrier—public perception of genetic status—will require new initiatives in large numbers. Public education for Tay-Sachs carrier screening worked because the target population was both defined and inclined to seek screening. In contrast, the potential target population for CF carrier screening is larger and more diffuse, with unknown attitudes toward carrier identification.

With respect to accessing health care coverage and CF carrier screening, OTA's survey found that the majority of third-party payors offering individual or medically underwritten group policies view genetic information as no different from other types of medical information. Genetic information is used in decisions determining risk classification and underwriting, but no blanket statement can be made as to the weight placed on it. Not surprisingly, respondents rank genetic information as relatively more important to individual policies than for medically underwritten groups. Medical directors and chief underwriters view personal and family medical histories as the most important determinants in classifying and rating candidates for individual or medically underwritten insurance. Whether a condition is genetically based or not is of less import. The increased availability of genetic information, however, adds to the amount of medical information that insurers can use for underwriting. Concern is expressed this additional information will lead to risk assessments that are so accurate on an individual level that they undermine the risk spreading function of insurance.

OTA's survey of health insurers illustrates why some claims of inappropriate or illegal health care coverage decisions based on genetic test results have occurred. These decisions might continue to arise, or they could disappear with time if such insurers become familiar with and educated about the significance of CF carrier screening. OTA's survey of genetic counselors and nurses reveals more than 14 percent of respondents have clients who reported difficulties in obtaining or retaining health insurance due to results from genetic tests.

Finally, Federal antidiscrimination law, particularly the Americans With Disabilities Act, clearly encompasses individuals who have a genetic condition that substantially limits one or more major life activities. According to regulations promulgated by the Equal Employment Opportunity Commission, however, ADA does not include predisposition to illness or disease if the individual is asymptomatic. Similarly, carriers of genetic disorders per se are not covered by ADA's provisions according to EEOC; genetic status is not a defining factor in determining disability or impairment under the ADA. Nor does the ADA restrict insurers from carrying out existing underwriting practices based on risk classification. Thus, if any health care reform is viewed as necessary because of the future of widespread carrier screening for CF, predictive testing for other disorders, or increased knowledge stemming from the Human Genome Project, it will necessarily, and probably appropriately, be done under the umbrella of general health care reform currently being debated in the United States.

## CHAPTER 8 REFERENCES

1. American Council of Life Insurance and Health Insurance Association of America, *Report of the ACLI-HIAA Task Force on Genetic Testing, 1991* (Washington, DC: American Council of Life Insur-

ance or Health Insurance Association of America, 1992).

2. Angastiniotis, M., Cyprus Thalassaemia Centre, Archbishop Makarios III Hospital, Nicosia, Cyprus, "Development of Genetics Services From Disease Oriented National Genetics Programs," presentation at the 8th International Congress of Human Genetics, Washington, DC, October 1991.
3. Asch, A., "Ethical Implications of Population Screening for Cystic Fibrosis: Disability and Stigmatization," contract document prepared for the U.S. Congress, Office of Technology Assessment, May 1991.
4. Beckwith, J., Harvard University, in N. Touchette, "Genes In Good Hands With Insurers?," *Journal of NIH Research* 2:37, 1990.
5. Berg, P., and Wolff, S., National Institutes of Health/Department of Energy Joint Subcommittee on the Human Genome, letter to Evan J. Kemp, Chairman, Equal Employment Opportunity Commission, July 10, 1991.
6. Billings, P.R., "Genetics and Insurance Discrimination," presentation before the U.S. Congress, Office of Technology Assessment advisory panel on *Genetic Monitoring and Screening in the Workplace*, March 1989.
7. Billings, P.R., Kohn, M.A., de Cuevas, M., et al., "Discrimination as a Consequence of Genetic Testing," *American Journal of Human Genetics* 50:476-482, 1992.
8. *Bowen* v. *American Hospital Association*, 476 U.S. 610 (1986).
9. Burris, S., "Rationality Review and the Politics of Public Health," *Villanova Law Review* 34:933-993, 1989.
10. Cao, A., Istituto di Clinica e Biologia dell'Età Evolutiva, Università Studi, Calgliari, Italy, "A Disease Oriented National Health Service," presentation at the 8th International Congress of Human Genetics, Washington, DC, October 1991.
11. Childs, B., "The Personal Impact of Tay-Sachs Carrier Screening," *Tay-Sachs Disease: Screening and Prevention*, M.M. Kaback (ed.) (New York, NY: Alan R. Liss, Inc., 1977).
12. Childs, B., Gordis, L., Kaback, M.M., et al., "Tay-Sachs Screening: Social and Psychological Impact," *American Journal of Human Genetics* 28:550-558, 1976.
13. Clarke, A., "Is Non-Directive Genetic Counselling Possible?," *Lancet* 338:998-1991.
14. Clow, C.L., and Scriver, C.R., "Knowledge and Attitudes Toward Genetic Screening Among High-School Students: The Tay-Sachs Experience," *Pediatrics* 59:86-91, 1977.
15. Clow, C.L., and Scriver, C.R., "The Adolescent Copes With Genetic Screening: A Study of Tay-Sachs Screening Among High-School Students," *Tay-Sachs Disease: Screening and Prevention*, M.M. Kaback (ed.) (New York, NY: Alan R. Liss, Inc., 1977).
16. Council for Responsible Genetics, Human Genetics Committee, "Position Paper on Genetic Discrimination," *Genewatch*, May 1990, p. 3.
17. Eckholm, E., "Health Benefits Found to Deter Job Switching," *New York Times*, Sept. 26, 1991, p. A1.
18. Eddy, D.M., "How To Think About Screening," *Common Screening Tests*, D.M. Eddy (ed.) (Philadelphia, PA: American College of Physicians, 1991).
19. Elias, S., and Annas, G.J. (eds.), *Reproductive Genetics and the Law* (Chicago, IL: Year Book, 1987).
20. Equal Employment Opportunity Commission, Washington, DC, letter to P. Berg and S. Wolff, Aug. 2, 1991.
21. Feldblum, C.R., "The Americans With Disabilities Act Definition of Disability," *The Labor Lawyer* 7:11-26, 1991.
22. Fernbach, S.D., Baylor College of Medicine, Houston, TX, personal communication, November 1991.
23. Fletcher, J.F., "Knowledge, Risk, and the Right to Reproduce: A Limiting Principle," *Genetics and the Law II*, A. Milunsky and G.J. Annas (eds.) (New York, NY: Plenum, 1980).
24. *Gerben* v. *Holsclaw*, 692 F. Supp. 557 (E.D. Pa. 1988).
25. Goffman, E., *Stigma: Notes on the Management of Spoiled Identity* (Englewood Cliffs, NJ: Prentice-Hall, 1963).
26. Gostin, L., "The Future of Public Health Law," *American Journal of Law and Medicine* 12:461-490, 1986.
27. Gostin, L., "Genetic Discrimination: The Use of Genetically Based Diagnostic and Prognostic Tests by Employers and Insurers," *American Journal of Law and Medicine* 17(1&2):109-144, 1991.
28. Green, E.D., and Waterston, R.H., "The Human Genome Project: Prospects and Implications for Clinical Medicine," *Journal of the American Medical Association* 266:1966-1975, 1991.
29. Hahn, H., "The Politics of Physical Differences: Disability and Discrimination," *Journal of Social Issues* 44:39-47, 1988.
30. Havighurst, C.C., Duke University School of Law, Durham, NC, personal communication, September 1991.
31. Health Insurance Association of America, *Source Book of Health Insurance Data, 1989* (Washington, DC: Health Insurance Association of America, 1989).
32. Holtzman, N.A., Johns Hopkins University, in N. Touchette, "Genes In Good Hands With Insurers?," *Journal of NIH Research* 2:37, 1990.

33. Holtzman, N.A., and Rothstein, M.A., "Invited Editorial: Eugenics and Genetic Discrimination," *American Journal of Human Genetics* 50:457-459, 1992.
34. *Jane Doe* v. *Region 13 Mental Health-Mental Retardation Commission*, 704 F.2d 1402 (5th Cir. 1983).
35. Kaback, M.M., Becker, M., and Ruth, V.M., "Sociologic Studies in Human Genetics: I. Compliance Factors in a Voluntary Heterozygote Screening Program," *Ethical, Social and Legal Dimensions of Screening for Human Genetic Disease*, D. Bergsma (ed.) (New York, NY: Stratton Intercontinental Medical Book Corp., 1974).
36. Kaplan, F., Clow, C., and Scriver, C.R., "Cystic Fibrosis Carrier Screening by DNA Analysis: A Pilot Study of Attitudes Among Participants," *American Journal of Human Genetics* 49:240-243, 1991.
37. Kass, N.E., Johns Hopkins University, personal communication, December 1991.
38. Kenen, R.H., and Schmidt, R.M., "Stigmatization of Carrier Status: Social Implications of Heterozygote Genetic Screening Programs," *American Journal of Public Health* 68:116-1120, 1978.
39. Kimura, R., "Genetic Screening and Testing: Report of a Working Group," *Genetics, Ethics, and Human Values: Human Genome Mapping, Genetic Screening, and Gene Therapy—Proceedings of the XXIVth CIOMS Round Table Conference*, Z. Bankowski and A.M. Capron (eds.) (Geneva, Switzerland: Council for International Organizations of Medical Sciences, 1991).
40. Kosterlitz, J., "Unrisky Business," *National Journal*, Apr. 6, 1991, pp. 794-797.
41. Lippman, A., "Prenatal Genetic Testing and Screening: Constructing Needs and Reinforcing Inequities," *American Journal of Law and Medicine* 17(1&2):15-50, 1991.
42. Mayerson, A., "The Americans With Disabilities Act—An Historic Overview," *The Labor Lawyer* 7:1-9, 1991.
43. Meyers, A.S., National Organization for Rare Disorders, "The Rising Cost of Health Insurance," testimony, U.S. House of Representatives, Committee on Energy and Commerce, Subcommittee on Commerce, Consumer Protection, and Competitiveness, Apr. 30, 1991.
44. Myers, S.T., and Grasmick, H.G., "The Social Rights and Responsibilities of Pregnant Women: An Application of Parson's Sick Role Model," *The Journal of Applied Behavioral Science* 26:157-172, 1990.
45. Murray, R.F., Jr., Howard University College of Medicine, Washington, DC, personal communication, March 1992.
46. Murray, T.H., "Genetics, Ethics, and Health Insurance," contract document prepared for the U.S. Congress, Office of Technology Assessment, July 1991.
47. National Center for Human Genome Research, National Institutes of Health, "Ethical, Legal and Social Issues Program, National Center for Human Genome Research: Grant-Making Status Report," August 1991.
48. Natowicz, M.R., Alper, J.K., and Alper, J.S., "Genetic Discrimination and the Law," *American Journal of Human Genetics* 50:465-475, 1992.
49. Northwestern National Life Insurance Co., "Americans Speak Out on Health Care Rationing" (Minneapolis, MN: Northwestern Life Insurance Co., 1990).
50. Parmet, W., "Discrimination and Disability: The Challenges of the ADA," *Law, Medicine, and Health Care* 18:331-344, 1990.
51. Pauly, M.V., Leonard Davis Institute of Health Economics, University of Pennsylvania, Philadelphia, PA, personal communication, September 1991.
52. Peterson, K.S., "Pregnancy Often Seen As Illness," *USA Today*, Mar. 12, 1992.
53. President's Commission for the Study of Ethical Problems in Medicine and Biomedical and Behavioral Research, *Securing Access to Health Care* (Washington, DC: U.S. Government Printing Office, 1983).
54. Purdy, L.M., "Genetic Diseases: Can Having Children Be Immoral?," *Ethical Issues in Modern Medicine*, 3rd ed., J.D. Arras and N.K. Rhoden (eds.) (Mountain View, CA: Mayfield, 1989).
55. Rapp, R., "Chromosomes and Communication: The Discourse of Genetic Counseling," *Medical Anthropology Quarterly* 2:143-157, 1988.
56. Raymond, H.E., Health Insurance Association of America, Washington, DC, personal communication, December 1991.
57. Richardson, S.A., "Attitudes and Behavior Toward the Physically Handicapped," *Birth Defects: Original Article Series* 12:15-34, 1976.
58. Rothstein, M.A., University of Houston Health Law and Policy Institute, Houston, TX, remarks at U.S. Congress, Office of Technology Assessment workshop, "Cystic Fibrosis, Genetic Testing, and Self-Insurance Under ERISA," Aug. 6, 1991.
59. Rothstein, M.A., "Genetic Discrimination in Employment and the Americans With Disabilities Act," *Houston Law Review*, vol. 29, in press, 1992.
60. Schneider, J., and Conrad, P., *Having Epilepsy: The Experience and Control of Illness* (Philadelphia, PA: Temple University Press, 1983).
61. *School Board of Nassau County, Florida* v. *Arline*, 480 U.S. 273 (1987).
62. Schulman, J.D., Genetics & IVF Institute, Fairfax, VA, personal communication, December 1991.
63. Seligmann, J., "Whose Baby Is It, Anyway?," *Newsweek*, Oct. 28, 1991.

64. Singer, E., "Public Attitudes Toward Genetic Testing," *Population Research and Policy Review*, 10:235-255, 1991.
65. Stone, D., Brandeis University, in S. Brownlee and J. Silberner, "The Assurances of Genes," *U.S. News & World Report*, July 23, 1990.
66. Susser, P., "The ADA: Dramatically Expanded Federal Rights for Disabled Americans," *Employee Relations Law Journal* 16(2):157-176, 1990.
67. *United Auto Workers* v. *Johnson Controls, Inc.*, 111 S. Ct. 1196 (1991).
68. U.S. Congress, Office of Technology Assessment, *Genetic Monitoring and Screening in the Workplace*, OTA-BA-455 (Washington, DC: U.S. Government Printing Office, October 1990).
69. U.S. Congress, Office of Technology Assessment, *Medical Monitoring and Screening in the Workplace: Results of a Survey—Background Paper*, OTA-BP-BA-67 (Washington, DC: U.S. Government Printing Office, October 1991).
70. U.S. Congress, Office of Technology Assessment, *Cystic Fibrosis and DNA Tests: Policies, Practices, and Attitudes of Genetic Counselors—Results of a Survey*, OTA-BP-BA-97 (Washington, DC: U.S. Government Printing Office, forthcoming 1992).
71. U.S. Congress, Office of Technology Assessment, *Cystic Fibrosis and Genetic Screening: Policies, Practices, and Attitudes of Health Insurers—Results of a Survey*, OTA-BP-BA-98 (Washington, DC: U.S. Government Printing Office, forthcoming 1992).
72. U.S. Congress, Senate, Committee on Labor and Human Resources, *Americans With Disabilities Act of 1990*, report to accompany S. 933, S. Rpt. 101-116 (Washington, DC: U.S. Government Printing Office, 1989).
73. Wertz, D.C., and Fletcher, J.C., "An International Survey of Attitudes of Medical Geneticists Toward Mass Screening and Access to Results," *Public Health Reports* 104:35-44, 1989.
74. Wexler, N.S., Beckwith, J., Cook-Deegan, R.M., et al., "Statement of Recommendations by the Joint Working Group on Ethical, Legal, and Social Issues (ELSI) Concerning Genetic Discrimination and ADA Implementation, Apr. 29, 1991," *Human Genome News*, September 1991.

Chapter 9

# Costs and Cost-Effectiveness

# Contents



### *Box*



### *Tables*

# Chapter 9
# Costs and Cost-Effectiveness

One of the least examined of the many issues surrounding carrier screening for cystic fibrosis (CF) is that of costs and cost-effectiveness. OTA found no comprehensive studies on the cost-benefit and cost-effectiveness of screening large numbers of people for CF carrier status, although one recent study examined the net economic benefit of prenatal screening for CF (11).

How much money might be involved in large-scale CF carrier screening? Under which, if any, conditions would it be cost-effective? What factors are important in optimizing cost-effectiveness? If large numbers of individuals are screened, would one strategy maximize cost-effectiveness and identify the highest number of carriers? This chapter first discusses costs associated with CF (medical and caregiving) and costs associated with carrier screening. It then analyzes the cost-effectiveness of widespread CF population carrier screening under varying assumptions and approaches. This analysis is necessarily based on *modeling*; experienced-based data are lacking.

## A CAUTIONARY NOTE

Examining potential costs and savings associated with routine CF carrier screening is fraught with technical and social pitfalls. Some data exist on attitudes of families with CF children or relatives toward CF carrier screening, prenatal diagnosis, and selective abortion (12,31,32). General agreement exists, however, that the perceptions of the general population (i.e., those without a relative or close friend with CF) might well differ from those of people with family histories of CF, but these perceptions are less well documented. Some survey data have been published on what people say they want and, theoretically, would do (6).

Since OTA found a paucity of experienced-based data on the attitudes of the general public toward key factors such as willingness to undergo CF carrier screening and to terminate CF-affected pregnancies, OTA calculated cost-effectiveness for several hypothetical scenarios. As explained later in this chapter, OTA attempted to construct the alternatives based on three sources: population survey data specific to CF carrier screening; survey data on the public's attitudes toward genetic tests, generally; and data from privately funded CF carrier screening pilots.

Aside from uncertainty about reproductive decisionmaking related to CF carrier screening, the costs of CF, itself, are uncertain and variable. As described in chapter 3, CF's clinical course varies widely from person to person; hence, so do medical costs. Even data on "average" medical costs are less than optimal. The effect of new pharmaceuticals, such as DNase, cannot be measured because data related to their cost or potential to extend median life expectancy do not exist. While new treatments might be expensive, they could be quite successful, with the percentage of women choosing to terminate affected pregnancies shifting dramatically. This chapter examines, to a limited extent, the effect on health care costs if lifespan is extended using current "average" expenses. It does not speculate on how much new pharmaceuticals or gene therapy might cost, nor the effect their availability might have on how individuals make decisions about screening and subsequent reproductive alternatives.

Most importantly, nearly 10 years ago, the President's Commission for the Study of Ethical Problems in Medicine and Biomedical and Behavioral Research concluded the fundamental value of CF carrier screening lies in its potential for providing people with information they consider beneficial for autonomous reproductive decisionmaking (25). Thus, while economic analyses can help inform resource allocation issues surrounding genetic screening, they have limits. In the context of public policy, the President's Commission articulated solid guidance about the benefits and limits of cost-effectiveness and cost-benefit analyses for genetic screening: These analyses are tools to be used within an overall policy framework, not solely as a method of making or avoiding judgment.

OTA concurs that the value of CF carrier screening is in information gained. No one can place a value on having information. Similarly, valuing births as parents likely value them is speculative, at best. Thus, personal considerations likely outweigh societal considerations of cost-effectiveness. Here an estimate of the impact of CF carrier screening on systemwide health costs is sought; there is no intimation that something that saves or costs money is more or less desirable from a welfare standpoint.

## COSTS OF CYSTIC FIBROSIS

The cost of any illness is the answer to the hypothetical question: If the disease disappeared and everything else stayed the same, how much more output (valued in dollars) would be available to the economy? One part of the calculation to answer this question involves the direct medical costs related to CF. Costs used in this chapter represent estimated charges for treating the average person with conventional strategies.

In addition to direct medical costs, disease has other manifestations. If the condition could be eliminated, then CF-specific costs of caring for persons with CF by parents, spouses, or friends could be avoided. Such nonmedical direct costs are included in this analysis.

OTA does not consider other dimensions of benefits and costs that might be included. For example, eliminating or ameliorating disease reduces premature death and permits people to work or carry on normal activities, producing a benefit in terms of future lifetime earnings. OTA does not recognize these benefits in this chapter, but some analysts would include this benefit to the economy in a cost-benefit analysis (the human capital approach) (11,22).

Conversely, anxiety and anguish for parents, friends, and patients could be counted a cost of illness, although it is almost never included in calculations because of difficulty in assigning a dollar value to such elements. Similarly, a decision to avoid childbearing as a result of CF carrier screening could be included as cost due to lost productivity from any unaffected children who might otherwise have been born (again, a human capital approach)—i.e., when couples choose CF carrier screening to avoid childbearing, society is made worse off by their decision to avoid conception because the output from unaffected children they might have had is lost. Yet, some parents will almost certainly choose to avoid conception, and a societal human capital approach is inappropriate for analyzing costs and savings of avoided births for individual families. It is far from obvious that the value associated with children born with or without CF provides any measure of the value of CF carrier screening to any real person.

Some estimates of the total cost of CF exist, but include indirect costs associated with lost productivity (10,11,21,22). In this chapter, OTA does not account for the present value of earnings that a (potential) child (with or without CF) might contribute to the economy. On a conceptual basis, OTA adopts a conservative approach, valuing only CF-related costs, rather than including future market earnings or non-CF medical costs for either affected or unaffected individuals.

### *Medical Direct Costs*

Estimating savings that might result from CF carrier screening requires an estimate of present value lifetime costs of the disorder. Several groups have compiled figures on the annual costs for medical services obtained by persons with CF (8,33). Hospitalization, CF clinic use, physical therapy, and drugs account for most CF-related medical costs. Calculations indicate annual costs of medical care, projected to 1989 using the Consumer Price Index, range from about $9,000 to nearly $14,000 (table 9-1).

OTA uses estimated annual costs for CF of $10,000, based on examining medical expense data from these other sources (22). Assuming an average life expectancy in 1990 of 28 years, an average expense of $10,000 per year, and using a 5 percent discount rate, the net present value of estimated lifetime medical expenses would be $148,981 (1990 dollars).

Using $10,000 for average annual medical costs might be an underestimate, but it is generally consistent with the estimates reported in table 9-1. For example, data from a 1989 Cystic Fibrosis Foundation (CFF) survey provide measures of hospital and clinic use (8). The average number of outpatient visits to CF centers was 3.9; at an assumed cost of $200 per visit to CF centers, such visits add up to $800 if four visits are made per year (33). Hospital stays for CF-related reasons averaged 8 days; at an average hospital cost of $700 per day, hospital costs for CF are about $5,600 per year. The final major expenses are drugs, including at-home intravenous antibiotics for about 10 percent of patients, and physical therapy. A reasonable estimate of such costs is approximately $5,000 per patient per year (33). Thus, average medical ex-

**Table 9-1—Estimated Average Annual Medical Expenses for Cystic Fibrosis Treatment**

| Average annual expense per patient | Source |
|---|---|
| $7,500 (1980)<br>13,870 (1989)[a] | M.V. Pauly "The Economics of Cystic Fibrosis," *Textbook of Cystic Fibrosis,* J.D. Lloyd-Still (ed.) (Boston, MA: PSG, Inc., 1983), using data from Cystic Fibrosis Foundation, *Cystic Fibrosis Patient Registry, 1980: Annual Data Report.* |
| 8,098 (1985)<br>9,220 (1989)[a] | Cystic Fibrosis Foundation, *Cystic Fibrosis Patient Registry, 1983: Annual Data Report.* |
| 12,300 (1989) | M.V. Pauly, "Cost-Effectiveness of Screening for Cystic Fibrosis," contract document prepared for the U.S. Congress, Office of Technology Assessment, August 1991, using data from Cystic Fibrosis Foundation, *Cystic Fibrosis Patient Registry, 1989: Annual Data Report.* |
| 10,885 (1989-90) | M.V. Pauly, "Cost-Effectiveness of Screening for Cystic Fibrosis," contract document prepared for the U.S. Congress, Office of Technology Assessment, August 1991, using data from Wilkerson Group, Inc., *Annual Cost of Care for Cystic Fibrosis Patients* (New York, NY: Wilkerson Group, Inc., 1991). |
| 11,400 (1989) | Office of Technology Assessment, 1992, based on M.V. Pauly, "Cost-Effectiveness of Screening for Cystic Fibrosis—Addendum" contract document prepared for the U.S. Congress, Office of Technology Assessment, February 1992. |

[a]Projected to 1989 dollars using the Consumer Price Index.

SOURCE: Office of Technology Assessment, 1992.

**Table 9-2—Annual Cost of Medical Care for Cystic Fibrosis Patients**

| Treatment | Mild | Moderate | Severe |
|---|---|---|---|
| Acute treatment | | | |
| Antibiotics | $ 2,000[a] | $6,000 | $12,000 |
| IV supplies | 300 | 500 | 900 |
| Hospitalization | 3,500 | 14,000 | 28,000 |
| Miscellaneous | 100 | 200 | 400 |
| Total cost acute | 5,900 | 20,700 | 41,300 |
| Chronic management | | | |
| Visits to CF Center | 600 | 800 | 1,200 |
| Medications | 2,000 | 3,000 | 4,000 |
| Total cost chronic | 2,600 | 3,800 | 5,200 |
| Total cost acute and chronic treatment | 8,500 | 24,500 | 46,500 |

[a]All values in 1989 dollars.

SOURCE: Wilkerson Group, Inc., *Annual Cost of Care for Cystic Fibrosis Patients* (New York, NY: Wilkerson Group, Inc., 1991).

penses in 1989 for CF-related care using these assumptions is about $11,400 per person with CF.

Another set of direct medical costs has been calculated based on interviews with CF patients' families and clinicians (33) (table 9-2). This estimate characterizes individuals with CF as "mild" (one inpatient episode every 2 years), "moderate" (two episodes per year), and "severe" (four or more episodes per year). Data from the 1989 CFF survey describe the distribution of hospital episodes, which are then used to categorize the number of persons in each of these categories. Similarly, average numbers of outpatient visits, pharmaceutical costs, and other medical expenses are estimated for each patient population. Examining average medical expenses based on data in table 9-2 requires one adjustment, however.

Even if all CF patients were "mild," the expected number of persons with no episodes would be no greater than 50 percent; in fact, it was 61 percent. Thus, there is another category, "submild," whose illness requires infrequent hospitalization. If approximately 40 percent of patients were submild and 40 percent were mild, about 60 percent of persons would not be hospitalized and 20 percent would be hospitalized once. About 13 percent of all patients had two or three episodes per year; this group represents the "moderate" portion. Finally, about 6 percent of all patients had four or more hospitalizations per year and comprise the "severe" patient group. Medical expenses in table 9-2 need to be adjusted to account for the "submild" group (22).

No data exist on the average expenses of "submild" persons with CF, but a reasonable assumption might be their expenses are about twice the average medical care cost of the average American under 65 years of age, or $2,000 (22). In fact, costs might be slightly higher. Actual costs for one submild case (parents providing physical therapy and no hospitalizations in 9 years) were approximately $4,700 in 1990; the cost of drugs alone was $1,900 (23). Nevertheless, erring on the conservative side ($2,000), the estimated average annual medical costs when the proportion of individuals with submild, mild, moderate, or severe cases of CF is accounted for yields a second estimate of annual medical costs at $10,885.

### *Nonmedical Direct Costs*

The chief nonmedical direct cost of CF is family caregiving time. CF centers estimate parents need to spend about 2 hours per day on therapy for a child with CF (730 hours per year) (22); many families have therapists provide chest physical therapy (23). In addition, parents, spouses, or other family members lose time from work or housework when an individual with CF misses school or work.

It is difficult to obtain direct estimates of the number of sick days for people with CF—and hence the nonmedical (caregiving) direct costs to family members—because the severity varies considerably across patients and over time. The OTA analysis assumes 20 sick days per CF patient per year, involving 8 hours per day of work or housework missed (160 hours per year). Nonmedical direct costs also must account for time spent taking the person with CF to medical appointments. The average number of hospital days is eight (8,33), and OTA assumes 4 hours of caregiving time are associated with each hospital day (32 hours per year). Time also is spent on physician and clinic visits not associated with an acute episode. The frequency of such visits is assumed to be four per year, at a time cost of 4 hours (16 hours per year). The total number of hours per year for CF-related caregiving is estimated to be 938.

OTA assumes caregiving time costs an estimated domestic/nursing wage of $10 per hour. While no empirical work supports this value, it is taken to be reasonable. Overall, then, the present value of lifetime nonmedical direct costs associated with CF is estimated at $139,744 (assuming an average life expectancy in 1990 of 28 years and using a 5 percent discount rate). This chapter presents only data using $10 per hour as the cost for time associated with nonmedical direct costs. Using a value that is 30 percent lower changes the relative cost-effectiveness less than changing other parameters, such as reproductive behavior or test sensitivity. It can be an important assumption in scenarios that are borderline net savings, however, since savings decrease.

## COSTS OF CARRIER SCREENING FOR CYSTIC FIBROSIS

Since CF is the most common recessive genetic disease among American Caucasians of European descent, there is intense commercial interest in marketing CF mutation assays. At least six commercial companies currently market CF carrier tests and at least 40 university and hospital laboratories conduct CF carrier assays. Table 9-3 lists a sample of prices for commercial facilities and university and hospital laboratories. The number of mutations assayed differs from facility to facility, and test costs reflect this variation. Additionally, some quoted prices include costs of pretest education and posttest counseling, while others do not. Costs of tests reflect differential royalty license fees among the facilities for patents related to the CF tests, as well (box 9-A). Nevertheless, the average charge for CF mutation tests among these facilities is approximately $170.

**Table 9-3—Costs for Cystic Fibrosis Carrier Tests At Selected Facilities**

| Institution | Price per sample |
|---|---|
| Baylor College of Medicine | $ 55 or 200 |
| Boston University | 170 |
| Collaborative Research, Inc. | 173 |
| Cornell University Medical Center | 75 |
| GeneScreen | 165 |
| Genetics & IVF Institute | 225 |
| Hahnemann University | 225 |
| Hospital of the University of Pennsylvania | 150 |
| Integrated Genetics | 150 |
| Johns Hopkins University Hospital | 270 |
| Mayo Medical Laboratories | 200 |
| St. Vincent's Medical Center | 150 |
| University of Minnesota | 136 |
| University of North Carolina | 150 |
| Vivigen, Inc. | 200 to 220 |

SOURCES: Office of Technology Assessment, 1992, and M.V. Pauly, "Cost-Effectiveness of Screening for Cystic Fibrosis," contract document prepared for the U.S. Congress, Office of Technology Assessment, August 1991.

*Photo credit: University of Kansas Medical Center*

H.C. Miller Building, University of Kansas Medical Center. University medical centers, as well as commercial companies, perform CF mutation analyses for testing and screening purposes.

### *Box 9-A—Licensing of Polymerase Chain Reaction*

The polymerase chain reaction (PCR) allows minute quantities of an identical sequence of DNA to be replicated millions of times (ch. 4); it is a critical tool for the CF mutation test and virtually all other new DNA-based diagnostic procedures. In 1987, Cetus Corp. received a patent for PCR. Currently, patent rights for PCR diagnostics are held by Hoffmann-La Roche, Inc. Because PCR is used in so many current and potential genetic tests, the terms of PCR licensing agreements are important determinants of future costs of tests to consumers.

Shortly after Hoffmann-La Roche acquired patent rights for PCR, commercial and hospital laboratories expressed concern that the proposed fee structure would discourage some laboratories from seeking agreements. For example, the terms for commercial facilities would have amounted to roughly 15 percent of the cost of each test that was performed. Any new applications—e.g., for a different disease—that a laboratory wanted had to be approved by Roche. Such licensing terms might have had a chilling effect on diagnostic companies and hospital laboratories that perform molecular diagnostics. Costs of DNA-based analyses, in general, would likely have risen rather than decreased as many expected they would with greater numbers of tests and increased volume.

Hoffmann-La Roche, Inc. announced in February 1992 that a new company, Roche Molecular Systems, had been formed to handle the development of PCR. The new licensing agreement announced includes permission to use PCR for a broad range of applications. Licenses will be available to all academic and commercial laboratories who request them, and academic and nonprofit licenses will require no down payment or minimum royalty payments and a royalty rate of less than 10 percent.

The saga of PCR licensing illustrates the importance of patents to future costs of DNA-based diagnostic tests, but the intellectual property issue is not solely confined to the PCR patent. Patents for the CF gene and its mutations, for example are pending. Thus, while automation will likely lower costs of DNA diagnostics, intellectual property protection to some extent might counter lower prices. Royalty licensing fees from patents will be reflected in charges for the tests to consumers. Resolving debates surrounding the Human Genome Project and intellectual property will have important consequences for ultimate cost—and hence, utilization and cost-effectiveness—of DNA-based diagnostics.

SOURCES: Office of Technology Assessment, 1992; based on M. Hoffmann, "Roche Eases PCR Restrictions," *Science* 225:528, 1992; and D. McQuilken, Roche Molecular Systems, Inc., personal communications, January 1992, February 1992.

OTA uses a lower cost per test—$100 per person—in its base case and most other analyses to reflect the expectation that test costs will likely decrease as CF carrier screening becomes routine; many believe economies of scale will be possible with larger volumes of screening assays. This cost includes the test itself and post-test counseling, but not pretest education or marketing. The $100 per individual cost ($200 per couple) also assumes all couples require the same amount of post-test counseling (and therefore incur an equivalent cost). In fact, couples who both screen negative (-/- couples) are likely to require far less post-test counseling than couples where one screens positive and one screens negative (+/- couples) or couples whose results are both positive (+/+ couples), but no assumptions are made about such variation. OTA also presents scenarios with higher and lower costs per test.

Three other components were included in test-related costs of CF carrier screening. First, some expenses will be incurred to inform people about the test's availability and to educate them about CF carrier screening. No estimates exist for costs related to providing such information and services. OTA uses a value of $25 per initial screening contact as the pretest cost of information and education (22).

Some argue $25 far exceeds what should be assigned to this cost—that most individuals will be visiting their physician for other purposes and that promotional pamphlets, videos, or mailings should constitute the sole cost of information services. OTA posits, however, that even if physicians include the cost of informing patients as part of their standard charge for a visit, this does not mean that providing information is costless—it consumes physician time and office space. Attaching a cost of $100 per hour to physician time and office use (equivalent to gross revenues of $200,000 per year), and assuming furnishing information takes 10 minutes, the cost per person informed would be about $17. To account for additional expenses related to nonphysician time and promotional materials, OTA uses a total cost of pretest information and education of $25. No charge is assessed for people who do not elect screening, although such individuals might well receive some pretest information. Similarly, although most scenarios considered here involve screening the woman first followed by the male only if she is positive, no charge is assessed for informational costs of screening the partner of those 1 in 25 women who test positive (would increase total cost of screening, but not significantly). Thus, others will argue OTA underestimates the total cost of pretest expenses. As results of the analysis show, the total cost associated with performing CF carrier assays is sensitive to assumptions about this charge; further research on actual expenditures related to it is warranted.

Second, because prenatal testing will be part of CF carrier screening when +/+ couples decide on childbearing, the cost of chorionic villus sampling (CVS) is relevant. CVS charges in two northeastern medical centers were investigated (22). As for other procedures, the specifics of what was included at what price varied. CVS sampling and the cost of a CF mutation analysis is priced at $1,200 in this analysis. Third, the cost of an abortion is priced at $900 (22).

## KEY VARIABLES AND ASSUMPTIONS

As just described, the cost of CF carrier screening includes providing pretest information and education, the actual test cost, post-test counseling, prenatal testing, and abortion. Beyond direct costs associated with performing CF carrier analyses, however, other key parameters that affect the economic analysis include:

- sensitivity of the CF test;
- percent of individuals who voluntarily elect to determine their CF carrier status;
- family size, percent who alter their reproductive behavior, and how the behavior is altered; and
- screening approach—e.g., preconception versus postconception, or women first, followed by men only for positive women, versus couples.

The following sections present the assumptions or values used for each of these factors in OTA's base case. Later sections describe how costs and savings were calculated, and how the assumptions or values for each were varied singly or in combination.

### *Test Sensitivity*

As described in detail in chapter 4, CF mutation analyses using delta F508 and an additional 6 to 12 mutations (ΔF508+6-12) detect about 85 percent of CF carriers, although depending on ethnic background and the battery of mutations used, test sensitivity can approach 90 percent (3,20). In Ashkenazic Jews, ΔF508+6 identifies nearly 95 percent of carriers (27). OTA uses 85 percent sensitivity in the base case and in most alternative scenarios, but varies the sensitivity in some scenarios to demonstrate its affect on costs and savings.

### *Participation*

What percent of eligible individuals will elect to be screened for their CF carrier status? Estimates are available for individuals with family histories of CF (12,18,31), as well as for one general population sample from a midwestern urban hospital and suburban health maintenance organization. A survey of this latter population found 84 percent of respondents had a strong interest in CF carrier screening before pregnancy and 69 percent would avail themselves during a pregnancy (6). Other surveys measure the acceptance of the general American populace toward prenatal genetic tests (28,30) and carrier screening (30), although not specific to CF. According to both these surveys, just over 80 percent of Americans say they would avail themselves of such tests.

To what extent, however, do such surveys represent real-life decisions? The key data would come from knowledge about what percent of individuals participate in CF carrier screening, but to date no published data exist. Early results from privately funded pilots, however, offer insights into what participation rates realistically might be expected. Through March 1992, 78 percent of participants (Caucasian Americans of European descent or Hispanic ethnicity) in a California pilot study have elected CF carrier screening (34); these individuals do not pay for their tests. Out-of-pocket costs dramatically affect the percent of people electing CF carrier screening: a Texas study reveals participation dropped from about 80 percent to 20 percent when a pilot ended and charges for screening began (3).

OTA uses 80 percent in the base case. Some might argue such a level is too high, despite experience from pilot projects. On the other hand, while the 80 percent figure will likely exceed participation in CF carrier screening's early phases of dissemination, it does not seem unreasonable in light of current steady-state data on the percentage of pregnant women who voluntarily elect prenatal testing for maternal serum alpha-fetoprotein or prenatal genetic analysis due to advanced maternal age (34): For both tests, 80 percent accept the procedure(s). OTA also examines, however, a participation rate of 20 percent, in light of OTA survey results that most third-party payors say they are unlikely to currently pay for CF carrier tests without a family history (ch. 7). Nevertheless, should CF carrier screening be incorporated into routine obstetric practice, third-party payment will likely increase, out-of-pocket expenses decrease, and participation increase. OTA also models 50 percent participation.

The population from which the 100,000 women or couples is drawn is assumed to be one for which the overall test sensitivity is at least 85 percent and the carrier frequency is 1 in 25. A more accurate approach might be to adjust the pool for racial and ethnic demographics. Different attitudes toward genetic screening and reproductive behavior prevail and would complicate the calculations. Further, the frequency of newborns with CF within other populations is so low—e.g., 1 in 17,000 (4) to 19,000 (16) African American newborns, 1 in 9,600 Hispanic, and 1 in 90,000 Asian American babies (16)—that weighting a random sample would have little net effect.

Finally, the pool is presumed to consist only of individuals who contemplate having children. Some argue CF mutation analysis would be of little interest to those choosing to be childless, although people who do not intend to have children sometimes do. Similarly, arguments can be made that some who do not contemplate children still might seek CF carrier screening solely for informational purposes or because of its impact in informing their relatives.

### *Reproductive Behavior*

Fundamental to the cost-effectiveness analysis of CF carrier screening is reproductive behavior. First, the analysis in this chapter assumes all couples (+/+, +/-, and -/-) seek the current U.S. average of 2.1 children per family (24). Some research indicates that families who have a child with CF alter their reproductive behavior by having fewer children (12,18,32), but these are retrospective analyses. No data exist for the total number of children that couples might ultimately have after identification as +/+ in the absence of a child with CF.

Second, only changes in the reproductive behavior of +/+ couples are modeled. Clearly, +/- couples might choose to avoid conception, seek prenatal testing, or consider pregnancy termination of carrier fetuses because current tests are not 100 percent sensitive, and hence there is some chance a fetus identified as a carrier actually has CF. Results from privately funded pilot studies, however, reveal no such decisions to date (2,26) (ch. 6).

Third, specific assumptions are necessary for precisely how people alter their reproductive behavior once they are identified as +/+ couples. The base case uses an infertility frequency in the general population of 8.4 percent (19,29). Of fertile couples, 10 percent is used as a reasonable estimate of the fraction of +/+ couples who will avoid conception (3). Avoiding conception obviously incurs costs, but fertile couples are also likely to incur contraception costs over their reproductive lives, and so contraception or sterilization costs are not included in this analysis.

Assumptions about the reproductive behavior of +/+ couples are critical to the cost analysis of population carrier screening for CF. The base case assumes all of the remaining 90 percent of +/+ fertile couples become pregnant and seek prenatal testing and that all couples with affected fetuses opt to terminate the pregnancy. Alternative scenarios vary the proportion of +/+ couples seeking or declining prenatal testing and the percentage electing abortion of CF-affected fetuses. Again, limited data exist on reproductive behaviors as they relate to CF carrier screening in the general population. Less than 20 CF-affected fetuses have occurred in pilot studies; in one study in Texas, 7 of 14 affected pregnancies were terminated (3). In the attitudinal survey of midwest urban and suburban women, 29 percent said they would terminate a pregnancy if the fetus were found to have CF (6). (As described in chapter 5, data from families of children or relatives with CF have been reported (18,31,32), but those data are not thought to be wholly representative of the general population.)

### *Screening Strategies*

Should both a man and woman be screened as a couple, or should the man be screened only if the woman's results are positive? Should the negative

327-046 - 92 - 8 : QL 3

partner in a +/- couple be screened for additional mutations to detect a higher proportion of carriers? Is there an effect on cost whether screening is preconception or postconception? The strategy employed in a CF carrier screening protocol affects costs and savings.

If the CF mutation assay detected 100 percent of mutations, it only would be necessary to screen one partner—usually the woman—because paternity is never assured. Placing primary focus on women is objectionable to some, however (14,15). Additionally, men and women are not always linked as a unit through their reproductive lives. Finally, limiting carrier screening to only men who are partners of positive women loses the opportunity to identify male carriers for whom CF carrier status might be of personal interest or future importance—to them and their relatives. On the other hand, from a cost perspective, it is clear the total number of individuals who will be screened in a "woman, then man" strategy will be less than in a "couple" strategy, and hence CF-related screening costs will be less—but to what extent?

Strategies for CF carrier screening can be preconception or postconception. The analysis in this chapter examines preconception screening. Modeling a postconception strategy is difficult because, although much CF carrier screening is of pregnant women and their fetuses, such screening is offered because the patients are being seen for other prenatal or genetic services—e.g., advanced maternal age or a family history of another disorder. Modeling reproductive behavior would be more complex because it becomes confounded by results for these other tests. Nonpaternity might be expected to be a greater factor in postconception CF carrier screening. Some fraction of postconception individuals will receive test results of an affected fetus in the late stages of pregnancy, where termination might not be feasible, which could then affect subsequent decisions about total family size.

Cost estimates for preconception strategies that screen only women first, as well as the strategy of screening couples, are presented. (The base case involves the former.) The analysis also examines the effect on costs and savings of screening the negative partner of +/- couples for additional mutations to detect a higher proportion of carriers. OTA assigns no additional cost for the followup analysis because at least three institutions follow such a protocol and do not charge extra for the additional mutations tested (3,9,34).

# COSTS AND SAVINGS

The analysis models a steady-state equilibrium, in which CF mutation analysis is available to all prospective parents at the outset of their reproductive planning—i.e., before the birth of any children. This assumption incorporates the fact that identification of CF carrier status affects all subsequent pregnancies.

### *Costs With No Carrier Screening*

At a carrier frequency of 1 in 25, 160 of 100,000 couples would be +/+, and be at 1 in 4 risk of having a child with CF in each pregnancy. Of these couples, 13.4 would be infertile (0.084 x 160). The remaining 146.6 couples each would have, on average, 2.1 children, of which 0.53 would theoretically have CF. Overall, the total lifetime CF-related medical costs without CF carrier screening are $11,575,536 (0.53 children/couple x $148,981 lifetime medical spending/child x 146.6 couples). Total nonmedical direct costs are $10,857,782 (0.53 children/couple x $139,744 lifetime CF-related caregiving costs x 146.6 couples). Total direct costs systemwide in the absence of screening are $22,433,318.

In fact, 36.7 couples of the 146.6 preconception couples will have a child with CF for their first pregnancy. As a result, some of these couples will alter their reproductive behavior (12,18,31,32) and avoid further conception (thereby saving potential costs associated with having an affected pregnancy in the future—i.e., cost offsets). Others will seek prenatal tests and consider abortion in subsequent pregnancies (thereby adding costs, but if abortion is chosen, contributing cost offsets). Overall, however, the net effect on total direct costs of altered reproductive behaviors for the 36.7 couples in the absence of DNA-based CF mutation analysis is negligible.

### *Calculating Costs*

The ways in which costs were calculated for the base case are presented in this section for illustrative purposes (table 9-4). Calculations for the alternative scenarios are not presented, but were performed in the same manner. Again, the base case involves the following:

**Table 9-4—Costs, Cost Offsets, and Net Savings for Base Case[a]**

| | | Costs | | Cost offsets (savings) | | |
|---|---|---|---|---|---|---|
| Description | Number | Cost of CF carrier screening per couple | Total CF carrier screening costs | Total medical savings | Total caregiving savings | Medical plus caregiving savings |
| Woman tests negative | 77,280 | $ 125 | $9,727,500 | 0 | 0 | 0 |
| Woman tests positive, man tests negative | 2,627.5 | 225 | 591,188 | 0 | 0 | 0 |
| Woman tests positive, man tests positive | 92.5 | — | — | — | — | — |
| Infertile | 7.8 | 225 | 1,755 | — | — | — |
| Fertile, voluntary childless | 8.5 | 225 | 1,913 | $670,415 | $628,848 | $1,299,263 |
| Fertile, prenatal testing, and abortion | 76.2 | 4,215 | 321,183 | 6,018,832 | 5,645,658 | 11,664,490 |
| Total cost of CF carrier screening | — | — | 10,643,539 | — | — | — |
| Total caregiving and medical offsets | — | — | — | — | — | 12,963,753 |
| Net savings per 100,000 couples | | | | | | |
| ***Net savings per 100,000 couples*** | — | — | — | — | — | 2,320,214 |

[a]Eighty percent elect to participate in screening; cost per test is $100; test sensitivity is 85 percent; all +/+ couples seek prenatal testing and terminate affected pregnancies.

SOURCE: Office of Technology Assessment, 1992.

- lifetime direct CF medical costs of $148,981 and lifetime CF-related caregiving costs of $139,744 per child with CF;
- carrier frequency of 1 in 25;
- 100,000 preconception women are screened, followed by screening the partner if the woman is positive;
- $25 cost per initial screening contact; test cost of $100 per test performed; and test sensitivity of 85 percent;
- 2.1 children per couple regardless of test results;
- 8.4 percent infertility and 10 percent of fertile +/+ couples choose to avoid conception; and
- 80 percent who are offered CF mutation analysis participate, all +/+ fertile couples seek prenatal testing (CVS at a cost of $1,200 per pregnancy) (22) and terminate all affected fetuses ($900 per pregnancy) (22).

Of 100,000 women offered screening, 80,000 elect to participate. Of these 80,000 women, 2,720 carriers are identified and so 2,720 men are tested. Of these, 92.5 males will be identified as carriers, and hence 92.5 +/+ couples are identified and 2,627.5 receive results indicating they are +/-. In fact, among these +/- couples are 16.3 +/+ couples who are missed, of whom 14.9 are fertile. Of the 92.5 +/+ couples who are identified, 7.8 are infertile (0.084 x 92.5) and 8.5 are voluntarily childless (0.10 x [92.5-7.8]). Thus, 76.2 fertile couples seek prenatal testing. Finally, 51.2 +/+ couples are missed (46.9 are fertile), among the 20,000 women who elected not to participate or who were undetected because the test is 85 percent sensitive, not 100 percent.

The cost per woman screened is $125 ($25 pretest for information and education + $100 for CF mutation analysis and post-test counseling). Women with negative results do not incur additional costs for screening. For identified carriers, there is an additional cost of $100 for CF mutation analysis and post-test counseling for the man, for a total of $225 per couple. This cost applies to couples who are identified as +/-, couples who are +/+, but infertile, and +/+ couples who decide not to have children. For +/+ couples who chose to conceive, costs are $225 plus the additional cost of prenatal screening and abortion. Since these couples seek a final family size of 2.1 children, a theoretical 2.8 pregnancies must be undertaken and 0.7 abortions per couple performed. Therefore prenatal screening and abortion costs add $3,990 per couple ($1,200 x 2.8 + $900 x 0.7). Screening related costs per +/+ couple, then, are $4,215 ($25 + $100 + $100 + $3,990).

The total cost related to performing CF carrier screening for this base case is $10,643,539.

### *Calculating Cost Offsets*

Two types of systemwide cost offsets (i.e., savings) flow from CF carrier screening: avoiding direct medical costs and avoiding nonmedical direct costs associated with time for caregiving. The benefit calculations that follow are a means to examine systemwide economic effects from CF carrier screening, not to positively or negatively reflect the intrinsic or extrinsic value to any individual or couple.

Neither medical nor caregiving cost offsets flow from infertile couples. The voluntarily childless couples, however, avoid medical and caregiving costs. Of 2.1 total expected children, 0.53 with CF would be expected per couple; overall, 4.5 CF-affected births would be avoided for this population (8.5 couples x 0.53 affected births/couple). Total medical cost offsets from those choosing to be childless are $670,415 ($148,981 per birth x 4.5 births). Savings from caregiving cost offsets for this group are $628,848 ($139,744 x 4.5).

Similarly, cost offsets arise from +/+ couples who use prenatal testing and, in the base case, terminate all fetuses diagnosed with CF. Some 40.4 affected births are avoided (76.2 couples x 0.53 affected births per couple). Total medical cost offsets for this group are $6,018,832 ($148,981 per birth x 40.4 births); total caregiving offsets are $5,645,658. Total medical and caregiving savings—costs avoided—are $12,963,753.

Because the test is less than 100 percent sensitive and because 20,000 women elect no screening, some +/+ couples are missed and some children with CF are born. Thus, the costs avoided fall short of the $22,433,318 spent in the absence of CF carrier screening. Overall, 77.7 babies with CF would be expected in the base case from the pool of 100,000 couples (146.6 fertile +/+ couples x 0.53 children with CF/couple), but 49.9 affected births are avoided (4.5 + 40.4). Systemwide, $12,963,753 are saved, but $10,643,539 are spent on screening, for a net savings over no screening of $2,320,214. That is, in the base case, sufficient savings accrue from avoided medical and caregiving costs to pay for costs associated with screening.

### *Alternative Scenarios*

Costs and savings for several alternative scenarios were developed. Table 9-5 presents the base case and 14 representative scenarios that demonstrate the effects of varying price, participation, test sensitivity, reproductive behavior, and screening strategy (a strategy that screens partners of positive women with a more sensitive test (90 percent) and a couples strategy).

Six alternative cases actually yield sufficient savings from avoided medical and caregiving costs to pay for all costs associated with screening—scenarios A, B, E, F, G, and J. All but scenario J, however, include the unlikely assumption of 100 percent termination of affected pregnancies. These scenarios are presented for illustrative purposes only or to examine the effect of other variables on cost-effectiveness, not as representations of likely occurrences or a goal to be achieved.

Scenario J also yields net economic savings over no screening from a systemwide perspective. It assumes 80 percent participation, a goal that might be achieved if CF carrier becomes as accepted as other prenatal tests (34) or as in the pilot studies in California and Texas described earlier (2,34). However, unlike in these pilots, a cost per test of $75 is assumed; participants in the pilots are/were not charged, and it is known that participation declines when out-of-pocket costs rise (2). Finally, scenario J assumes 64 percent[1] of affected fetuses detected are terminated, which could well be more frequent than will actually occur.

The remaining eight scenarios (C, D, H, I, K, L, M, N) do not yield net savings on a systemwide basis—i.e., they cost more than if no screening exists. As discussed below, several factors account for why population CF carrier screening is not cost-effective under assumptions used in these scenarios.

[1] Eighty percent of +/+ couples undergo prenatal diagnosis, and the remaining 20 percent choose no prenatal diagnosis because they would not consider terminating an affected pregnancy, regardless of outcome. The parents of 80 percent of affected fetuses diagnosed through the prenatal test elect to terminate, but 20 percent do not. This latter group might have sought prenatal testing with the thought of terminating, but then chose not to. Or they might have chosen prenatal testing knowing they would not terminate, but wanted information to prepare for the birth of a child with CF. The net frequency of abortion is 64 percent after combining the split decision process—i.e., that some will not seek any prenatal testing, but others might (which costs the system), but then opt not to terminate an affected pregnancy (also adds costs).

**Table 9-5—Effect of Assumptions on Net Savings Over No Screening**

| | Test sensitivity (%) | Cost per test | Participation | Percent +/+ who seek prenatal testing | Percent +/+ who abort affected pregnancies | Net systemwide savings compared to no screening[a] |
|---|---|---|---|---|---|---|
| Base case | 85 | $100 | 80% of 100,000 women; man only if woman positive | 100 | 100 | $2,320,214 |
| A | 100 | 100 | 100% of 100,000 women; man only if woman positive | 100 | 100 | 9,007,651 |
| B | 85 | 100 | 100% of 100,000 women; man only if woman positive | 100 | 100 | 2,977,225 |
| C | 85 | 100 | 80% of 100,000 women; man only if woman positive | 100 | 50 | (3,456,025)[b] |
| D | 85 | 100 | 80% of 100,000 women; man only if woman positive | 80 | 64[c] | (1,786,030) |
| E | 85 | 100 | 20% of 100,000 women; man only if woman positive | 100 | 100 | 589,498 |
| F | 85 | 100 | 50% of 100,000 women; man only if woman positive | 100 | 100 | 1,474,375 |
| G | 100 | 100 | 80% of 100,000 women; man only if woman positive | 100 | 100 | 7,188,877 |
| H | 85 | 100 | 20% of 100,000 women; man only if woman positive | 100 | 50 | (840,078) |
| I | 85 | 100 | 20% of 100,000 women; man only if woman positive | 80 | 64[c] | (430,182) |
| J | 85 | 75 | 80% of 100,000 women; man only if woman positive | 80 | 64[c] | 338,729 |
| K | 85 | 100 | 50% of 100,000 women; man only if woman positive | 80 | 64[c] | (1,075,074) |
| L | 85/90 | 100 | 50% of 100,000 women; if the woman is positive, the man is tested with additional mutations at 90% sensitivity | 80 | 64[c] | (765,919) |
| M | 85 | 100 | 100% of couples participate | 100 | 100 | (6,682,775) |
| N | 85 | 50 | 100% of couples participate | 100 | 50 | (4,077,693) |

[a]Per 100,000 eligibles.
[b]( ) denotes a negative value.
[c]Eighty percent choose prenatal diagnosis. Of these, 80 percent terminate affected pregnancies and 20 percent do not, either because when faced with the choice they decide against it or because they did not plan to, but sought prenatal diagnoses to prepare for the birth of a child with CF. Twenty percent do not undergo prenatal diagnosis because they will not terminate regardless of the outcome.

SOURCE: Office of Technology Assessment, 1992.

## Effect of Varying Rates of Selective Termination

Varying the ratio of those who elect to continue pregnancies diagnosed with CF to those who elect termination exerts a significant effect on net savings over no screening. The base case and scenarios D have similar assumptions except the fraction of affected pregnancies terminated varied—100 percent and 50 percent, respectively. Net savings per 100,000 women screened compared to no screening declines from $2,320,214 to -$1,786,030. Scenarios E and H are likewise identical except for the fraction of affected pregnancies terminated, and cost savings are eliminated when the frequency is halved.

### Effect of Test Cost

Reduced test costs appear as important as reproductive behavior, especially when participation is high. Not surprisingly, decreasing the cost of CF mutation analysis results in increased net savings (compare scenarios D and J)—to a point where cost-effectiveness can be achieved when the price per test is dropped 25 percent under the assumptions used. Net savings of -$1,786,030 in scenario D rise to $338,729 in scenario J when the cost per test drops from $100 to $75 per individual.

### Effects of Test Sensitivity or Screening Strategy

Increasing test sensitivity results in net system savings, if reproductive behavior and other variables are constant (scenarios A versus the base case, and scenarios L versus K). And as expected, the strategy of screening both individuals (scenario M), rather than only the male when the woman is positive (scenario B), decreases savings from $2,977,225 per 103,400 total individuals screened for the woman first strategy ($28.79 saved per person screened) to - $6,682,775 per 200,000 individuals screened in the couples strategy (cost of $33.41 per person). Decreasing the cost of screening by 50 percent in the couples strategy (scenario M versus N) is still not cost-effective under the assumptions used. But again, since the primary value of CF carrier screening is information, for which no cost can be assigned, the informational value to +/- couples where the male is positive and the female is negative would be entirely missed in the woman-first strategy.

### Effect of Participation

The percent of individuals electing CF carrier mutation analysis has an effect on system savings, all other factors being equal. Participation rate is less important, however, than test cost or reproductive behavior. In fact, lower participation is more cost-effective, depending on test cost or selective termination, because costs associated with screening are not incurred, although high participation and a high frequency of termination or low test cost is most cost-effective. Scenarios D, I, K, and J best illustrate this point for test cost. The first three scenarios assume a test cost of $100, 85 percent sensitivity, and the bifurcated reproductive decision option that results in a net termination of 64 percent of affected pregnancies; scenario J assumes the same at a test cost of $75. What differs among the first three is agreement to be screened—80 percent, 20 percent, and 50 percent, respectively; scenario J assumes 80 percent participation to compare it to scenario D. At 80 percent acceptance and $100 test cost per 100,000 eligibles, CF carrier screening is not cost-effective (net cost of $1,786,030), but when test cost is reduced to $75, CF carrier screening is cost effective (net savings of $338,729). In contrast, when participation falls to 50 percent and test cost is $100, net cost drops to $1,075,074; when acceptance is only 20 percent the net cost is $430,182. Scenarios C, E, and H demonstrate the effect of participation versus selective termination).

## QUALIFICATIONS ON THE ANALYSIS

Results of this analysis are highly dependent on the assumptions made. Some of the more critical or controversial assumptions are highlighted, including:

- **Agreement to be screened.** The results are sensitive to the proportion of the eligible population who consent to initial screening, but less so than reproductive behavior or test sensitivity. Even with 20 percent participation, screening can be cost-effective, depending on the other variables.
- **Reproductive behavior.** Assumptions about reproductive behavior are the most important factor in the analysis, but experience-based data are sparse. How completed fertility is affected by the occurrence of CF (in the absence of screening), how the availability of CF carrier screening affects potential parents' choices between remaining childless versus prenatal screening with selective termination, and how both CF carrier screening and prenatal testing availability affect final average family size are critical.
- **Pretest costs.** A cost for marketing CF carrier screening and for pretest education is assessed as a screening expense. Although this cost is small per eligible individual, it is an important part of the total cost because it is incurred for all first contacts who elect screening. No data empirical data exist to support the $25 used. OTA estimated the value by analogy. While it might be such costs are higher for those who

elect screening, balanced against higher pretest costs for these individuals is the likelihood of lower or no costs for those who do not participate.

- **Cost of the CF carrier assay.** A cost of $100 per test is assumed in most scenarios, which is lower than current charges. Using costs of $75 and $50 per test increases net savings. Ultimately, the assay's cost might be less important overall to net savings than it is to its impact on the willingness of individuals to elect screening if they must pay the cost themselves. Nevertheless, should the cost reach $50 to $75 per individual, savings are achieved, depending on the frequency of pregnancy termination.
- **Median life expectancy.** A median life expectancy of 28 years in 1990 is used. Using a longer life expectancy increases net savings if CF-affected births are avoided, but also increases costs of the disease when +/+ couples are missed because of test sensitivity, when people elect no screening, and when reproductive behavior is not altered. Overall, the effect of median life expectancy on costs and savings, however, is negligible if the other variables in the analysis are held constant. More importantly, as median life expectancy increases, individuals might be less likely to alter their reproductive behavior, which will likely result in fewer CF-affected births avoided and lower net savings.
- **Present value of avoided CF-related costs.** An estimate of $10,000 as the annual medical cost per person with CF is used, which is at the low end of annual cost estimates compiled by several sources. Higher estimates, however, are based on calculations that appear to conflict with actual data on use of care by CF patients (21,22). Another potentially important adjustment to this estimate is the assumption that the $10,000 occurs uniformly every year over the person's lifetime. As CF patients live longer and are treated more effectively, the period of high medical costs likely will be pushed further into the future. Even if those costs eventually are substantial, however, they are discounted back to the present, which means they add little to the present value of CF-related medical costs. Conversely, some new treatments and technologies likely will be expensive. Greater use of heart-lung transplants or even gene therapy, for example, will increase CF-related medical costs—significantly if the therapies become a common option. The choice of discount rate is also important.
- **Paternity.** The analysis assumes the prospective father can be identified and screened and that each woman has the same partner for all births, so that CF carrier screening need be done only once per couple. Uncertainty of paternity already confounds real-world use of genetic tests, with frequencies of nonpaternity reported or estimated from 2 to 15 percent (1,17). People also may have multiple partners over their reproductive years. Both behaviors reduce net savings.
- **Test precision.** Any costs (e.g., anxiety) that theoretically might be imposed by a less that 100 percent sensitive test were unaccounted for, although post-test costs related to counseling were included. Similarly, psychological costs associated with false positive findings—most likely from laboratory handling error—were also unaccounted for, since evidence indicates DNA tests per se yield accuracy greater than 99 percent (5,13). False positive findings would not result in a CF-affected pregnancy, but could result in the cost of a prenatal test for a fetus not actually at risk.
- **Preferential screening of relatives of carriers.** Although no specific scenario is examined, close relatives of carriers might seek and use CF mutation analysis at a greater frequency, at least initially. The effect of preferential screening of relatives would be to enhance the efficiency of CF carrier identification. For example, in a population of 100,000 individuals with 50 percent participation, 75.2 percent of +/+ couples would be identified. If the population were relatives of previously identified CF carriers, 50 percent participation identifies 88.3 percent of +/+ couples (7). If the reproductive behavior were similar, then preferential screening would increase net savings. On the other hand, if acceptance of CF carrier screening approaches the 80 percent level that exists for similar tests, preferential screening of relatives will have little net economic effect.
- **Subsequent generations.** Future offspring of +/- couples might preferentially seek CF mutation analysis because their carrier risk will be 1 in 2. In contrast, offspring of -/-couples will have a very low risk of being carriers and might be less likely to utilize CF tests. Such a scenario

would decrease costs of screening and increase net savings.

## SUMMARY AND CONCLUSIONS

One of the least examined of the many issues surrounding CF carrier screening is the potential systemwide savings or costs if large numbers of individuals are screened. Examining cost-effectiveness for CF carrier screening, however, is fraught with technical and social pitfalls. In 1983, the President's Commission for the Study of Ethical Problems in Medicine and Biomedical and Behavioral Research concluded the fundamental value of CF carrier screening rests in its potential for providing people with information they consider beneficial for autonomous reproductive decision-making. In short, personal considerations of having information—as well as societal considerations of avoiding eugenics, stigmatization, and discrimination—outweigh considerations of cost-effectiveness. There is no intimation in OTA's analysis that something that saves or costs money is more or less desirable from a welfare standpoint. Nevertheless, while a cost-effectiveness analysis of CF carrier screening is useful to examine issues of resource allocation, some will find it offensive that such calculations are even performed in the context of genetic screening, since at its core it involves the potential to terminate affected pregnancies.

Overall, whether CF carrier screening can be paid for on a population basis through savings accrued by avoiding CF-related medical and caregiving costs depends on the assumptions used—including how many children people will have, average CF medical costs, and average time and cost devoted to caring for a child with CF, as well as variations in reproductive behaviors, costs of CF mutation analyses, and screening participation rates. Eight of 14 scenarios examined result in a net negative output to the economy over no screening. In the remaining scenarios, CF carrier screening is cost-effective, but most of these scenarios involve 100 percent participation, test sensitivity, or selective termination—all unlikely to be realized in the near term, if ever. Nevertheless, CF carrier screening can save money compared to no screening even under less absolute circumstances. The balance between net savings versus net costs in nearly all scenarios is fine. How many individuals participate in screening is relatively unimportant to cost-effectiveness, but it is clear the frequency of affected pregnancies terminated and the assay's price will ultimately affect this balance.

## CHAPTER 9 REFERENCES


1. Ashton, G.C., "Mismatches in Genetic Markers in a Large Family Study," *American Journal of Human Genetics* 32:601-613, 1980.
2. Beaudet, A.L., Howard Hughes Medical Institute, Houston, TX, remarks at the 8th International Congress of Human Genetics, Washington, DC, October 1991.
3. Beaudet, A.L., Howard Hughes Medical Institute, Houston, TX, personal communications, November 1991, February 1992.
4. Boat, T.F., "Cystic Fibrosis," *Textbook of Respiratory Medicine*, J.F. Murray and J.A. Nadel (eds.) (Philadelphia, PA: W.B. Saunders, 1988).
5. Boehm, C.D., and Kazazian, H.H., Jr., "Prenatal Diagnosis by DNA Analysis," *The Unborn Patient: Prenatal Diagnosis and Treatment*, 2nd ed., M.R. Harrison, M.S. Golbus, and R.A. Filly (eds.) (Philadelphia, PA: W.B. Saunders Co., 1991).
6. Botkin, J.R., and Alemagno, S., "Carrier Screening for Cystic Fibrosis: A Pilot Study of the Attitudes of Pregnant Women," *American Journal of Public Health* 82:723-725, 1992.
7. Cox, T.K., and Chakravarti, A., "Detection of Cystic Fibrosis Gene Carriers: Comparison of Two Screening Strategies by Simulations," *American Journal of Human Genetics* 49(Supp.):327, 1991.
8. Cystic Fibrosis Foundation, *Cystic Fibrosis Patient Registry* (Bethesda, MD: Cystic Fibrosis Foundation, 1989).
9. Davis, J.G., New York Hospital, New York, NY, personal communication, March 1992.
10. Fenerty, J.P., and Garber, A.M., *Costs and Benefits of Prenatal Screening for Cystic Fibrosis*, Working Paper No. 2749 (Cambridge, MA: National Bureau of Economic Research, 1988).
11. Garber, A.M., and Fenerty, J.P., "Costs and Benefits of Prenatal Screening for Cystic Fibrosis," *Medical Care* 29:473-491, 1991.
12. Kaback, M., Zippin, D., Boyd, P., et al., "Attitudes Toward Prenatal Diagnosis of Cystic Fibrosis Among Parents of Affected Children," *Cystic Fibrosis: Horizons. Proceedings of the 9th International Cystic Fibrosis Congress*, D. Lawson (ed.) (New York, NY: John Wiley & Sons Inc., 1984).
13. Lebo, R.V., Cunningham, G., Simons, M.J., et al., "Defining DNA Diagnostic Tests Appropriate or Standard of Clinical Care," *American Journal of Human Genetics* 47:583-590, 1990.
14. Lippman, A., "Prenatal Genetic Testing and Screening: Constructing Needs and Reinforcing Inequities," *American Journal of Law and Medicine* 17:15-50, 1991.

15. Lippman, A., ''Mother Matters: A Fresh Look at Prenatal Diagnosis and the New Genetic Technologies,'' *Reproductive Genetic Testing: Impact on Women*, proceedings of a conference, Nov. 21, 1991.
16. MacLusky, I., McLaughlin, F.J., and Levinson, H.R., ''Cystic Fibrosis: Part 1,'' *Current Problems in Pediatrics*, J.D. Lockhart (ed.) (Chicago, IL: Year Book Medical Publishers, 1985).
17. MacIntyre, S., and Sooman, A., ''Non-Paternity and Prenatal Genetic Screening,'' *Lancet* 338:869-871, 1991.
18. Miller, S.R., and Schwartz, R.H., ''Attitudes Toward Genetic Testing of Amish, Mennonite, and Hutterite Families With Cystic Fibrosis,'' *American Journal of Public Health* 82:236-242, 1992.
19. Mosher, W.D., and Pratt, W.R., ''Fecundity and Infertility in the United States, 1965-88,'' *Advance Data From Vital and Health Statistics, No. 192* (Hyattsville, MD: National Center for Health Statistics, 1990).
20. Ng, I.S.L., Pace, R., Richard, M.V., et al., ''Methods for Analysis of Multiple Cystic Fibrosis Mutations,'' *Human Genetics* 87:613-617, 1991.
21. Pauly, M.V., ''The Economics of Cystic Fibrosis,'' *Textbook of Cystic Fibrosis*, J.D. Lloyd-Still (ed.) (Boston, MA: PSG, Inc., 1983).
22. Pauly, M.V., ''Cost-Effectiveness of Screening for Cystic Fibrosis,'' contract documents prepared for the U.S. Congress, Office of Technology Assessment, August 1991, November 1991, February 1992.
23. Poling, S., Parent, Silver Spring, MD, personal communication, December 1991.
24. Population Reference Bureau, *1991 World Population Data Sheet* (Washington, DC: Population Reference Bureau, 1991).
25. President's Commission for the Study of Ethical Problems in Medicine and Biomedical and Behavioral Research, *Screening and Counseling for Genetic Conditions: The Ethical, Social, and Legal Implications of Genetic Screening, Counseling, and Education Programs* (Washington, DC: U.S. Government Printing Office, 1983).
26. Schulman, J.D., Genetics & IVF Institute, Fairfax, VA, personal communication, December 1991.
27. Shoshani, T., Augarten, A., Gazit, E., et al., ''Association of a Nonsense Mutation (W1282X), the Most Common Mutation in Ashkenazi Jewish Cystic Fibrosis Patients in Israel, With Presentation of Severe Disease,'' *American Journal of Human Genetics* 50:222-228, 1992.
28. Singer, E., ''Public Attitudes Towards Genetic Testing,'' *Population Research and Policy Review* 10:235-255, 1991.
29. U.S. Congress, Office of Technology Assessment, *Infertility: Medical and Social Choices*, OTA-BA-358 (Washington, DC: U.S. Government Printing Office, 1988).
30. U.S. Congress, Office of Technology Assessment, *New Developments in Biotechnology: Public Perceptions of Biotechnology*, OTA-BP-BA-45 (Washington, DC: U.S. Government Printing Office, May 1987).
31. Wertz, D.C., Janes, S.R., Rosenfield, J.M., et al., ''Attitudes Toward the Prenatal Diagnosis of Cystic Fibrosis: Factors in Decision Making Among Affected Families,'' *American Journal of Human Genetics* 50:1077-1085, 1992.
32. Wertz, D.C., Rosenfield, J.M., Janes, S.R., et al., ''Attitudes Toward Abortion Among Parents of Children With Cystic Fibrosis,'' *American Journal of Public Health* 81:992-996, 1991.
33. Wilkerson Group, Inc., *Annual Cost of Care for Cystic Fibrosis Patients* (New York, NY: Wilkerson Group, Inc., 1991).
34. Witt, D., Kaiser Permanente Medical Group, San Jose, CA, personal communication, March 1992.

Chapter 10

# Cystic Fibrosis Carrier Screening in the United Kingdom

# Contents



## *Boxes*



## *Figures*



## *Tables*

# Chapter 10
# Cystic Fibrosis Carrier Screening in the United Kingdom

Three pilot projects to explore the implications of population screening for cystic fibrosis (CF) carriers are under way in the United Kingdom, funded by the Cystic Fibrosis Research Trust (CF Trust), a private, nonprofit philanthropic organization. The goals of the programs, which began in 1990, are to identify the most appropriate populations for screening, evaluate various test protocols and techniques, and understand the psychosocial consequences of carrier identification. In addition to pilot programs supported by the CF Trust, at least two other ongoing CF carrier screening pilots are supported by private and various research funds.

This chapter discusses the structure of CF carrier screening programs in the United Kingdom—the most extensive, comprehensive, and advanced pilots currently under way. It analyzes the strategies being used and reports on results recorded to date—both of which could bear on how CF carrier screening is approached in the United States. This chapter is based, in part, on interviews conducted by OTA during June 1991 visits with pilot project staff and staff of the Medical Research Council (MRC).

## GENETIC SERVICES IN THE UNITED KINGDOM

Through the British National Health Service (NHS), all citizens receive medical care. Individuals self-assign themselves to a general practitioner (GP) within close geographic proximity. Private, for-pay medical care is also available to those willing and able to pay. Through the NHS, the British government informs GPs of new tests and medical practices, although some believe that such information does not always get communicated in a timely fashion (12). As in the United States, GPs are likely to learn about new developments in diagnosis and therapy through continuing education.

The British health care and legal systems have protected medical professionals from malpractice suits because there is no contingency fee arrangement as in the United States. Moreover, in general, patients can only sue for actual cost. Because health care is free, the actual costs are likely to be low.

In the United Kingdom, individuals are usually referred to a genetics unit by their GP, who is responsible for primary health care screening and prevention and is the usual means by which individuals are introduced to the need for genetic information. Family planning and prenatal clinics are also sources of genetics information. Ideally, when a woman tests positive for pregnancy, she is booked with a hospital prenatal clinic for management of her pregnancy, where she may receive relevant genetic information. The provision of routine genetic counseling is increasingly offered through primary care. Unusual or difficult cases are referred to a genetics specialty unit.

Medical genetics is a rapidly developing specialty that has been widely introduced in the United Kingdom. Population screening and prenatal diagnosis have been available for groups at risk for Tay-Sachs disease and thalassemias. Genetics services routinely deliver neonatal screening for phenylketonuria and congenital hypothyroidism, maternal serum alpha-fetoprotein screening, and fetal karyotyping in women of advanced maternal age (12,18). A national network of regional genetic services exists but is currently threatened by recent governmental changes in the NHS (2,10,12).

### *Cystic Fibrosis in the United Kingdom*

In Great Britain, about 300 babies a year are born with CF. The average annual cost of treating someone with CF is estimated to be £5,000 (20) to £10,000 (6). The Caucasian population in South East England has undergone extensive genetic analysis for CF carrier status (8,9), and the frequency of the ΔF508 mutation in this part of England has been variously reported at a minimum of 70 percent in adult CF populations (21), 71.5 percent (8,9), and 80 percent (27). In the Scottish population, ΔF508 represents about 71 percent of all CF mutations (15,22). (See app. A for international distribution of mutation frequencies.)

Buccal mouthwashes are increasingly used as the source of DNA for screening in the United Kingdom. The mouthwash technique is inexpensive and considered safe, and the DNA can be extracted rapidly and reliably in sufficient quantity for amplification by polymerase chain reaction (PCR; ch. 4). Welsh investigators have compared mouthwash, buccal scrapes, and finger pricks as the methods for sample collection, and determined that the mouthwash is the most desirable in terms of patient acceptability, successful DNA extraction, and cost (12). Among pregnant women, blood—already being drawn for other diagnostic tests—is used rather than mouthwash.

With births numbering 700,000 annually, an estimated 1.4 million screening assays would have to be performed if all couples were screened prenatally. About 56,000 carriers would be detected from this cohort, requiring counseling and the option of prenatal diagnosis. The identification of 56,000 carriers annually would overwhelm the 157 full-time doctors and clinical coworkers in clinical genetics centers in the United Kingdom, excluding laboratory scientists (10). One survey of health professionals showed that approximately 75 percent of both GPs and family planning clinic staff thought that the introduction of CF carrier screening was appropriate; less than 10 percent opposed it (27).

In addition to staffing difficulties, it is widely acknowledged in Britain (as it is in the United States) that there are serious deficiencies in the teaching of clinical genetics in medical schools, which makes experts reluctant to rely on primary care doctors to provide genetics advice (13). In contrast to the United States, genetic counseling in Britain is most frequently offered by M.D. and Ph.D. clinicians. Counselors trained at the level of a master's degree are rare; nurses trained in genetics are more common. As a result, genetic services tend to be provided through highly specialized, highly trained individuals. A survey of health visitors (the U.K. equivalent of a U.S. visiting nurse) showed that while generic health visitors (i.e., those not working in genetics) had a reasonable knowledge of the more obvious aspects of genetic services, there were a number of areas about which they were unsure. Furthermore, they viewed their own knowledge of genetics as poor (7).

### *The Role of the Medical Research Council in Cystic Fibrosis Carrier Screening*

The MRC is a public agency of the British Government. "The MRC has a long history of supporting research of direct relevance to health services, in addition to, and often built around, the biomedical and clinical research that forms the bulk of its work" (16). Much of the MRC's health services research is conducted by its units (comparable to the intramural programs of the U.S. National Institutes of Health). In 1981, the MRC established a Health Services Research Panel (HSRP) as an advisory body to the Council's Boards and Grants Committees. In 1986, the HSRP was reconstituted as a committee, the Health Services Research Committee (HSRC), with powers to promote the development of health services research and to make funding recommendations (16). In general, HSRC concentrates on: research into the effectiveness and efficiency of health services in implementing medical knowledge to improve health; and needs for which effective medical interventions exist or could be developed in the future. Design and delivery of health services is integral to the research interests of the HSRC; thus CF carrier screening projects fall squarely within its domain (24).

Involvement of the MRC with CF began in 1989 when it received outline proposals for CF carrier screening pilot studies. At the same time, the CF Trust invited applications for pilot studies. The outlines and the proposals funded by the CF Trust had a broad focus, whereas the MRC was interested in studies that would address the costs, benefits, disbenefits, and acceptability of screening programs. The MRC, therefore, convened a workshop for geneticists and social scientists to discuss the research questions raised by CF carrier screening (24). After the workshop, pilot proposals were considered. Based on the conclusions of the workshop, the MRC placed a high priority on studies to:

- evaluate the objectives of screening programs (e.g., identify carriers to reduce birth incidence, aid reproductive choice, or provide information);
- address the acceptability and participation rate of carrier screening for different target groups and in different settings;
- compare screened and nonscreened groups;
- clarify what information should be provided about carrier screening to the general public and to individuals or couples to whom screening might be offered, and evaluate ways of presenting that information;
- assess anxiety specific to carrier screening in addition to general anxiety levels;
- measure short- and long-term outcomes of carrier status identification;
- follow up couples at low risk who subsequently produce an affected child; and

- measure the costs of providing a screening program (17).

In addition to the preceding research issues, the MRC considers the cost-effectiveness of screening programs an important factor and encourages researchers to address this issue (24). This could prove difficult: Because of the way the NHS operates, cost accounting has always been elusive. Few providers of genetic services would likely document evidence of beneficial outcome, workload, or costs for each of the districts they serve and would have difficulty assigning a figure to the cost of tests.

The MRC currently supports two projects. A pilot study at St. Mary's Hospital Medical School will produce a videotape for CF carriers and monitor and evaluate the use of the videotape (£19,000). An evaluation of carrier screening for CF in couples at the Human Genetics Unit in Edinburgh also receives MRC funds (£97,000) (3,24). Some staff working on the pilot projects funded by the CF Trust also receive money through the "Genetic Approach to Human Health" project. Through this project, the MRC aims to encourage more applications for studies of the health services research issues associated with carrier screening.

## THE CYSTIC FIBROSIS RESEARCH TRUST PILOT PROGRAMS

The CF Trust, analogous to the Cystic Fibrosis Foundation in the United States, funds research related to improved diagnosis, understanding, and treatment of the disorder. The CF Trust has been the lead organization in the United Kingdom's investigation of carrier screening programs. Acknowledging the complexity of a national screening program, the CF Trust sought advice from an expert panel, the MRC, and the Department of Health before deciding to fund three pilot projects for evaluation of heterozygote screening. The three sites to receive funds were:

- University of Wales College of Medicine, Cardiff, Wales, for 3 years, £197,540;
- University of Edinburgh, Western General Hospital, Edinburgh, Scotland, for 3 years, £197,540; and
- Guy's Hospital, London, England, for 2 years, £177,775.

In addition to performing their own tests, all three pilots are simultaneously using Cellmark Diagnostics' amplification refractory mutation system (ARMS) multiplex kit. The ARMS multiplex assays $\Delta$F508, 621+1G$\rightarrow$T, G551D, and G542X, with a sensitivity of 83 percent. Cellmark is not yet charging for the use of the kits because it has not yet been licensed to do so.

*Photo credit: Peter Harper, University of Wales College of Medicine*

Institute of Medical Genetics, University of Wales College of Medicine, Cardiff, Wales.

### *University Hospital of Wales, Cardiff, Wales*

The 3-year pilot project under way at the Institute of Medical Genetics, University Hospital of Wales in Cardiff has the following aims:

- evaluate the attitudes of GPs and primary care staff to CF carrier screening;
- evaluate the feasibility of screening for CF carriers in general practices;
- evaluate clients' interest in and reactions to CF carrier screening;
- develop accurate, informative, and acceptable educational material for the public and health professionals; and
- consider the possibility of a pilot prenatal carrier screening project.

In addition to these clinical objectives, the Cardiff group has established an effective PCR multiplex system to test for the major CF mutations. The Institute of Medical Genetics of the University Hospital of Wales is not a basic CF research group, but a comprehensive genetics center. As a result, the clinic assumes the need for well-prepared GPs in order to make new screening programs work (11). Much of the first year of this pilot was spent evaluating attitudes about screening in urban and rural/industrial mining-town general practices. Although geneticists view CF as a relatively common genetic disorder, GPs do not; the Cardiff pilot group did not want to mistakenly assume that GPs would give the necessary time to ensure a successful screening program. In addition, the pilot coordinator has been working with health professionals such as nurses and midwives to educate them about CF carrier screening and prepare them for questions that may arise.

Opportunistic screening refers to the practice of approaching individuals either in person or via brochure while they are waiting for appointments for other reasons at either their GP or family planning clinic. Screening is offered opportunistically to all individual adults between the ages of 16 and 45 who are registered with two general practices in Wales (each approximately 1,500 individuals). The Welsh plan for screening and followup of individuals is depicted in figure 10-1.

Prenatal screening is not offered at this time, but will be offered in 1993, the third year of the pilot. Known CF families have been contacted, and family testing has been completed for that group. Screening of couples is under consideration (box 10-A). This project, if implemented, will offer CF carrier screening to all couples (but not individuals) in a separate general practice (approximately 1,000 individuals) who are in the reproductive age group and are planning a pregnancy (figure 10-2). The attitudes of participating couples will be evaluated. This protocol, as with all protocols, must be approved by the Ethical Committee of the Division of Child Health of the University Hospital of Wales.

Mouthwashes are used for collection of DNA. The Cardiff multiplex tests for ΔF508, ΔI507, G551D, R553X, and 621+1G→T, providing 81.5 percent sensitivity. As with the other two pilots, Cellmark's ARMS is run concurrently, with a sensitivity of 83 percent.

The project is designed to compare the differences of participation in screening following invitation by letter, opportunistic invitation, and self-referred requests for screening. In addition, socioeconomic variables will be compared, including social class, age, sex, and marital status. As with the London and Edinburgh pilots, the Cardiff pilot will also evaluate the psychological sequelae to screening for CF. Participants will be asked to complete questionnaires at a minimum of three junctures: prior to screening, after being told of their carrier status, and at a 3-month followup. The questionnaires will evaluate perceived health through the perceived health measure (14), a shortened State of Anxiety Inventory (23), a reproductive intentions questionnaire, and a knowledge-of-CF questionnaire. Box 10-B describes the protocol.

Those who choose not to be screened will be contacted to determine their reasons for refusing screening. Finally, the Cardiff team will calculate the costs and time implications of screening for general practices and laboratory staff.

### *Western General Hospital, Edinburgh, Scotland*

The CF pilot at Western General Hospital in Edinburgh, Scotland is modeled after existing β-thalassemia and Tay-Sachs programs offered through prenatal clinics in the United Kingdom. The long-term goal is to introduce prepregnancy CF carrier screening. The philosophy of this pilot is markedly different from that of the other two CF Trust pilots and the program at St. Mary's Hospital in London. The pilot director believes that CF mutation analysis should first be offered to individuals with the most limited choices (i.e., women who are pregnant) while simultaneously initiating screening in the broader preconception population (3). This philosophy is based on the assumption that pregnant couples coming through prenatal clinics are more motivated, more in need of this type of information, and more likely to make immediate use of the information provided through screening (4).

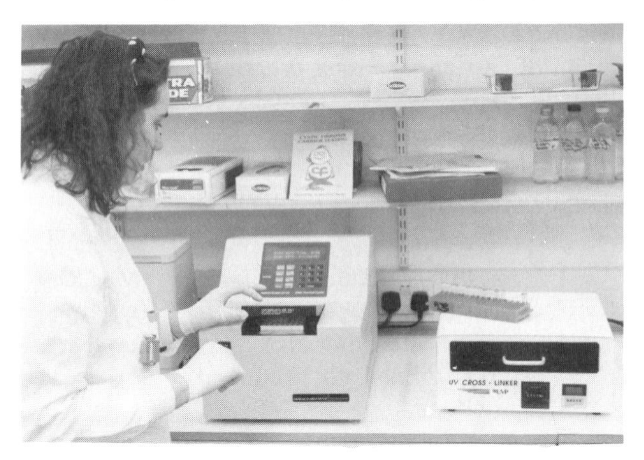

*Photo credit: David J.H. Brock, University of Edinburgh*

CF mutation analysis of samples in a laboratory at Western General Hospital, University of Edinburgh, Edinburgh, Scotland.

**Figure 10-1—Plan for Screening and Followup in West Glamorgan, Wales (Individuals)**

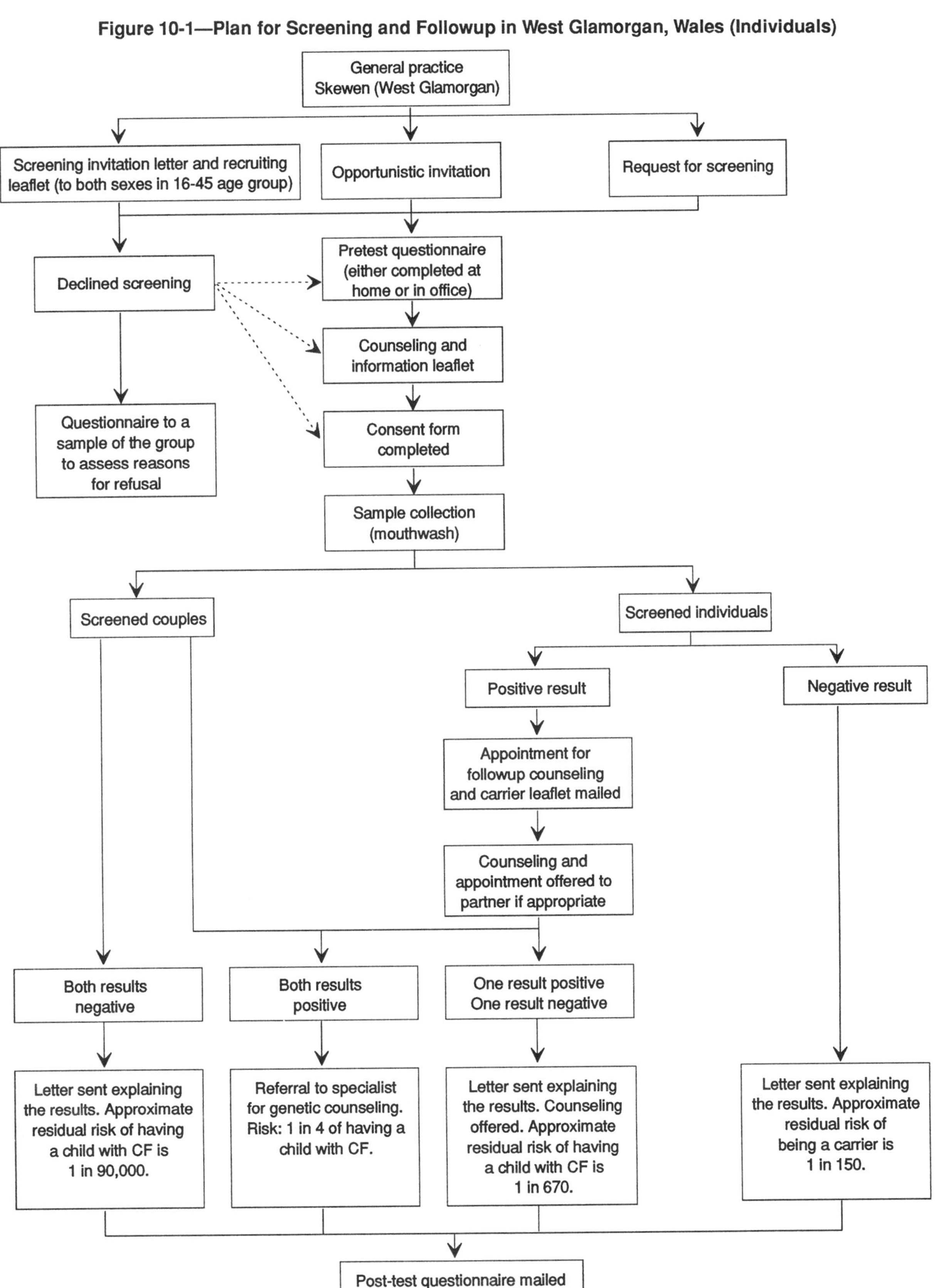


SOURCE: P. Harper, University of Wales College of Medicine, Cardiff, Wales, personal communication, 1991.

### *Box 10-A—The Couples Approach to Carrier Screening*

One approach currently used in London and under consideration elsewhere in the United Kingdom is the "couples" protocol for carrier screening. This protocol aims to identify high risk couples (both partners are carriers, thereby each pregnancy has a 1 in 4 risk of having an affected child). Using this protocol, couples are screened as a unit and receive their results as either high or low risk of producing an affected child. Individual carrier status is not discussed. Even if the geneticist determines that one of the members of the couple is a carrier (a +/- couple), that couple is grouped with negative-screening couples (-/-) in terms of risk. (See figure 10-2 for protocol.)

Several medical geneticists in the United Kingdom are disturbed by this approach as it involves a failure to disclose known information. At a meeting convened by the Medical Research Council in May 1991, participants considered the concept of blind testing. If test results could not be attributed to an individual but rather the couple, then, the group concluded, information was not being concealed and the protocol could be considered ethically acceptable.

Approximately 3 percent of couples will test +/-, and their risk of producing an affected child is 1 in 600. This compares to the lower risk group of -/- couples, whose chances of producing an affected child are 1 in 50,000. The residual risk to the first group presents ethical dilemmas currently being sorted out in the United Kingdom.

Proponents of this approach feel it is more economical and reduces the anxiety associated with knowing one's carrier status. In addition, both partners are screened simultaneously. If the woman is pregnant, the amount of time needed to identify a couple at risk is reduced.

Just how to obtain informed consent in this process remains a problem, although proponents claim that if the contractual agreement is to identify only high risk couples, then those who find such an approach disturbing could opt out. In addition, the implications for liability should a couple deemed "low risk" give birth to a child with cystic fibrosis have not been addressed.

SOURCE: Office of Technology Assessment, 1992.

When a woman is booked for her first appointment through a family planning service, she also receives a CF carrier screening booklet. Upon her arrival at the clinic, a genetic nurse or midwife (who has been previously trained to handle CF questions) asks her if she wants to be screened. Women who are late in their pregnancy (beyond 16 weeks' gestation) are excluded, as are those with no partners, since prenatal screening is most informative using results from both parents. The goal of the program is to reach at least 80 percent of patients coming through a major maternity hospital with the aim to expand to two other hospitals in the near future. In addition to the prenatal program, the Edinburgh pilot study is going back to CF families and offering the test. Offering screening to school-age individuals has been explored but not initiated. In a study of 14- to 16-year-old school children in Edinburgh, investigators found a positive attitude to carrier screening for CF and to the offer of prenatal tests of those couples shown both to be carriers (5).

The Edinburgh pilot employs a two-step model, first screening the pregnant woman for carrier status using assays for three mutations (ΔF508, G551D, and G542X, providing a predictive value of 85 percent in the population) and then testing her partner for 15 mutations if she is diagnosed as a carrier (evaluating 12 more mutations provides a predictive value of 92 percent in the population). The Edinburgh laboratory is currently running 50 to 60 samples a week and running the ARMS test kit simultaneously. By December 1991, more than 2,000 samples had been run, detecting 74 carriers (table 10-1).

Like the other pilot programs, Edinburgh is devoting considerable effort to examining the acceptability of offering the test, in this case prenatally. Maternal and parental anxiety is the focus of the assessments—in specific, determining to what extent offering the test to pregnant women raises anxiety in preconception and prenatal populations. Consulting psychologists and psychiatrists are developing a survey instrument that will measure anxiety, self-esteem, and perceptions of stigma in those screened.

As noted earlier, the program received funds in January 1992 from the MRC to carry out a separate

**Figure 10-2—Plan for Screening in a Cardiff, Wales General Practice (Couples)**

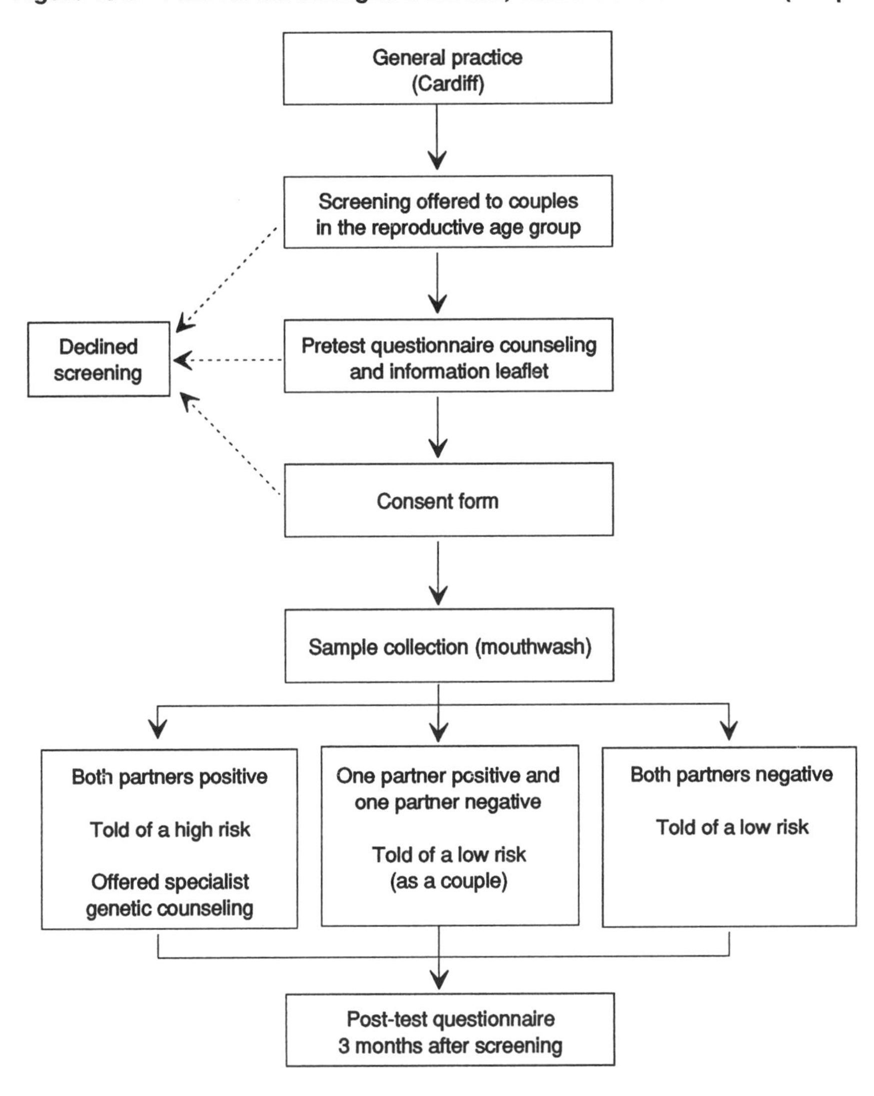


SOURCE: P. Harper, University of Wales College of Medicine, Cardiff, Wales, personal communication, 1991.

## *Box 10-B—Cardiff Protocol for Evaluating Psychological Sequelae*

*Step 1*

Participants are invited either through the mail or directly through their general practitioner. Prior to screening they are asked to complete the perceived health measure questionnaire and the State of Anxiety Inventory. Immediately prior to counseling, they are asked to complete the reproductive intentions questionnaire and the knowledge-of-CF questionnaire.

*Step 2*

Within 1 week of notification of their carrier status, all participants are asked to complete the questionnaires again.

*Step 3*

Three months after screening, participants are again sent all the questionnaires.

SOURCE: P. Harper, Institute of Medical Genetics, University Hospital of Wales, Cardiff, personal communication, 1991.

**Table 10-1—Results of the Edinburgh Pilot Program (as of December 1991)**

| | |
|---|---|
| Number of women approached | 2,780 |
| Ineligible for screening | |
| late gestation | 245 |
| abnormal pregnancy | 50 |
| no partner | 31 |
| other | 35 |
| Total | 361 (13 percent) |
| Eligible for screening | 2,419 (87 percent) |
| Declined screening | 331 (12 percent) |
| Screened | 2,088 (75 percent of all; 86 percent of eligible) |
| Carrier women | 74 |
| Carrier partners | 3 |
| Carrier frequency found | 1 in 28 |
| Carrier frequency expected | 1 in 30 |

SOURCE: D.J.H. Brock, Western General Hospital, Edinburgh, Scotland, personal communication, 1991.

trial of couples screening in another maternity hospital in Scotland.

### *Guy's Hospital, London, England*

Guy's Hospital in London offers CF carrier screening to adults of reproductive age (18 to 45 years old) through a general practice. Screening prenatally is not the goal of the program. The philosophy of this pilot is to introduce screening through the GP so that after the pilot programs expire, screening can be run by the GPs.

The predictive value of the test when used in the Guy's population is approximately 80 percent using the four mutations detected with the Cellmark ARMS kit. Through mid-1991, approximately 200 samples had been run at a rate of 10 to 20 per week (1).

As with the Cardiff and Edinburgh pilot programs, participants are asked to complete the psychological questionnaires (which were developed in collaboration with a Guy's psychologist). At the end of the pilot programs, all three groups will compare results.

## OTHER PILOT PROGRAMS IN THE UNITED KINGDOM

In addition to the pilot programs funded by the CF Trust, two other programs—funded privately or by a variety of public and private funding mechanisms—have also begun to evaluate the implications of population carrier screening for CF. The most active non-CF-Trust program in the United Kingdom is at St. Mary's Hospital Medical School, Imperial College, London. The St. Mary's program is similar to those in Cardiff and at Guy's Hospital, focusing on screening nonpregnant adults of reproductive age.

Another program, funded entirely through private sources, is at St. Bartholomew's Hospital (Bart's) in London. The Bart's program is unique, in that it offers only couples screening. In fact, the director of the Bart's program has been the most outspoken advocate in the United Kingdom of the couples screening approach (box 10-A) (4). Pilots are also under way in Italy, Denmark (box 10-C), and Austria (table 10-2).

**Box 10-C—Carrier Screening in Denmark**

The carrier frequency for the ΔF508 mutation is nearly 88 percent in Denmark. This makes screening simpler because of the higher sensitivity of the test. This factor, combined with a national health program, made the introduction of CF carrier screening more feasible. Using national health service funds, carrier screening is offered to all pregnant women seen at an out patient clinic at the Rigshospitalet in Copenhagen, as well as to families of CF patients. Blood samples are drawn and tested for the presence of the ΔF508 mutation. Among 3,664 women tested as of September 1991, 91 were found to be carriers of the ΔF508 mutation (1 in 40). When the partners of these women were tested, only one was found to also be a carrier of the ΔF508 mutation. Prenatal diagnosis revealed that the fetus was homozygous for ΔF508 and the pregnancy was terminated. In addition, prenatal diagnosis was performed on an additional 42 fetuses. Of these, 24 were found to be carriers (heterozygotes) for the ΔF508 mutation.

SOURCE: M. Schwartz and N.J. Brandt, Rigshospitalet, Copenhagen, Denmark, personal communication, 1991.

### *St. Mary's Hospital Medical School, London*

The pilot study operated out of St. Mary's Hospital Medical School in London is evaluating preconception CF carrier screening of males and females of reproductive age through three GPs and three family planning clinics (figure 10-3). This genetic service opted to offer the test to nonpregnant individuals of reproductive age, because the director believes screening this population maximizes reproductive choice and autonomy.

**Table 10-2—International Cystic Fibrosis Carrier Screening Pilot Programs**

| Country | Institution | Target population |
|---|---|---|
| Austria | University of Vienna | Newborns |
| Denmark | Rigshospitalet | Prenatal |
| England | St. Mary's | Adult preconception |
| | Guy's | Adult preconception |
| | St. Bartholomew's | Adult preconception |
| Italy | University of Padua | Newborns |
| Wales | University of Wales College of Medicine | Adult preconception |
| Scotland | University of Edinburgh | Prenatal |

SOURCE: Office of Technology Assessment, 1992.

Using mouthwash samples, the St. Mary's group is presently looking for the ΔF508, G551D, and R553X mutations, which together should detect 82 percent of the CF carriers in their population. Through mid-1991, St. Mary's had screened about 1,600 individuals at the rate of approximately 50 samples a week. Rough estimates of the cost of screening run about £1.75 per sample for laboratory costs alone. Estimating total costs, however, is difficult as some laboratory staff are paid out of a research grant, and others are paid through the NHS.

**Figure 10-3—Carrier Screening in Primary Care (St. Mary's Hospital)**

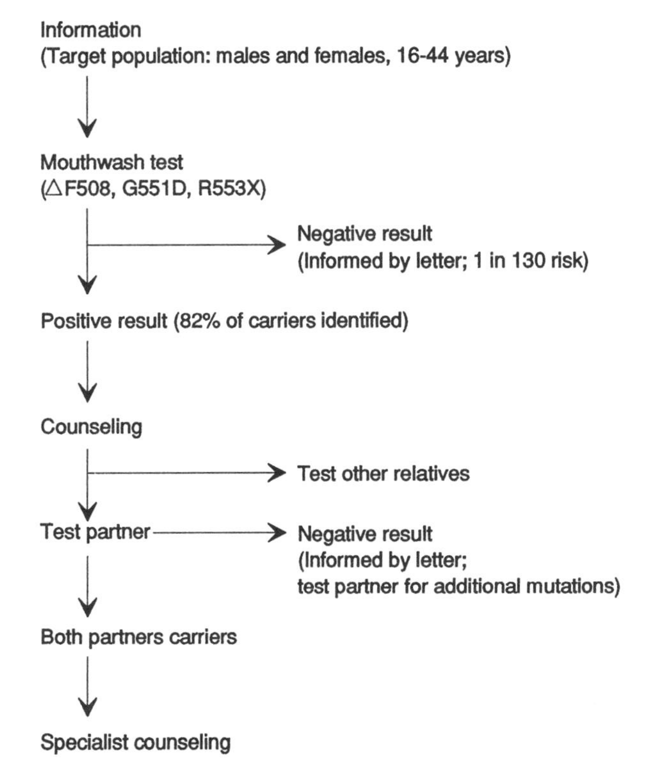


SOURCE: R. Williamson, St. Mary's Hospital, London, England, personal communication, 1991.

Counseling is not a major cost as it is, for the most part, carried out by GPs or practice nurses who are already employed by the NHS (28). Because the tests are part of a research protocol, for which the laboratory does not charge, they do not pay PCR royalties.

One approach used by this group is opportunistic screening. The St. Mary's group has found that approximately 66 percent of individuals approached through their GP eventually request screening, while 87 percent of individuals in the family planning clinics request the test (25,28). Another approach being tested by this pilot group involves solicitation by invitation letter. Individuals receive a letter offering the test on Saturday mornings at their GP's office. The response rate of this method approximates 10 percent (25).

With both approaches, each person is given a leaflet that explains the test. Those who opt for screening are told about the limited sensitivity of the assay. The results are sent through the mail. The letter to those screening negative reemphasizes the sensitivity of the test, informs the individual that his or her risk of being a carrier has been reduced to 1 in 130, and offers screening for partners or spouses. Carriers are invited to attend a counseling session where risks are explained and the testing of partners or relatives is discussed. The partners of identified carriers are screened for several additional mutations, which brings the detection rate up to around 86 percent (26,29).

Participants in this screening program were asked how they thought their future reproductive plans might be affected if both they and their partners were found to be carriers. For those with no experience with CF, 38 percent felt they might choose not to have children, 78 percent would request prenatal diagnosis should they become pregnant, and 16 percent would not consider terminating an affected pregnancy. For those who had a relative or knew someone with CF, 45 percent felt they might choose not to have children, 82 percent would opt for prenatal diagnosis in pregnancy, and 20 percent would not consider terminating an affected pregnancy (25).

This study has also begun offering screening at selected work sites, such as police barracks or the Royal Mail. There are also plans to mail the mouthwash kits to the homes of relatives of carriers.

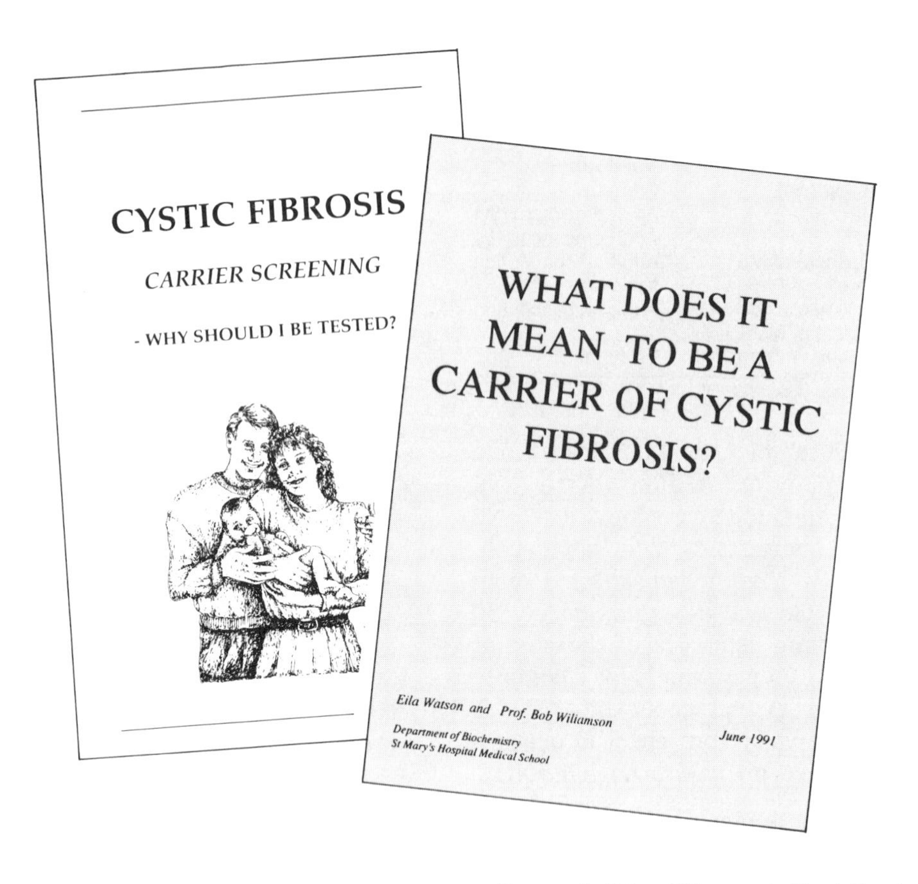


*Photo credit: Robert Williamson, St. Mary's Hospital*

Sample educational materials used by the CF carrier screening pilot study at St. Mary's Hospital Medical School in London, England.

## PROSPECTUS

The future of broad population carrier screening for CF in the United Kingdom is yet to be decided. Many interviewees were skeptical that widescale screening will be pursued by the NHS once the pilot programs are completed. Moreover, even if the NHS takes up carrier screening, there is no guarantee that it would last. It is generally easier to retract programs in the United Kingdom than in the United States.

As in the United States, there is speculation about whether the necessary infrastructure exists to proceed with broad based screening. In the United Kingdom, it will be necessary to provide:

- information to the population;
- a system for collecting samples from a cohort of the population at some point before reproduction, and delivering the samples to the laboratory;
- a network of diagnostic laboratories with a quality control system;
- a system for reporting the results to doctors and the people concerned;
- an information storage and retrieval system;
- information and counseling for carriers;
- adequate expert centers for counseling couples at risk and providing prenatal diagnosis; and
- a system for monitoring the service (18).

There is a general consensus in the United Kingdom that newborn screening would be an inefficient approach to reducing the incidence of CF. One of four carrier couples would already have an affected child before being identified as carriers, and an additional 25 percent would not be identified because their child had inherited neither CF mutation. Ensuring that this information follows the carrier child into adulthood is also problematic (27). Legal, ethical, and logistical problems also make school-based screening programs difficult to implement in the United Kingdom (4).

There remains some disagreement as to whether prenatal CF carrier screening unduly raises maternal anxiety in the approximately 24 of 25 women who will test negative. Proponents of prenatal screening feel that pregnant couples are most in need of and most likely to use this information. Furthermore, the infrastructure already exists for working with these individuals. Those opposed to prenatal screening feel that it raises anxiety at an already anxious time and leaves little time for reflection (18,27). People of this view tend to believe that screening should be offered preconceptionally, when carrier couples will have a maximum range of reproductive options. Carrier screening offered when pregnancy is known has the advantage of a captive population, but the disadvantage of limited time for screening and decisionmaking, as well as eliminating the option of avoiding conception if both partners are carriers. The pilot studies will help clarify some of these issues as they encompass both prenatal and preconception populations.

If the pilots are successful and the NHS embraces the notion of a widescale screening program, the laboratory service would be provided by centralized regional DNA laboratories run by the NHS. One area in which the United Kingdom lags behind the United States is in the area of quality control and assurance. Lacking regulatory agencies comparable to the U.S. Health Care Financing Administration and the U.S. Food and Drug Administration, laboratories need only voluntarily comply with quality standards.

In the late 1980s, the European Concerted Action on Cystic Fibrosis was formed for the purposes of data coordination, information exchange, and establishment of international standards of quality control (table 10-3). Organized out of St. Mary's Medical School in London, the consortium supplies every participating laboratory with oligonucleotides for CF analysis in exchange for data. The group interacts with the Genetic Analysis Consortium in North America, but does not consider itself a "Euro-equivalent," in that it is not primarily a research group, but a clinical assistance group. A newsletter published six times a year serves as a point of exchange for information about ongoing work and technical advances.

Coded samples and a list of mutations to be tested were distributed in June 1991 to 35 voluntary hospital-based research groups. A database program for the collection of mutation and patient data is available through the European Concerted Action on Cystic Fibrosis (19). The program will allow laboratories to computerize their records and analyze data according to mutations, clinical details, and ethnic groups. Each of the pilot projects is participating in this quality control exercise. The MRC is seeking additional information regarding quality control and assurance.

**Table 10-3—Members of the European Concerted Action on Cystic Fibrosis (as of February 1991)**

| | |
|---|---|
| Australia | New Zealand |
| Austria | Northern Ireland |
| Belgium | Norway |
| Bulgaria | Poland |
| Cuba | Portugal |
| Czechoslovakia (former) | Republic of Ireland |
| Denmark | Scotland |
| England | South Africa |
| Finland | Spain |
| France | Sweden |
| Germany | Switzerland |
| Greece | USSR (former) |
| Israel | Wales |
| Italy | Yugoslavia (former) |
| The Netherlands | |

SOURCE: R. Williamson, St. Mary's Hospital Medical School, London, England, personal communication, 1991.

Establishing the infrastructure necessary for CF carrier screening, if it is done, could lay the ground work for other forms of genetic screening in the future. Current reform of the NHS might well be the best predictor of the future of genetic screening programs in the United Kingdom. Under the new reforms, health authorities will assess the health needs of their resident populations and then negotiate contracts to purchase the services that they expect will achieve the most improvements in health (10). With the reforms, hospitals can function outside health authority control as "self-governing hospital trusts," and it is not yet clear how specialized, preventive, genetic services will be administered under this new plan. They may be deemed too costly. Results of the pilot programs in the United Kingdom will provide valuable information on the ability of primary care providers to assist in screening, the acceptability by the public of screening, and the most appropriate population for screening.

# SUMMARY AND CONCLUSIONS

The private CF Trust is largely responsible for the existence of CF carrier screening pilot projects in the United Kingdom. Were the availability of government funding the determinant of pilot project initiation, it is uncertain any pilots would exist. Results of pilot studies in the United Kingdom will be directly relevant to consideration of population screening in the United States because of similar concerns about the appropriate target population, levels of anxiety, and the role of primary care providers. Although the latter concern is of a different nature in the United Kingdom because of the role of the general practitioner and the National Health Service retain, the ability of GPs to participate in screening in the United Kingdom will be of significant interest in the United States.

The role of the British GP as the likely first point of contact for CF carrier screening makes preconceptional carrier screening of adults more easily achieved than in the United States, where primary care physicians are less likely to refer individuals for screening in the absence of a positive family history. Targeting GPs as important collaborators and resources in CF carrier screening is done by nearly every British pilot program. GPs are actively recruited to participate in the development and implementation of screening.

Prenatal clinics provide another population easily targeted for screening. Yet there is no consensus on the appropriateness of targeting pregnant women for CF carrier screening. Only the Edinburgh pilot project is actively recruiting pregnant women. Concerns about raising anxiety in pregnant women and the logistical restrictions to offering first, rather than second, trimester prenatal diagnosis, are the impetus for screening programs aimed at preconceptional individuals. Anxiety levels are being followed by the CF Trust pilot projects and the results of these analyses should shed light on the validity of those concerns.

Debate over couples screening has focused on the ethics of not informing carriers of their status in couples in which one partner is a CF carrier and the other has a negative test result (an informing practice that would be considered legally and ethically dangerous in the United States). In addition, those opposed to the concept of couples screening find the treatment of individuals as reproductive units unsettling, given the possibility of nonpaternity or new partners in the future. Unlike the United States, the British medical community does not operate under the fear of malpractice or litigation. Because the British pay for their health care indirectly, through taxation, they do not view themselves as consumers or buyers of services.

While screening cost is a major consideration in the United States, it is viewed differently in the United Kingdom, as the total cost is likely to be borne by the NHS. New programs, such as routine CF carrier screening, must compete with other desirable projects for available funds. In addition, services are more centralized, lowering overall costs. Reform of the NHS, however, will likely alter the manner in which genetic services are offered and made available.

Except for samples from pregnant women, investigators in the United Kingdom rely on mouthwash for their DNA extractions. This approach, seldom used in the United States, is thought to be as effective for DNA extraction as other sources and is less costly. In addition, investigators feel that the use of this noninvasive procedure contributes to higher rates of participation in screening programs. Quality control and assurance, currently conducted on an informal basis throughout the United Kingdom, would have to be addressed.

The British health care system is significantly different from that found in the United States. The existence of preliminary carrier screening pilot projects in no way commits the system to sustain the programs. It is generally easier in the United Kingdom than in the United States to retract policies or cease offering services if they are deemed unnecessary or inappropriate. Results of the pilot programs in the United Kingdom will provide valuable information on the ability of primary care providers to assist in screening, the acceptability of screening by the public, and the most appropriate population for genetic carrier screening.

## CHAPTER 10 REFERENCES

1. Bobrow, M., Guy's Hospital, London, personal communication, June 1991.
2. Brock, D.J.H., "A Consortium Approach to Molecular Genetic Services," *Journal of Medical Genetics* 27:8-13, 1990.
3. Brock, D.J.H., University of Edinburgh, Western General Hospital, Edinburgh, Scotland, personal communications, June 1991, December 1991.
4. Brock, D.J.H., Mennie, M.E., McIntosh, I., et al., "Heterozygote Screening for Cystic Fibrosis," *Antenatal Diagnosis of Fetal Abnormalities*, J.O. Drife and D. Donnai (eds.) (London, England: Springer-Verlag, 1991).
5. Cobb, E., Holloway, S., Elton, R., et al., "What Do Young People Think About Screening for Cystic

Fibrosis?," *Journal of Medical Genetics* 28:322-324, 1991.

6. Dodge, J.A., "Implications of the New Genetics for Screening for Cystic Fibrosis," *Lancet* 2(8612):672-674, 1988.
7. Guilbert, P., and Cheater, F., "Health Visitor's Awareness and Perception of Clinical Genetic Services," *Journal of Medical Genetics* 27:508-511, 1990.
8. Harris, A., "DNA Markers Near the Cystic Fibrosis Locus: Further Analysis of the British Population," *Journal of Medical Genetics* 27:39-41, 1990.
9. Harris, A., Beards, F., and Mathew, C., "Mutation Analysis at the Cystic Fibrosis Locus in the British Population," *Human Genetics* 85:408-409, 1990.
10. Harris, R., "Genetic Services in Britain: A Strategy for Success After the National Health Service and Community Care Act 1990," *Journal of Medical Genetics* 27:711-714, 1990.
11. Harper, P., "Genetic Services in the Community," *Journal of Medical Genetics* 27:473-474, 1990.
12. Harper, P., Institute of Medical Genetics, University of Wales College of Medicine, Cardiff, Wales, personal communications, June 1991, December 1991.
13. Johnston, A.W., "Teaching of Clinical Genetics in Britain: A Report From the Royal College of Physicians of London," *Journal of Medical Genetics* 27:707-709, 1990.
14. Marteau, T.M., Johnston, M., Shaw, R.W., et al., "The Impact of Prenatal Screening and Diagnostic Testing Upon the Cognitions, Emotions, and Behaviour of Pregnant Women," *Journal of Psychosomatic Research* 33:7-16, 1989.
15. McIntosh, I., Curtis, A., Lorenzo, M-L., et al., "The Haplotype Distribution of the ΔF508 Mutations in Cystic Fibrosis Families in Scotland," *Human Genetics* 85:419-420, 1990.
16. Medical Research Council, *Health Services Research*, brochure available from the MRC, Park Crescent, London, England.
17. Medical Research Council, *Report on the HSRC Workshop to Discuss Research in Cystic Fibrosis Carrier Screening* (91/HSRC 163, file No. A861/132), 1991.
18. Modell, B., "Cystic Fibrosis Screening and Community Genetics," *Journal of Medical Genetics* 27:475-479, 1990.
19. Newsletter of the European Concerted Action on Cystic Fibrosis, vol. 4, February 1991.
20. Royal College of Physicians, *Prenatal Diagnosis and Genetic Screening: Community and Service Implications* (London: Royal College of Physicians, 1989).
21. Santis, G., Osborne, L., Knight, R., et al., "Cystic Fibrosis Haplotype Association and the ΔF508 Mutation in Adult British CF Patients," *Human Genetics* 85:424-425, 1990.
22. Shrimpton, A.E., McIntosh, I., and Brock, D.J.H., "The Incidence of Different Cystic Fibrosis Mutations in the Scottish Population: Effects on Prenatal Diagnosis and Genetic Counseling," *Journal of Medical Genetics* 28:317-321, 1991.
23. Spielberger, C.S., Gorsuch, R.L., and Lushene, R.E., *The State Trait Anxiety Inventory* (Palo Alto, CA: Consulting Psychologists Press, 1970).
24. Vickers, M., Medical Research Council, London, England, personal communications, June 1991, December 1991.
25. Watson, E.K., Mayall, E., Chapple, J., et al., "Screening for Carriers of Cystic Fibrosis Through Primary Health Care Services," *British Medical Journal* 303: 504-507, 1991.
26. Watson, E.K., Mayall, E.S., Simova, L., et al., "The Incidence of ΔF508 CF Mutation and Associated Haplotypes in a Sample of English Families," *Human Genetics* 85:435-436, 1990.
27. Watson, E.K., Williamson, R., and Chapple, J., "Attitudes to Carrier Screening for Cystic Fibrosis: A Survey of Health Care Professionals, Relatives of Sufferers, and Other Members of the Public," *British Journal of General Practice* 41:237-240, 1991.
28. Williamson, R., St. Mary's Hospital Medical School, London, England, personal communications, June 1991, December 1991.
29. Williamson, R., Allison, M.E.D., Bentley, S.M.C., et al., "Community Attitudes to Cystic Fibrosis Carrier Testing in England: A Pilot Study," *Prenatal Diagnosis* 9:727-734, 1989.

# Appendixes

# Appendix A
# Epidemiology of Mutations for Cystic Fibrosis

The differential distribution of mutations causing cystic fibrosis (CF) has clear implications for carrier screening. Besides ΔF508, which accounts for about 70 percent of CF chromosomes in Caucasians of Northern and Central European descent, more than 170 mutations have been identified, and the number increases steadily. The vast majority of these are rare mutations present in only a few individuals or families.

Not only are CF mutations heterogeneous, they are distributed with varying frequencies among populations according to geographic, ethnic, and racial distinctions; regional differences within countries also exist. In Europe, ΔF508 occurs along a decreasing gradient from north to south (figure A-1); other mutations also occur differentially among racial and ethnic groups. Denmark has the most homogeneous CF mutation population in Northern Europe, with ΔF508 accounting for about 88 percent of mutations.

International studies continue to document the frequency distribution of ΔF508 and other mutations (table A-1). Results are coordinated through international collaborative efforts, including the Cystic Fibrosis Genetic Analysis Consortium, the European Working Group on Cystic Fibrosis, and the European Concerted Action on Cystic Fibrosis.

In the United States and Canada, diverse heritages are reflected genetically (table A-2): The distribution of ΔF508, as an aggregate, is an average of European values. For example, in North American populations from mixed European descent, ΔF508 accounts for 68 to 76 percent of CF mutations, while the combined European data also average 68 percent (table A-3). Preliminary studies of some distinct ethnic groups in North America have similar profiles to areas of origin in Europe.

By studying markers surrounding the CF locus, researchers believe that the ΔF508 deletion most likely resulted from a single mutational event that was subsequently passed along by invading peoples. Some explanations of the distribution patterns of ΔF508 attempt to correlate waves of invasions with distribution of the mutation, although this remains speculation.

**Figure A-1—Occurrence of ΔF508 in Europe**

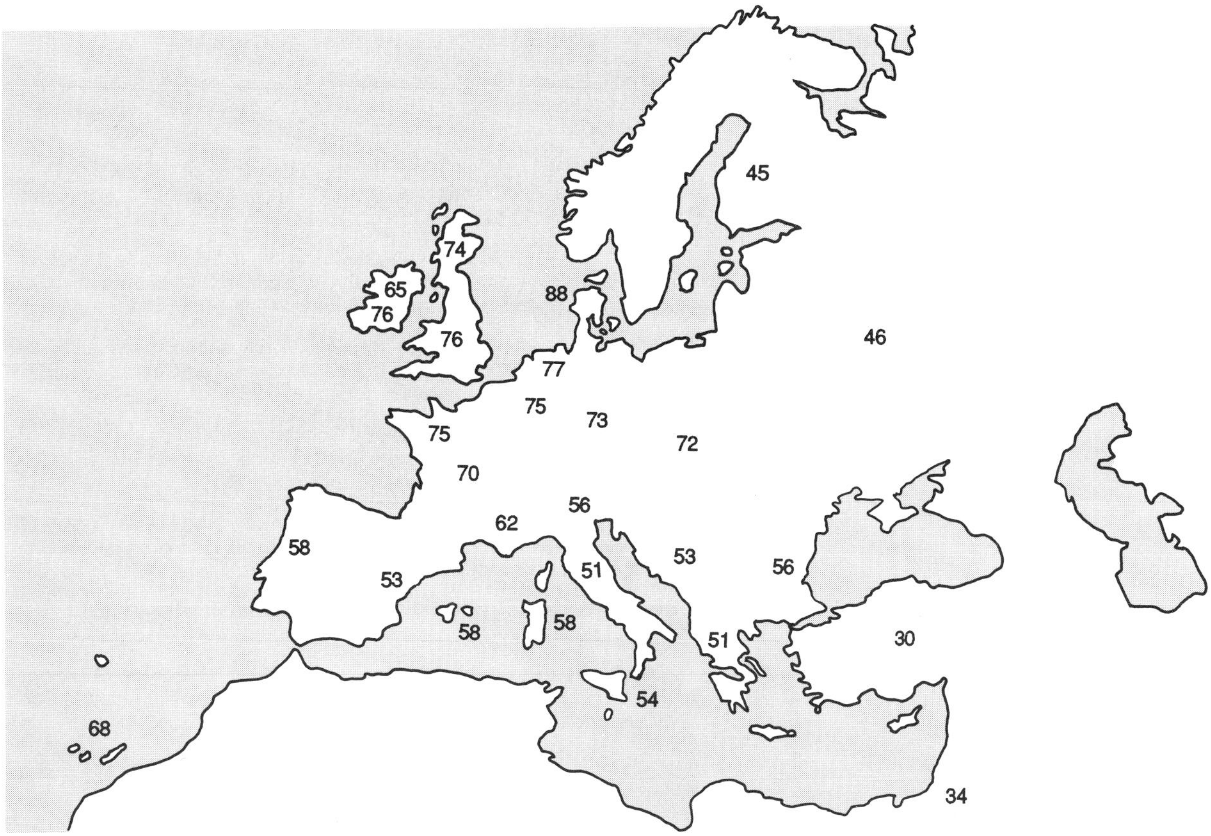

SOURCE: European Working Group on Cystic Fibrosis Genetics, "Gradient of Distribution in Europe of the Major CF Mutation and of Its Associated Haplotype," *Human Genetics* 85:436-445, 1990.

## Table A-1—Distribution of ΔF508 in Europe

| Country | Frequency of ΔF508 (percent of cystic fibrosis chromosomes) | Study size (number of cystic fibrosis chromosomes) | Source |
|---|---|---|---|
| **Albania** | 75 | 92 | G. Novelli, F. Sangiuolo, V. Mokini, et al., "The Cystic Fibrosis ΔF508 Mutation in the Albanian Population," *American Journal of Human Genetics* 50:875-876, 1992. |
| **Belgium** | 59.5 | 116 | H. Cuppens, E. Legius, P. Cabello, et al., "Association Between XV2c/CS7/KM19/D9 Haplotypes and the ΔF508 Mutation," *Human Genetics* 85:402-403, 1990. |
| | 65<br>67<br>77 | 124<br>60<br>150 | The Cystic Fibrosis Genetic Analysis Consortium, "Worldwide Survey of the ΔF508 Mutation—Report From the Cystic Fibrosis Genetic Analysis Consortium," *American Journal of Human Genetics* 47:354-359, 1990. |
| | 76 | 146 | M. Bonduelle, W. Lissens, A. Malfroot, et al., "The Deletion F508 is the Major Gene Mutation in a Representative Belgian Cystic Fibrosis Population," *Human Genetics* 85:395-396, 1990. |
| | 78 | 214 | P. Cocheaux, R. Van Geffel, D. Baran, et al., "Prevalence of the ΔF508 Deletion of the Cystic Fibrosis Gene in Belgian Patients," *Human Genetics* 85:400, 1990. |
| Antwerp | 80 | 71 | J.G. Wauters, J. Hendrickx, P. Coucke, et al., "Frequency of the Phenylalanine Deletion (ΔF508) in the CF Gene of Belgian Cystic Fibrosis Patients," *Clinical Genetics* 39:89-92, 1991. |
| **Bulgaria** | 56 | 96 | L. Kalaydjieva, J. Antov, J. Bronzova, et al., "Molecular Data on Cystic Fibrosis Data in Bulgaria," *Human Genetics* 85:412-413, 1990. |
| | 58 | 110 | The Cystic Fibrosis Genetic Analysis Consortium, "Worldwide Survey of the ΔF508 Mutation—Report From the Cystic Fibrosis Genetic Analysis Consortium," *American Journal of Human Genetics* 47:354-359, 1990. |
| **Czech and Slovak Federal Republic** | 68 | 354 | The Cystic Fibrosis Genetic Analysis Consortium, "Worldwide Survey of the ΔF508 Mutation—Report From the Cystic Fibrosis Genetic Analysis Consortium," *American Journal of Human Genetics* 47:354-359, 1990. |
| Bohemia/Moravia | 67 | NA | M. Macek, V. Vavrová, I. Böhm, et al., "Frequency of the ΔF508 Mutation and Flanking Marker Haplotypes at the CF Locus From 167 Czech Families," *Human Genetics* 85:417-418, 1990. |
| Slovakia | 63 | 46 | The Cystic Fibrosis Genetic Analysis Consortium, "Worldwide Survey of the ΔF508 Mutation—Report From the Cystic Fibrosis Genetic Analysis Consortium," *American Journal of Human Genetics* 47:354-359, 1990. |
| **Denmark** | 87 | 304 | M. Schwartz, H.K. Johansen, C. Koch, et al., "Frequency of the ΔF508 Mutation on Cystic Fibrosis Chromosomes in Denmark," *Human Genetics* 85:427-428, 1990. |
| | 88 | 423 | The Cystic Fibrosis Genetic Analysis Consortium, "Worldwide Survey of the ΔF508 Mutation—Report From the Cystic Fibrosis Genetic Analysis Consortium," *American Journal of Human Genetics* 47:354-359, 1990. |
| **Finland** | 45 | 40 | The Cystic Fibrosis Genetic Analysis Consortium, "Worldwide Survey of the ΔF508 Mutation—Report From the Cystic Fibrosis Genetic Analysis Consortium," *American Journal of Human Genetics* 47:354-359, 1990. |
| | 45 | 40 | J. Kere, E. Savilahti, R. Norio, et al., "Cystic Fibrosis Mutation ΔF508 in Finland: Other Mutations Predominate," *Human Genetics* 85:413-415, 1990. |
| **France** | 66<br>73<br>79 | 268<br>271<br>332 | The Cystic Fibrosis Genetic Analysis Consortium, "Worldwide Survey of the ΔF508 Mutation—Report From the Cystic Fibrosis Genetic Analysis Consortium," *American Journal of Human Genetics* 47:354-359, 1990. |
| | 67 | 248 | J.C. Chomel, A. Haliassos, L. Tesson, et al., "Frequency of the Major CF Mutation in French CF Patients," *Human Genetics* 85:397-398, 1990. |
| | 72.5 | 258 | B. Simon-Bouy, E. Mornet, J.L. Serre, et al., "The Cystic Fibrosis ΔF508 Mutation in the French Population," *Human Genetics* 85:431-432, 1990. |

**Table A-1—Distribution of ΔF508 in Europe—Continued**

| Country | Frequency of ΔF508 (percent of cystic fibrosis chromosomes) | Study size (number of cystic fibrosis chromosomes) | Source |
|---|---|---|---|
| | 75 | 422 | M. Vidaud, C. Ferec, O. Attree, et al., "Frequency of the Cystic Fibrosis ΔF508 Mutation in a Large Sample of the French Population," *Human Genetics* 85:434-435, 1990. |
| Britanny (Celtic) | 81 | 224 | The Cystic Fibrosis Genetic Analysis Consortium, "Worldwide Survey of the ΔF508 Mutation—Report From the Cystic Fibrosis Genetic Analysis Consortium," *American Journal of Human Genetics* 47:354-359, 1990. |
| Lyon | 74 | 230 | The Cystic Fibrosis Genetic Analysis Consortium, "Worldwide Survey of the ΔF508 Mutation—Report From the Cystic Fibrosis Genetic Analysis Consortium," *American Journal of Human Genetics* 47:354-359, 1990. |
| Paris | 70 | 102 | The Cystic Fibrosis Genetic Analysis Consortium, "Worldwide Survey of the ΔF508 Mutation—Report From the Cystic Fibrosis Genetic Analysis Consortium," *American Journal of Human Genetics* 47:354-359, 1990. |
| Southern | 64 | 98 | M. Claustres, M. Desgeorges, H. Kjellberg, et al., "Cystic Fibrosis Typing With DNA Probes and Screening for ΔF508 Deletion in Families From Southern France," *Human Genetics* 85:398-399, 1990. |
| **Germany** | | | |
| Berlin | 70 | 290 | The Cystic Fibrosis Genetic Analysis Consortium, "Worldwide Survey of the ΔF508 Mutation—Report From the Cystic Fibrosis Genetic Analysis Consortium," *American Journal of Human Genetics* 47:354-359, 1990. |
| East (former) | 60 | 388 | The Cystic Fibrosis Genetic Analysis Consortium, "Worldwide Survey of the ΔF508 Mutation—Report From the Cystic Fibrosis Genetic Analysis Consortium," *American Journal of Human Genetics* 47:354-359, 1990. |
| | 60 | 518 | Ch. Coutelle, K. Grade, R. Bruckner, et al., "CF DNA-Diagnosis and Gene Mutation Analysis: Data From East Germany," *Pathologie Biologie* 6:585-586, 1991. |
| | 62 | 314 | K. Grade, K. Will, R. Szibor, et al., "First Analysis of the F508 Deletion in Cystic Fibrosis Patients From the GDR," *Human Genetics* 85:406-408, 1990. |
| West (former) | 77<br>77 | 244<br>234 | The Cystic Fibrosis Genetic Analysis Consortium, "Worldwide Survey of the ΔF508 Mutation—Report From the Cystic Fibrosis Genetic Analysis Consortium," *American Journal of Human Genetics* 47:354-359, 1990. |
| | 77 | 400 | A. Reis, S. Bremer, M. Schlösser, et al., "Distribution Patterns of the ΔF508 Mutation in the CFTR Gene on CF-Linked Marker Haplotypes in the German Population," *Human Genetics* 85:421-422, 1990. |
| | 80 | 186 | C. Aulehla-Scholz, R. Kaiser, J. Weber, et al., "The Frequency of the ΔF508 Deletion in CF Chromosomes of Different Ethnic Origin," *Human Genetics* 85:392-393, 1990. |
| **Greece** | 54 | 194 | A. Balassopoulou, D. Loukopoulos, P. Kollia, et al., "Cystic Fibrosis in Greece: Typing With DNA Probes and Identification of the Common Molecular Defect," *Human Genetics* 85:393-394, 1990. |
| | 54 | 194 | The Cystic Fibrosis Genetic Analysis Consortium, "Worldwide Survey of the ΔF508 Mutation—Report From the Cystic Fibrosis Genetic Analysis Consortium," *American Journal of Human Genetics* 47:354-359, 1990. |
| **Hungary** | 64 | 66 | M. Nemeti, E. Louie, Z. Papp, et al., "Molecular Analysis of Cystic Fibrosis in the Hungarian Population," *Human Genetics* 87:511-512, 1991. |
| **Ireland, Republic of** | 76 | 120 | The Cystic Fibrosis Genetic Consortium, "Worldwide Survey of the ΔF508 Mutation—Report From the Cystic Fibrosis Genetic Analysis Consortium," *American Journal of Human Genetics* 47:354-359, 1990. |
| | 76 | 88 | M.A. De Arce, D. Mulherin, P. McWilliam, et al., "Frequency of Deletion 508 Among Irish Cystic Fibrosis Patients," *Human Genetics* 403-404, 1990. |

*(Continued on next page)*

**Table A-1—Distribution of ΔF508 in Europe—Continued**

| Country | Frequency of ΔF508 (percent of cystic fibrosis chromosomes) | Study size (number of cystic fibrosis chromosomes) | Source |
|---|---|---|---|
| **Israel** | 32 | 113 | I. Lerer, S. Cohen, M. Chemke, et al., "The Frequency of the ΔF508 Mutation on Cystic Fibrosis Chromosomes in Israeli Families: Correlation to CF Jewish Haplotypes in Jewish Communities and Arabs," *Human Genetics* 85:416-417, 1990. |
| Arab | 22 | 23 | |
| Ashkenazic Jewish | 32 | 40 | |
| Non-Ashkenazic Jewish | 38 | 29 | |
| Arab | 25 | 40 | T. Shosani, A. Augarten, E. Gazit, et al., "Association of a Nonsense Mutation (W1282X), the Most Common Mutation in the Ashkenazi Jewish Cystic Fibrosis Patients in Israel, With Presentation of Severe Disease," *American Journal of Human Genetics* 50:222-228, 1992. |
| Ashkenazic Jewish | 23 | 95 | |
| Sephardic Jewish | 35 | 51 | |
| Unclassified | 25 | 8 | |
| Jewish (mixed) | 34 | 127 | I. Lerer, M. Sagi, G.R. Cutting, et al., "Cystic Fibrosis Mutations ΔF508 and G542X in Jewish Patients," *Journal of Medical Genetics* 29:131-133, 1992. |
| Ashkenazic Jewish | 30 | 84 | |
| Non-Ashkenazic Jewish | 42 | 43 | |
| **Italy** | 43 | 54 | The Cystic Fibrosis Genetic Analysis Consortium, "Worldwide Survey of the ΔF508 Mutation—Report From the Cystic Fibrosis Genetic Analysis Consortium," *American Journal of Human Genetics* 47:354-359, 1990. |
| | 47 | 122 | |
| | 55 | 284 | |
| | 43 | 348 | X. Estivill, M. Chillón, T. Casals, et al., "ΔF508 Gene Deletion in Cystic Fibrosis in Southern Europe," *Lancet* II:1404-1405, 1989. |
| | 50 | 35 | G. Restagno, S. Garnerone, C. Gennaro, et al., "ΔF508 Deletion in Cystic Fibrosis in Italian Families," *Human Genetics* 85:422-423, 1990. |
| | 53 | 350 | L. Cremonesi, L. Ruocco, M. Seia, et al., "Frequency of the ΔF508 Mutation in a Sample of 175 Italian Cystic Fibrosis Patients," *Human Genetics* 85:400-402, 1990. |
| | 53 | 624 | M. Devoto, P. Ronchetto, P. Fanen, et al., "Screening for Non-DeltaF508 Mutations in Five Exons of the Cystic Fibrosis Transmembrane Conductance Regulator (CFTR) Gene in Italy," *American Journal of Human Genetics* 48:1127-1132, 1991. |
| Campania | 54 | 102 | G. Sebastio, O. Castiglione, B. Incerti, et al., "The ΔF508 Mutation in Cystic Fibrosis Patients of Southern Italy," *Human Genetics* 85:430-431, 1990. |
| Central/ Southern | 45 | 350 | The Cystic Fibrosis Genetic Analysis Consortium, "Worldwide Survey of the ΔF508 Mutation—Report From the Cystic Fibrosis Genetic Analysis Consortium," *American Journal of Human Genetics* 47:354-359, 1990. |
| Northern | 40 | 218 | The Cystic Fibrosis Genetic Analysis Consortium, "Worldwide Survey of the ΔF508 Mutation—Report From the Cystic Fibrosis Genetic Analysis Consortium," *American Journal of Human Genetics* 47:354-359, 1990. |
| Rome/Verona | 42 | 424 | G. Novelli, P. Gasparini, A. Savoia, et al., "Polymorphic DNA Haplotypes and ΔF508 Deletion in 212 Italian CF Families," *Human Genetics* 85:420-421, 1990. |
| Sardinia | 57 | 42 | The Cystic Fibrosis Genetic Analysis Consortium, "Worldwide Survey of the ΔF508 Mutation—Report From the Cystic Fibrosis Genetic Analysis Consortium," *American Journal of Human Genetics* 47:354-359, 1990. |
| **Netherlands** | 75 | 166 | The Cystic Fibrosis Genetic Analysis Consortium, "Worldwide Survey of the ΔF508 Mutation—Report From the Cystic Fibrosis Genetic Analysis Consortium," *American Journal of Human Genetics* 47:354-359, 1990. |
| | 79 | 235 | |
| | 76 | 152 | H. Scheffer, D.J. Bruinvels, G.J. te Meerman, et al., "Frequency of the ΔF508 Mutation and XV2c, KM19 Haplotypes in Cystic Fibrosis Families From The Netherlands: Haplotypes Without ΔF508 Still in Disequilibrium," *Human Genetics* 85:425-427, 1990. |
| | 77 | 190 | D.J.J. Halley, H.J. Veeze, and L.A. Sandkuyl, "The Mutation ΔF508 on Dutch Cystic Fibrosis Chromosomes: Frequency and Relation to Patients' Age at Diagnosis," *Human Genetics* 85:407-408, 1990. |

**Table A-1—Distribution of ΔF508 in Europe—Continued**

| Country | Frequency of ΔF508 (percent of cystic fibrosis chromosomes) | Study size (number of cystic fibrosis chromosomes) | Source |
|---|---|---|---|
| **Poland** | 55 | 22 | The Cystic Fibrosis Genetic Analysis Consortium, "Worldwide Survey of the ΔF508 Mutation—Report From the Cystic Fibrosis Genetic Analysis Consortium," *American Journal of Human Genetics* 47:354-359, 1990. |
| **Portugal** | 52 | 82 | A. Duarte, C. Barreto, L. Marques-Pinto, et al., "Cystic Fibrosis in the Portuguese Population: Haplotype Distribution and Molecular Pathology," *Human Genetics* 85:404-405, 1990. |
| | 54 | 84 | The Cystic Fibrosis Genetic Analysis Consortium, "Worldwide Survey of the ΔF508 Mutation—Report From the Cystic Fibrosis Genetic Analysis Consortium," *American Journal of Human Genetics* 47:354-359, 1990. |
| **Spain** | 49 | 388 | X. Estivill, M. Chillón, T. Casals, et al., "ΔF508 Gene Deletion in Cystic Fibrosis in Southern Europe," *Lancet* II:1404-1405, 1989. |
| | 50 | 388 | M. Chillón, V. Nunes, T. Casals, et al., "Distribution of the ΔF508 Mutation in 194 Spanish Cystic Fibrosis Families," *Human Genetics* 85:396-397, 1990. |
| | 51<br>65 | 466<br>142 | The Cystic Fibrosis Genetic Analysis Consortium, "Worldwide Survey of the ΔF508 Mutation—Report From the Cystic Fibrosis Genetic Analysis Consortium," *American Journal of Human Genetics* 47:354-359, 1990. |
| Basque country | | | |
| Basque | 87 | 30 | T. Casals, C. Vázquez, C. Lázaro, et al., "Cystic Fibrosis in the Basque Country: High Frequency of Mutation ΔF508 in Patients of Basque Origin," *American Journal of Human Genetics* 50:404-410, 1992. |
| Mixed Basque | 58 | 60 | |
| Central/ Southern | 61 | 120 | B. Peral, C. Hernández-Chico, J.L. San Millán, et al., "The ΔF508 Mutation and RFLP-linked Loci in Spanish Cystic Fibrosis Families," *Human Genetics* 87:516-517, 1991. |
| Continental | 66 | 45 | B. Jaume-Roig, B. Simon-Bouy, A. Taillandier, et al., "Genotyping of the Spanish Cystic Fibrosis Population at the ΔF508 Mutation Site and RFLP Linked Loci," *Human Genetics* 85:410-411, 1990. |
| Balearic Islands | 58 | 13 | |
| Balearic Islands | 58 | 13 | |
| **Switzerland** | 69 | 334 | The Cystic Fibrosis Genetic Analysis Consortium, "Worldwide Survey of the ΔF508 Mutation—Report From the Cystic Fibrosis Genetic Analysis Consortium," *American Journal of Human Genetics* 47:354-359, 1990. |
| **Turkey** | 27 | 30 | J. Hundrieser, S. Bremer, F. Peinemann, et al., "Frequency of the ΔF508 Mutation in the CFTR Gene in Turkish Cystic Fibrosis Patients," *Human Genetics* 85:409-410, 1990. |
| Turkish population in Germany | 27 | 30 | The Cystic Fibrosis Genetic Analysis Consortium, "Worldwide Survey of the ΔF508 Mutation—Report From the Cystic Fibrosis Genetic Analysis Consortium," *American Journal of Human Genetics* 47:354-359, 1990. |
| **Union of Soviet Socialist Republics (former)** | 36 | 25 | The Cystic Fibrosis Genetic Analysis Consortium, "Worldwide Survey of the ΔF508 Mutation—Report From the Cystic Fibrosis Genetic Analysis Consortium," *American Journal of Human Genetics* 47:354-359, 1990. |
| | 49 | 58 | P. Ronchetto, M. Devoto, A. Puliti, et al., "Preliminary Results on the Frequency of the ΔF508 Mutation in Cystic Fibrosis Patients From the USSR," *Human Genetics* 85:423-425, 1990. |
| Moscow/Odessa | 45 | 58 | The Cystic Fibrosis Genetic Analysis Consortium, "Worldwide Survey of the ΔF508 Mutation—Report From the Cystic Fibrosis Genetic Analysis Consortium," *American Journal of Human Genetics* 47:354-359, 1990. |
| **United Kingdom** | 77 | 39 | The Cystic Fibrosis Genetic Analysis Consortium, "Worldwide Survey of the ΔF508 Mutation—Report From the Cystic Fibrosis Genetic Analysis Consortium," *American Journal of Human Genetics* 47:354-359, 1990. |

*(Continued on next page)*

327-046 - 92 - 9 : QL 3

**Table A-1—Distribution of ΔF508 in Europe—Continued**

| Country | Frequency of ΔF508 (percent of cystic fibrosis chromosomes) | Study size (number of cystic fibrosis chromosomes) | Source |
|---|---|---|---|
| England | 70<br>70<br>75 | 210<br>150<br>180 | The Cystic Fibrosis Genetic Analysis Consortium, "Worldwide Survey of the ΔF508 Mutation—Report From the Cystic Fibrosis Genetic Analysis Consortium," *American Journal of Human Genetics* 47:354-359, 1990. |
| | 75<br>79<br>80 | 186<br>252<br>600 | |
| | 70 | 108 | G. Santis, L. Osborne, R. Knight, et al., "Cystic Fibrosis Haplotype Association and the ΔF508 Mutation in Adult British CF Patients," *Human Genetics* 85:424-425, 1990. |
| | 71.5 | 144 | A. Harris, F. Beards, and C. Mathew, "Mutation Analysis at the Cystic Fibrosis Locus in the British Population," *Human Genetics* 85:408-409, 1990. |
| | 78.5 | 214 | M.J. Schwarz, M. Super, C. Wallis, et al., "ΔF508 Testing of the DNA Bank of the Royal Manchester Children's Hospital," *Human Genetics* 85:428-430, 1990. |
| | 80 | 195 | E.K. Watson, E.S. Mayall, L. Simova, et al., "The Incidence of ΔF508 CF Mutation, and Associated Haplotypes, in a Sample of English CF Families," *Human Genetics* 85:435-436, 1990. |
| Northern Ireland | 54 | 204 | The Cystic Fibrosis Genetic Analysis Consortium, "Worldwide Survey of the ΔF508 Mutation—Report From the Cystic Fibrosis Genetic Analysis Consortium," *American Journal of Human Genetics* 47:354-359, 1990. |
| Scotland | 71 | 361 | A.E. Shrimpton, I. McIntosh, and D.J.H. Brock, "The Incidence of Different Cystic Fibrosis Mutations in the Scottish Population: Effects on Prenatal Diagnosis and Genetic Counselling," *Journal of Medical Genetics* 28:317-321, 1991. |
| | 74 | 238 | The Cystic Fibrosis Genetic Analysis Consortium, "Worldwide Survey of the ΔF508 Mutation—Report From the Cystic Fibrosis Genetic Analysis Consortium," *American Journal of Human Genetics* 47:354-359, 1990. |
| | 74 | 215 | I. McIntosh, A. Curtis, and M.-L. Lorenzo, "The Haplotype Distribution of the ΔF508 Mutation in Cystic Fibrosis Families in Scotland," *Human Genetics* 85:419-420, 1990. |
| **Yugoslavia (former)** | 92 | 12 | The Cystic Fibrosis Genetic Analysis Consortium, "Worldwide Survey of the ΔF508 Mutation—Report From the Cystic Fibrosis Genetic Analysis Consortium," *American Journal of Human Genetics* 47:354-359, 1990. |
| Macedonia | 39.5 | 38 | L. Simova, C. Williams, G.D. Efremov, et al., "ΔF508 Frequency and Associated Haplotypes Near the Cystic Fibrosis Locus in the Yugoslav Population," *Human Genetics* 85:432-433, 1990. |
| Slovenia | 26 | 34 | The Cystic Fibrosis Genetic Analysis Consortium, "Worldwide Survey of the ΔF508 Mutation—Report From the Cystic Fibrosis Genetic Analysis Consortium," *American Journal of Human Genetics* 47:354-359, 1990. |
| Southern (mixed population) | 38 | 39 | The Cystic Fibrosis Genetic Analysis Consortium, "Worldwide Survey of the ΔF508 Mutation—Report From the Cystic Fibrosis Genetic Analysis Consortium," *American Journal of Human Genetics* 47:354-359, 1990. |

NA=Not available

SOURCE: Office of Technology Assessment, 1992.

### Table A-2—Distribution of ΔF508 in North America

| Population | Frequency of ΔF508 (percent) |
|---|---|
| Caucasian, mixed European ancestry | 68 to 76 |
| Hispanic | 71 |
| Louisiana French Acadian | 69 |
| French Canadian | 54 to 69 |
| Ashkenazic Jewish | 26 to 50 |
| African American | 37 |
| Hutterite | 35 |

SOURCE: Office of Technology Assessment, 1992.

### Table A-3—Frequencies of Common Cystic Fibrosis Mutations

| Mutation | European frequency[a] (aggregate data; percent) | North American frequency (percent Caucasian population) |
|---|---|---|
| ΔF508 | 68.0 | 68.0 to 76.0 |
| G551D | 4.4 | 3.2[b] |
| G542X | 6.0 | 2.7[b] to 4.6[c] |
| R553X | 3.4 | 0.8[c] to 1.4[b] |
| N1303K | 4.2 | 1.4[b] |
| W1282X | NA | 0.01[d] to 0.9[e] |
| R560T | 6.0 | 0.01[f] to 0.6[b,e] |
| ΔI507 | 1.0 | 0.01[f] to 0.6[b,e] |
| S549N | 0.7 | 0.01[f] |

[a] *Newsletter of the European Concerted Action on Cystic Fibrosis*, R. Williamson (ed.), vol. 3, 1990.

[b] A.L. Beaudet, Howard Hughes Medical Institute, Baylor College of Medicine, Houston, TX, personal communication, 1991.

[c] GeneScreen, "Two Years Later . . . Seven Mutations in the CF Panel," *In the Genome*, Addendum, summer 1991.

[d] S. Curristin, B.J. Rosenstein, and G.R. Cutting, "North American Caucasian and African-American Cystic Fibrosis (CF) Patients Have Different Distributions of CF Gene Mutations," *American Journal of Human Genetics* 49 (Supp.): 490, 1991.

[e] P. Kristidis, D. Bozon, M. Corey, et al., "Genetic Determination of Exocrine Pancreatic Function in Cystic Fibrosis," *American Journal of Human Genetics* 50: 1178-1184, 1992.

[f] IG Laboratories, Inc., "CF/12 Mutation Test for Cystic Fibrosis," brochure, fall 1991.

NA=Not available.

SOURCE: Office of Technology Assessment, 1992.

# Appendix B
# Case Studies of Other Carrier Screening Programs

Carrier screening programs have historically been focused within a particular group—e.g., Tay-Sachs among Ashkenazic Jews and sickle cell anemia among African Americans. With cystic fibrosis (CF), the potential target population is larger and less defined, which may introduce both technical and organizational complexity not present in past carrier screening. This appendix describes past carrier screening programs for Tay-Sachs, sickle cell anemia, and β-thalassemia. It also delineates the similarities and differences of these programs to reveal considerations for CF carrier screening.

## SCREENING FOR TAY-SACHS DISEASE

Tay-Sachs is a lethal genetic disorder that predominantly affects Jews of Eastern and Central European descent (Ashkenazic Jews) and populations in the United States and Canada descended from French Canadian ancestors; it also occurs infrequently in the general population. The disorder affects the central nervous system, resulting in mental retardation and death within the first 3 to 6 years of life. Unlike for CF, the means to screen for Tay-Sachs carrier and affected status have existed for more than two decades.

Shortly after accurate tests were developed, a number of large-scale carrier screening programs were initiated, first in the United States, and then throughout the world. Carrier screening programs for Tay-Sachs are frequently considered models for other genetic screening endeavors (54,64,66). While some of the experiences can be generalized to other genetic disorders, other facets of Tay-Sachs carrier screening programs derive from specific aspects of the condition and the populations targeted.

### *The Disease*

Tay-Sachs (also called $G_{M2}$ gangliosidosis, Type I) is a progressive, degenerative disorder of the central nervous system. Due to the absence of a particular enzyme, hexosaminidase A (Hex A), fatty substances—specifically $G_{M2}$ gangliosides—accumulate in cells throughout the central nervous system, eventually destroying them. The process begins prenatally, though the infant can appear normal for the first months of life. Symptoms usually begin between 4 and 12 months with an exaggerated startle response, followed by increasing motor weakness. Blindness, seizures, and continual decline into a vegetative state ensue until the child dies. No cure exists (19).

### *Genetics of Tay-Sachs Disease*

Like CF, Tay-Sachs is a single gene disorder inherited in an autosomal recessive pattern; that is, only a child receiving a defective copy of the gene from each parent will be affected; carriers have no clinical symptoms. Among the general population, Tay-Sachs mutations are rare, occurring with a carrier frequency of 1 in 167, giving rise to 1 in 112,000 affected births. The disease is significantly more common, however, among Ashkenazic Jews: the carrier frequency is 1 in 31 and birth incidence is 1 in 3,900 (77). Prevalence in Ashkenazic Jews is a function of ethnic background, and is unrelated to religious practice. Elevated incidence of variant alleles are also found among Moroccan Jews (97), French-Canadians (1), and a distinct population of Louisiana French Acadians (41).

In 1969, scientists identified the lack of Hex A as the underlying metabolic defect in Tay-Sachs disease (60). A year later, an enzymatic assay was developed to measure Hex A activity. Application of this test to known carriers revealed intermediate activity of the enzyme in these individuals, which led to the assay for carrier identification among those without family histories (59).

Tay-Sachs carrier screening is performed on blood serum, although for pregnant women, women using oral contraceptives, and persons with liver disease, the blood serum test does not yield accurate results. Instead, white blood cells, platelets, or tears are used; such methods are more time-consuming and costly. The biochemical assay also can be used to detect affected fetuses through either amniocentesis (79) or chorionic villus sampling. This assay has had significant false positive carrier rates but lower false negative rates (80).

In the mid-1980s, characterization of the Hex A gene structure and mutations led to the possibility of DNA-based Tay-Sachs screening and diagnosis. Three distinct mutations account for 90 to 98 percent of Tay-Sachs chromosomes in this population (28,62,94); one—a 4 base pair insertion—is present in over 70 percent of Tay-Sachs Ashkenazic alleles. In contrast, among non-Ashkenazic carriers, the three mutations comprise about 20 percent of defective mutations (28,62,94). For Moroccan Jews, a single base pair deletion predominates (56), while in French-Canadians, multiple mutations are present (29). Preliminary results in the Louisiana French Acadian population indicate the presence of two alterations correlated with Ashkenazic carrier alleles (50). Additionally, one of the prevalent Ashkenazic mutations results in an ''adult-onset'' form of the disease (94).

DNA analysis is useful in confirmatory and family studies and is being tested as a carrier screening method in some centers. Currently, DNA assays allow confirmation of diagnosis, reducing levels of both false positives and false negatives (62,94). In the future they might replace the enzymatic assay.

### *Screening Programs*

Initial awareness of Tay-Sachs centered in Jewish communities; hence, screening programs (using biochemical assays) for carrier status originated there. After these efforts were launched, similar programs spread to areas where French-Canadians clustered and, most recently, to parts of Louisiana. The almost immediate implementation of carrier screening programs among Ashkenazic Jews was facilitated by a number of factors:

- occurrence of the disease is concentrated in a defined population;
- determination of the carrier state is easy and relatively inexpensive;
- at-risk pregnancies can be monitored through prenatal diagnosis of the disease (34); and
- public funding or subsidies were available—e.g., through the National Sickle Cell Anemia, Cooley's Anemia, Tay-Sachs, and Genetic Diseases Act (Public Law 94-278).

## Tay-Sachs Carrier Screening Programs

Tay-Sachs carrier screening began in 1971 in the Baltimore, MD-Washington, DC Jewish populations at the behest of the Jewish communities (34). The pilot program involved community outreach: 14 months of organizational planning, technical preparation, and education of medical and religious leaders preceded massive public education campaigns. Every aspect was carried out within the community, from planning to sample collections, which were held in synagogues, high school gymnasiums, and Jewish community centers (34). Eighteen hundred people showed up in one day for the first screening; in under a year, 5,600 individuals were screened, and 245 carriers identified (34,35).

Following the success of this effort, Jewish communities throughout the United States and Canada implemented similar endeavors. By 1976, 52 cities in the United States, and others in Canada, England, Israel, South Africa, and Japan were conducting carrier screening programs (34). Many centers followed the initial protocol, adapting specifics to individual locations. Most changes were made in the target population group and in

WHAT DO YOU KNOW ABOUT YOUR GENETIC MAKEUP?

TAY-SACHS FACTS:

- TAY-SACHS DISEASE IS AN INHERITED GENETIC DISORDER OF INFANCY
- A CHILD WITH THE DISEASE CAN BE BORN TO HEALTHY PARENTS WHO ARE CARRIERS OF THE TAY-SACHS GENE
- ANYBODY CAN BE A CARRIER - CARRIER RATE IS 1:150 IN THE GENERAL POPULATION AND 1:30 IN THE JEWISH POPULATION
- EARLY CARRIER DETECTION CAN PROTECT FUTURE GENERATIONS

BE TESTED AT: STANFORD UNIVERSITY

WED., MAY 16 11:00am–2:00pm Tresidder Union, Oak Lounge West
5:30pm–7:30pm Business School, Room 54

THURS. MAY 17 8:00am–10:00am Med. Center, M106 (Med. Stu. Lounge)
11:00am–2:00pm Tresidder Union, Oak Lounge West

(over 17 and non-pregnants only, please)

Sponsored by: AEPi; Hillel; Stanford Genetic Counseling, Dept of Gyn/OB; Cowell Health Promotion Program

— BE SAFE — BE TESTED —

*Take the Carrier Detection Test*

For Additional Information Call
NORTHERN CALIFORNIA TAY-SACHS PREVENTION PROGRAM — (415) 658-5568

This is a public service program supported by the State of California Department of Health

*Photo credit: Alpha Epsilon Pi*

An advertisement for Tay-Sachs carrier screening.

emphasis on public or physician education. Three main approaches evolved:

- educating and screening the entire community (mass screening);
- targeting couples of reproductive age, usually with the involvement of physicians; and
- screening high school and college students.

**Mass Screening.** Mass screening programs essentially followed the prototype of the Baltimore, MD-Washington, DC endeavor. In a south Florida area with a high concentration of Ashkenazic Jews, for example, a program involving leaders of the local community, parents of children with Tay-Sachs, and the Department of Pediatrics at the University of Miami School of Medicine was organized in 1973. In 1974, the steering committee overseeing the Florida effort decided to reemphasize mass screening sessions, and established small, local clinics (including synagogues and colleges) where screening sessions could be coordinated with local blood banks' visits to the community. Though screening was available to all individuals, young, single persons were counseled that it might be more effective to delay screening until after marriage, when reproductive decisions would be more pertinent (93).

**Couples of Reproductive Age.** In 1977, clinical chemists in Akron, OH designed a program that focused on educating family physicians and minimizing public education (63). Information was disseminated to rabbis and Jewish community centers advising them to refer married couples, either pregnant or planning a pregnancy, to physicians for screening. In 1983, a unique program, Chevra Dor Yeshorim, in New York City's Orthodox Jewish population began screening people prior to arranged marriages to ensure that no marriages would occur between carriers (box B-1).

**Students.** A different approach was launched in Montreal, Canada in 1974. A school-based screening program, operated by physicians and paramedical personnel, garnered participation of 75 percent (48). An 8-year followup study indicated that Canadians screened in high school had largely positive attitudes toward genetic screening long after the experience, and made appropriate use of the results (82,83,106). Others criticize the followup study for underestimating the forgetfulness of the adolescents screened (91). Though voluntary genetic screening of high school students is considered acceptable and successful by many in the Montreal community (83), some outside the community accuse the program of coercion (30).

## Tay-Sachs Screening Today

Through 1987, 600,000 people had been screened through voluntary programs in the United States (55). Almost 22,000 heterozygotes had been identified, and more than 1,400 pregnancies were monitored. Over 200 affected fetuses were detected and more than 1,000 healthy babies were born from at-risk pregnancies (55).

### *Box B-1—Chevra Dor Yeshorim Tay-Sachs Program*

Chevra Dor Yeshorim, which originated in New York City's Orthodox Jewish community, is a unique Tay-Sachs screening program. The elevated incidence of Tay-Sachs in this population, combined with religious beliefs opposing abortion and contraception, led community leaders to conclude that preventing Tay-Sachs was possible only by avoiding reproduction between carriers (52). Such reproduction could be avoided because this Orthodox Jewish community arranges marriages through matchmakers.

In 1983, the Committee for Prevention of Jewish Genetic Diseases formed Chevra Dor Yeshorim, which still operates today. All marriageable people are assigned a number and screened for Tay-Sachs carrier status at a local center (52). No one is informed of his or her result, which is filed by number only. When a marriage is proposed, the matchmaker calls the center with the prospective couple's numbers and is informed whether the proposed match involves two Tay-Sachs carriers; if the match does not, marriage plans proceed. If the match involves two carriers, the matchmaker tells the two families to contact the center, where the families are informed that both children are carriers and referred to counseling. Carriers thus learn their status only if they match with another carrier; families can report the match has failed for other reasons and look for new matches. This system of anonymity is employed because of historical stigmatization of carriers (52).

Chevra Dor Yeshorim has eventually become an adolescent rite of passage. Similar programs have been developed in California, Tennessee, Michigan, Florida, Maryland, Massachusetts, Illinois, Montreal, Israel, Europe, and other communities (45). By 1992, more than 30,000 people have been screened through such programs. More than 7,000 inquiries have been made and 47 prospective matches between Tay-Sachs carriers identified and halted (45).

SOURCE: Office of Technology Assessment, 1992.

Today, the focus has changed somewhat from the early mass education and screening approach, with over 100 hospitals and clinics nationwide offering screening on a continual basis. According to the National Tay-Sachs and Allied Diseases Association (NTSAD), a major aim "continues to be the promotion of genetic screening programs nationally through affiliated hospitals and medical centers, and through [NTSAD'S] local chapters and community organizations" (55).

A program in Toronto, Canada illustrates some general trends in Tay-Sachs carrier screening. Begun in 1972 as a community outreach program along the lines of the Baltimore, MD-Washington, DC prototype, it faced lagging community attendance by the end of the decade. By 1978, the orientation changed from mass screening to one of case finding, where physicians referred patients for screening on an individual basis (14,15,47). In this manner, 600 to 700 individuals are screened each year. Nearly two-thirds are pregnant at the time of screening, suggesting that determination of carrier status is being viewed as part of prenatal care rather than preconception planning.

Though outreach efforts are directed beyond the Jewish community, knowledge of Tay-Sachs is generally confined to this population. When multiple cases of Tay-Sachs surfaced in a rural, Catholic community in Louisiana, NTSAD and Tulane University organized a mass education and screening program (41). NTSAD continues public education efforts through nationwide mailings to community and religious organizations and to college, high school, and grammar school libraries.

**Figure B-1—Normal and Sickled Red Blood Cells**

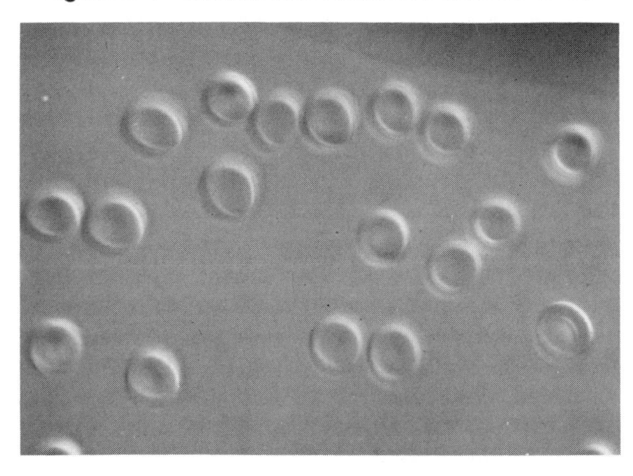

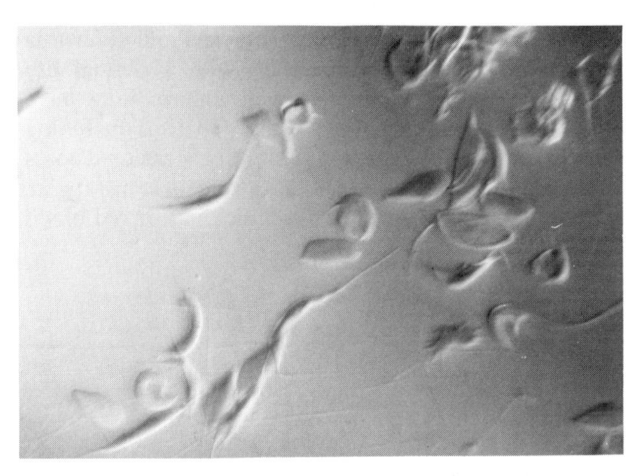

In sickle cell anemia, many red blood cells are distorted from their normal round shape (top), into the shape of a crescent or sickle (bottom). These distorted cells can obstruct smaller blood vessels or be removed too rapidly by the spleen.

SOURCE: M. Murayama, National Institute of Arthritis, Musculoskeletal, and Skin Diseases, National Institutes of Health, Bethesda, MD, 1992.

# SCREENING FOR SICKLE CELL ANEMIA

If Tay-Sachs screening is often held as an example of a successful screening campaign, early sickle cell screening efforts are frequently cited as "screening gone wrong." Like Tay-Sachs, sickle cell anemia is an autosomal recessive condition generally affecting a particular population—for sickle cell anemia, people of African descent. Sickle cell anemia impairs red blood cell flow through the circulatory system, causing complications in organ systems throughout the body.

As for Tay-Sachs, massive screening programs for sickle cell were undertaken in the United States in the 1970s. Sickle cell programs, however, differed immensely from the Tay-Sachs program in a variety of ways. Screening was mandatory in some States, and there was neither treatment nor prenatal diagnosis. Early programs also suffered from misinformation and discrimination against carriers.

## *The Disease*

The sickle cell mutation affects hemoglobin (Hb), the oxygen-carrying molecule in the blood stream. Hb is found in the red blood cells and can be of a variety of types. Hb A is found in healthy red blood cells; Hb S occurs in the red blood cells of sickle cell anemia patients. Hb S causes the cells to become deformed and sickle shaped (figure B-1). Sickled red blood cells become trapped, decreasing red blood cell survival and, therefore, oxygen transport. Individuals with sickle cell anemia are susceptible to episodes of extreme pain in the limbs, back, abdomen, or chest, which are thought to be caused by blockages of the deformed red blood cells in the circulatory system. These crises occur as often as once a month to once every other year and can last from 3 days to more than 3 weeks. Lack of oxygen in the spleen, kidneys, bones, and joints can also damage these organs and tissues. Eight to 30 percent of children with sickle cell anemia die in the first years of life from complications of severe bacterial infection (25). However, intensive antibiotic treatment—if begun early enough—imparts some resistance to these conditions (25). No cure exists for sickle cell anemia. Unlike Tay-Sachs, sickle cell anemia does not involve mental retardation.

## *Genetics of Sickle Cell Anemia*

Like CF and Tay-Sachs, sickle cell anemia is a genetic disorder inherited in an autosomal recessive pattern. A person with two copies of the sickle cell mutation is homozygous and has sickle cell anemia. A person with a single copy is heterozygous and is said to be a sickle cell carrier or to have sickle cell trait. The red blood cells of people with the sickle cell trait contain both Hb A and Hb S. These individuals do not have sickle cell anemia and are considered to be healthy with a normal life expectancy (88,101). Some clinical abnormalities have been found in people with sickle cell trait including defects in urine concentrating ability and occasional bouts of blood in the urine (88). Under extremely low-oxygen, high-exertion conditions, minimal sickling of red blood cells can occur (36).

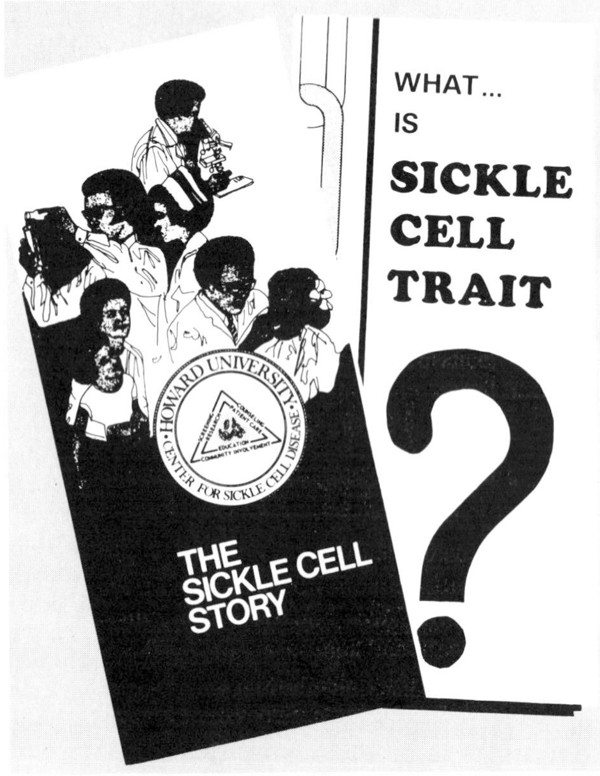


*Photo credit: Howard University*

Patient education brochures describing sickle cell anemia and sickle cell trait.

The precise difference between Hb A and Hb S was elucidated in 1956 (57); since then, the specific mutation causing these mutations has been found. The genetic basis of sickle cell is rare: One base change accounts for all cases of the disorder (57). A base change of an adenine for a thymine in the β-globin gene produces a single amino acid difference of the 146 amino acids in the β-chain—one of the two protein components of Hb.

Although the incidence of the sickle cell mutation is high in Greeks, Italians (particularly Sicilians), Eti-Turks, Arabs, southern Iranians, and Asian Indians, the highest frequency occurs in Africans and their descendants (7). One in 400 African American newborns has sickle cell anemia (32), and 1 in 10 or 11 has sickle cell trait (3,57,96).

Diagnostic procedures for sickle cell anemia evolved as knowledge of the biochemistry of Hb increased. Discovery of the differential electrophoretic mobility of Hb S and Hb A permitted testing of blood samples. The presence of only Hb A indicated noncarrier status, only Hb S indicated sickle cell anemia, and the combination of Hb A and Hb S in an individual indicated sickle cell trait. Restriction fragment length polymorphism analysis has had an important impact on diagnosis of sickle cell disease and trait, especially in the prenatal diagnosis of affected fetuses by amniocentesis and chorionic villus sampling. In the early 1980s, a restriction enzyme, *Mst* II, that recognizes a sequence at the site of the sickle cell mutation itself was found (61). Since 1982, researchers have performed this procedure with 100 percent informativeness and rare errors (39). Diagnosis no longer depends on having informative family members. DNA diagnosis to detect sickle cell has been shown to be 100 percent informative and 99 percent reliable (43). Additionally, polymerase chain reaction can now be used with nonradioactive, enzyme-labeled, allele-specific oligonucleotide probes in dot-blot format (76).

## *Screening Programs*

In the early 1970s, 16 States and the District of Columbia enacted laws to identify people with sickle cell trait and sickle cell anemia so that carrier couples could be informed of their risks of having affected children. Most laws were drafted and promoted by African American legislators at the height of the civil rights movement. At first glance, they offered an inexpensive benefit to African American citizens (65). Depending on the States, screening was mandated for newborns, preschool children, pregnant women, couples applying for marriage licenses, inmates of State institutions, or some combination of these groups. Some States mandated that tests be offered, while others mandated screening.

Many aspects of State sickle cell screening laws generated public critique. Statutes consistently contained blatant medical and scientific errors including calling for immunization for sickle cell anemia. Some States classified sickle cell carrier status with having sexually transmissible diseases on marriage licenses; others called it a communicable disease (65). Almost every State law failed to insist on using the most sensitive assay available. Controversy also focused on the racial distribution of sickle cell mutations and the target screening population. Sickle cell anemia was considered a "Black Disease" (32). The laws were seen by many citizens as racist eugenic measures aimed at reducing the number of marriages between carriers and decreasing the number of pregnancies at risk for affected children of a minority population. The fact that the programs were largely designed and operated by Caucasians fueled proclamations of genocide.

Most State laws failed to provide adequate education and counseling for persons with sickle cell anemia and trait. The most common error conferred disease status on those who were carriers. Those diagnosed with sickle cell trait were often told they could not have children, that childbirth would be hazardous, or other untruths. Only 4 of 13 programs mentioned counseling of any sort. People were often confused about which condition they had—sickle cell disease or sickle cell trait—which led to increased anxiety.

State laws also failed to provide public education to guard against discrimination and stigmatization. By 1972, for example, at least one flight attendant had been grounded (32). Stories of job and insurance discrimination multiplied as screening programs proliferated. Additionally, too little concern was expressed over confidentiality of results. Some States required that positive test results be filed with the State's public health entity.

### National Legislation

In 1971, President Nixon spoke of the problem of sickle cell anemia and the need for more research and education. The 1972 National Sickle Cell Anemia Control Act authorized $85 million over 3 years for the "establishment and operation of voluntary sickle cell anemia screening and counseling programs," and to "develop information and educational materials relating to sickle cell anemia and to disseminate such information and materials to persons providing health care and to the public generally" (Public Law 92-294). Some $30 million was allocated for research on the disease and for development of educational, screening, and counseling programs. Applicants for Federal funds were required to meet certain standards and to ensure confidentiality of all medical and counseling records. The law also promised community participation: African Americans delivering the service to African Americans.

### Sickle Cell Screening Today

By 1973, laws in eight States had been repealed (57). By 1977, more than one-third of States had enacted laws under the National Sickle Cell Anemia Control Act (65). Mandatory screening was eliminated, and the need for adequate genetic counseling, public education, and confidentiality of test results was recognized (65). Nevertheless, discrepancy between the ideal and actual practice existed, as criticism focused, in the late 1970s, on a continued lack of "community participation" and accusations of Federal money granted disproportionately to white institutions (65).

Today, sickle cell screening is often done routinely on pregnant women as part of routine blood workup and on newborns (3,23). Prenatal sickle cell screening infrequently results in selective pregnancy termination (103). Forty States screen newborns for sickle cell anemia (103), detecting babies with sickle cell anemia and babies who are carriers. Some States screen all newborns, some target a particular population, and some have voluntary screening that varies from hospital to hospital (103). Some newborn screening is done without informed consent, but with counseling of the parents if carrier status is detected (3).

## SCREENING FOR β-THALASSEMIA

β-thalassemia, also known as Cooley's anemia, thalassemia major, target cell anemia, and Mediterranean anemia, is an autosomal recessive condition for which carrier screening has been undertaken. β-thalassemia clusters in particular ethnic groups—people of Mediterranean, Middle Eastern, Asian Indian, Chinese, Southeast Asian, and African descent (7). Legislative action concerning β-thalassemia has occurred both in the United States and abroad. Screening for α-thalassemia, a related disorder, was first introduced in 1975 and continues today in some parts of the world (box B-2). Many European programs emphasize hemoglobinopathy screening encompassing all clinically significant α and β alleles and their combinations (103).

### *The Disease*

β-thalassemia is one of the most common genetic disorders in the world (92). Like sickle cell anemia, the condition affects the Hb of red blood cells. In β-thalassemia, the amount of Hb is diminished, causing red blood cells to be smaller and have a "target" appearance. Severe anemia, frequent infections, spleen enlargement, growth retardation, and marrow hypertrophy are all characteristics of the disorder (57). Death usually results from iron toxicity, most often in childhood. Therapy for β-thalassemia includes transfusions, folic acid supplementation, and intensive treatment of infections. Transfu-

### Box B-2—α-Thalassemia

α-thalassemia is another genetic condition often included in carrier screening programs. Like β-thalassemia, it is an autosomal recessive disorder affecting hemoglobin (Hb) production. The level of α-globin, a component of Hb, is decreased, resulting in abnormal Hb molecules: Hb H and Hb Bart. It is the most common genetic disease in some Chinese provinces and is found throughout Southeast Asia and the Mediterranean. According to the World Health Organization, 10,000 babies are born with α-thalassemia each year in Asia (42). α-thalassemia also affects people of African descent (7).

α-thalassemia results from mutations in the α-globin genes (87). While there is one β-globin gene in the human genome, there are two α-globin genes. The most severe type of α-thalassemia is caused by homozygous deletion of both α-globin genes. This condition, called hemoglobin Bart's hydrops fetalis, generally leads to intrauterine death (57,98) and is the primary cause of stillbirth in Southeast Asia (57). Three to five percent of African Americans are heterozygous carriers of mutations for both α-thalassemia genes, and about 26 percent are heterozygous carriers for mutations in one of the α-thalassemia genes (6).

Prior to DNA analysis, diagnosis relied on assaying the presence of Hb Bart in cord blood of babies suspected to have α-thalassemia. Identifying carriers by determination of mean corpuscular volume (MCV) and globin chain electrophoresis was inexact. Today, DNA technologies have increased the speed and efficacy of prenatal and carrier screening for α-thalassemia (44).

Carrier screening for α-thalassemia occurs worldwide (100). One program is currently underway in Hong Kong, where 98 percent of the population are ethnic Chinese from South China, with a carrier frequency of 3 percent (12). Public education, and formal organization of screening, counseling, and referral were required to make the program successful (12). Today, both carrier screening and prenatal diagnosis appear to be well accepted in Hong Kong (12). A similar effort is advancing in Guangdong Province in southern China (107), where the carrier frequency is about 5 percent.

α-thalassemia carrier screening has also been conducted in the United States. In the 1980s, researchers in southern California educated a Southeast Asian population about screening through the community's churches. Thalassemia screening was performed on a cross-section of the population and on newborns using MCV indices and Hb electrophoresis. Approximately 8 percent of more than 600 individuals were found to be carriers. Today—although the program is no longer operating because Federal funding expired—overall awareness of α-thalassemia appears to have been maintained among physicians, and screening now targets pregnant women in obstetricians' offices (21).

One study in Hawaii assessed consumer attitudes towards thalassemia screening (105). As part of the Hawaii Hereditary Anemia Project, 862 α- and β-thalassemia carriers who were identified through the project were surveyed about feelings of stigmatization and attitudes toward knowledge gained. Researchers concluded that learning carrier status provoked little anxiety (except in less educated populations) and provided meaningful benefits (105). Researchers in Rochester, NY have found that the percentages of people seeking an explanation of the test result, having the partner screened, and undergoing prenatal diagnosis are all higher among Southeast Asians than non-Southeast Asians (72).

SOURCE: Office of Technology Assessment, 1992.

sions, however, can result in hemochromatosis (iron overload), which can be improved slightly by using chelating agents to reduce excess iron. Splenectomy—removal of the spleen—is indicated if signs of hypersplenism exist and is performed preferably after age 5 (57).

## Genetics of β-Thalassemia

Like sickle cell, β-thalassemia is an autosomal recessive disease caused by mutations in the β-globin gene. Unlike sickle cell, however, several mutations lead to β-thalassemia (37). Depending on the number and type of mutations, symptoms range from mild to severe (9). Some mutations in the β-globin gene render the β-globin gene product useless and lead to severe β-thalassemia. Other mutations reduce the output of β-globin, but have relatively little clinical effect.

Each ethnic group has a unique distribution of β-thalassemia mutations (37). At least 91 β-globin mutations leading to β-thalassemia have been characterized, and common mutations exist in a variety of ethnic groups including Mediterranean, African American, Southeast Asian, Indian, and Middle-Eastern populations (38).

The advent of DNA technology has enhanced diagnostic testing for β-thalassemia and β-thalassemia trait. Prior to DNA assays, numerous methods were used to diagnose β-thalassemia (72). Biochemical assays continue to be adequate for carrier screening, but DNA assays allow for mutation identification and improved prenatal diagnosis (5,8,10,76). β-thalassemia diagnosis using DNA analysis is 95 percent informative and 99 percent reliable, comparing favorably to Hb electrophoresis and other biochemical methods (43).

## *Screening Programs*

β-thalassemia is considered a major world health problem, and a number of carrier screening programs have been established both within the United States and abroad. Each program has distinct characteristics and serves specific populations within its society.

### United States

β-thalassemia occurs most often in Mediterranean American, African American, and Asian American populations (69). Though 4 percent of the Italian American population has β-thalassemia (70), most Americans have never heard of the condition (75). In the late 1970s and early 1980s, adults in a health maintenance organization (HMO) in Rochester, NY were screened for β-thalassemia trait as part of general health care or multiphasic screening. There was no informed consent, and screening was done by Hb electrophoresis. Patients identified as β-thalassemia carriers received genetic counseling by a physician or watched a videotape that presented basic information about the disorder and the meaning of carrier status. Both methods of providing information were shown to be equally effective (24). Compared to noncarriers (who were not counseled), carriers demonstrated increased knowledge about β-thalassemia and its genetics (70), which was retained at 2 and 10 months postcounseling (71).

β-thalassemia was also incorporated in a 5-year comprehensive prenatal hemoglobinopathies (genetic disorders of Hb) carrier screening program in Rochester, NY, in the early 1980s (74). Pregnant women were screened for β-thalassemia and other hemoglobinopathies as part of routine prenatal screening (73). Every blood sample drawn was screened with no separate informed consent, since providers felt they had their patients' implicit consent for relevant diagnostic blood tests (74). Screening included examining red cells to determine mean corpuscular volume (MCV) and Hb $A_2$ determination.

In this study, 18,907 pregnant women were screened—about 35 percent of pregnancies in the Rochester metropolitan area. Over 800 women were identified as carriers or as homozygous for a hemoglobinopathy, of whom 92 were β-thalassemia carriers (73,74). Twenty-two percent of these β-thalassemia carriers were not Mediterranean American, African American, or Asian American (74), those populations generally considered at high risk. This finding argues in favor of screening all pregnant women, not only those usually regarded as at elevated risk based on ethnicity or race. Eighty-six percent of women who were counseled about their hemoglobinopathy carrier status said they wanted their partners screened, 55 percent had their partners screened, and 47 percent of carrier couples identified underwent prenatal diagnosis. Thus, unselected patients in a primary care setting in the Rochester region, even though pregnant, were receptive to and utilized genetic information (74). The researchers concluded that genetic screening in such a setting has many advantages over that in non-health-care settings, including comprehensive coverage of the population, screening at a time appropriate or relevant for the individual, and increased likelihood of appropriate medical followup (69).

The Federal Government has also been involved with β-thalassemia screening. In 1972, the National Cooley's Anemia Control Act (Public Law 92-414) was introduced and sponsored by legislators of Mediterranean heritage. It authorized $3 million for β-thalassemia screening, treatment, and counseling programs, $3 million for public education, and $5.1 million for disease research over a 3-year period (65).

### Canada

A program for β-thalassemia prevention comprising education, population screening for carriers, and reproductive counseling has been carried out in Canada (81). In a 25-month period (1979-81), 6,748 persons, including 5,117 high school students, were screened for β-thalassemia trait using MCV indices; the participation rate was 80 percent (81). Researchers surveyed 60 carriers and 120 noncarriers among the high school student population, and most carriers told parents (95 percent) and friends (67 percent) the test result. Most carriers (91 percent) reported they would ascertain their spouses' genotype; 95 percent approved of the screening effort (81).

### Sardinia

β-thalassemia is a significant health problem. One couple in 80 is a carrier couple—i.e., at 1 in 4 risk of an affected pregnancy. Prior to screening, 1 in 250 live births had β-thalassemia. Most β-thalassemia cases in Sardinia are severe—i.e., β-globin chains are absent. About 96 percent of the β-thalassemia mutations in this population are of a single type, and another mutation accounts for 2.1 percent of the remaining mutations (11).

In 1976, the region's Department of Public Health initiated a voluntary program to control β-thalassemia in Sardinia based on carrier screening, genetic counseling, and prenatal diagnosis. General practitioners, obstetricians, and paramedics were trained at specific educational meetings about the philosophical and technical aspects of the program. The program was targeted toward young unmarried men and women, married couples, and pregnant women. When a carrier was identified, the professionals urged that other family members also consider testing (11).

Prior to implementing the program, a mass media educational campaign geared to the general public was launched. About 47 percent of couples and 65 percent of singles screened said they learned of the program through this campaign. Screening was performed first on one member of a couple. If that individual was positive, his or her partner was screened. Prenatal diagnosis was also performed (11).

To date, 24 percent of Sardinia's population has been screened, and 85 percent of the theoretical number of carrier couples (+/+) in the population have been identified. In large part, this high efficiency results from followup screening of carriers' family members (11). Sixty-five percent of those screened in 1980 were pregnant women. By 1990, only 30 percent were pregnant, demonstrating a trend in increased knowledge—i.e., individuals came in for screening prior to conception. The incidence of β-thalassemia in Sardinia has declined from 1 in 250 live births in 1974 to 1 in 1,200 in 1991, an effective prevention of 90 percent of predicted cases (11). Of babies born with β-thalassemia, 67 percent were born to parents who were unaware of the disease and of carrier screening, 20 percent were to parents who, for ethical reasons, decided against abortion after prenatal diagnosis, and 13 percent were cases of false paternity (11).

An important early result of the program was the impact on carriers. Many had difficulty finding jobs, chiefly because the national army is the predominant employer in Sardinia and refuses to employ β-thalassemia carriers due to a misunderstanding about carrier status. Ongoing efforts to educate the army appear likely to reverse this practice in the near future (11). β-thalassemia screening in Sardinia is currently moving to encompass the school-aged population (11).

### Cyprus

In Cyprus, 1 in 7 individuals is a carrier, and 1 in 1,000 Cypriots is a patient under treatment for β-thalassemia (2). In the early 1970s, the realization that demand for treating this single disease could outstrip resources for all health care led to the development of a national carrier screening effort. Undertaken with World Health Organization support, the program consisted of public education, population screening for carriers, genetic counseling for carrier couples, prenatal diagnosis, and premarital screening (2).

An unusual compliance mechanism is one facet of the Cyprus program. The Greek Orthodox Church recognized the problems of the high incidence of β-thalassemia births, but was reluctant to endorse pregnancy termination to prevent such births. In 1983, however, the Church agreed that, before the Church would bless a marriage or engagement, a couple would be required to present a certificate proving they had been screened and had received genetic counseling. With the Church's cooperation, the number of people screened skyrocketed from 1,785 in 1977 to 18,202 in 1983. Today, 10,000 to 11,000 people are screened annually. Since most couples of reproductive age have been screened, attention now focuses on single people. Since 1985, a 20 percent fall in the expected proportion of +/+ couples has been noted, perhaps indicating that carriers are avoiding marriage to other carriers (2).

As in Sardinia, the number of newborns with β-thalassemia has fallen dramatically. In 1974, 53 of 8,594 births were affected: From 1986 through 1990, there were five, for a prevention rate of 97 percent. Ninety percent of the 69 affected babies born between 1978 and 1980 resulted from lack of knowledge among the public and the medical profession about the Cyprus carrier screening effort. Two births were due to refusal of prenatal diagnosis on religious grounds, two were laboratory errors, and one was parental choice after positive prenatal diagnosis.

The program's success is attributed to several factors, including extensive and continuous public education, small population size, and homogeneity of the population (2). Time spent on counseling is much less today than when the program was initiated—a direct result of increased public education. A newer program is under way in Turkish Cyprus.

### Other β-thalassemia Screening Programs

Screening programs for β-thalassemia in the United Kingdom (58) incorporate treatment and prevention into general health care. Primary health care, peripheral centers, and reference centers coordinate the program's efforts. Individuals learn about the screening through schools and public information posters. Community involvement, particularly through parents' and patients' associations is an important component, and a detailed education program provides effective information about genetic risks and services. Fewer β-thalassemia births have been avoided in the United Kingdom than in Cyprus or Sardinia, chiefly because of the large population size, the occurrence of β-thalassemia in ethnic minorities scattered throughout an indigenous population, and the beliefs of many of those at risk that prenatal diagnosis and abortion are socially and religiously unacceptable (58). Southeast Asia, Hong Kong, East Mediterranean countries, and South China have β-thalassemia carrier screening programs (12,46,107).

# LESSONS FOR CYSTIC FIBROSIS CARRIER SCREENING

The Tay-Sachs, sickle cell anemia, β-thalassemia, and α-thalassemia experiences offer lessons that might be applicable to routine carrier screening for CF. These lessons involve five principal aspects of the programs described in this appendix:

- nature of participation,
- setting and fiscal support,
- target population,
- public education, and
- counseling carriers and at-risk couples.

## *Nature of Participation*

Participation in the programs described in this appendix was voluntary or mandatory. The Tay-Sachs experience offers a model of voluntary participation, which contributed to its efficacy and success—both in reducing disease and in avoiding stigmatization (54,66). It is noteworthy, nonetheless, that one survey found that nearly half of participants in one Tay-Sachs program felt screening should be mandatory, although the breakdown of carriers and noncarriers expressing those opinions was not measured (13).

The sickle cell experience, on the other hand, included some mandatory State screening of people in schools or correctional institutions, or of couples applying for marriage licenses. Such mandatory participation was viewed as a contributing factor in the widespread failure of these efforts (65,66). β-thalassemia programs in the United States and abroad have used the voluntary approach and have been generally successful (2,11,74,105). β-thalassemia programs in the Mediterranean that use a quasi-mandatory approach through the involvement of the Greek Orthodox Church also have been successful in reducing disease. Although stigmatization appeared to be an issue initially, it seems to now be obviated (2). The success of these mandatory programs, however, can be attributed mainly to the cultures of Cyprus and Sardinia and is likely not applicable to carrier screening in the United States.

Thus, success of voluntary initiatives and the failure of mandatory programs (excluding the Cyprus example) lead many to conclude that voluntary participation for CF carrier screening is essential. The National Institutes of Health (NIH) Workshop on Population Screening for the Cystic Fibrosis Gene recommended that screening be voluntary (4).

## *Setting and Fiscal Support*

The environment in which a genetic screening program is administered also affects its success and acceptance. Tay-Sachs carrier screening was community-based. Leaders within the Jewish community became leaders of the screening programs, which were offered through synagogues, storefronts, and community centers. The community was involved in all aspects of the program. For CF, however, the target population is larger and more diffuse, with a wide range of religious beliefs. Thus, it would be an ill fit to organize CF carrier screening solely through religious centers.

In stark contrast, the sickle cell anemia programs were run largely by the State and Federal governments, their contractors, and grant recipients. During a time of tension over civil rights, Caucasians from outside of the community generally controlled screening, which was viewed extremely negatively by the African American community. The resulting lack of support by the targeted population doomed sickle cell screening programs to failure.

β-thalassemia programs in the United States have been more diverse, involving primary care physicians in neighborhood health centers, a large HMO, hospital prenatal clinics, a family medicine program, and private practice. The particular setting seems less important to success than other factors, (e.g., patient and provider education)—although a primary care clinical setting is desirable (69)—and likely will prove most appropriate to CF carrier screening. (β-thalassemia experiences in Europe revolve around national health care systems, making it difficult to compare setting and funding to U.S. experiences.)

Beyond the particular setting is the issue of funding for organizing and maintaining the effort. Historical perspectives probably apply least in this area. Most efforts launched in the 1970s and 1980s involved some degree of public funding. With respect to routine CF carrier screening in the United States, the funding issue is complex, but direct funding in the manner of Tay-Sachs, sickle cell, or the thalassemias is unlikely to materialize beyond the current NIH pilot projects (ch. 6). Thus, to the extent that direct public subsidy contributed to the success of past screening, routine CF carrier screening will unlikely realize such a benefit.

## *Target Population*

Tay-Sachs programs targeted a narrow slice of the U.S. population. The population was easily identifiable, largely urban, educated, and receptive to screening and reproductive options; all are population characteristics often cited as beneficial to a program's success (34). Sickle cell programs targeted African American communities at the height of the civil rights movement, causing considerable controversy about segregation, discrimination, and stigmatization. Racial tensions created by singling out a population already historically discriminated against stress the importance of prior awareness of the implications of focusing on members of a particular group and their potential reactions to screening.

The age and sex of the screened population also have implications for CF carrier screening. With the ability to perform prenatal diagnosis, much attention focuses on screening pregnant women (95). Some argue that screening during pregnancy offers the advantages of immediate interest in knowledge of genetic information and comprehensive coverage of the population (75). Disadvantages of screening during pregnancy include eliminating the option to avoid conception, so that prenatal fetal testing might be perceived as necessary. Or, it might be too late in a pregnancy for prenatal diagnosis even if desired. Some also believe that carrier identification can be too anxiety-producing during pregnancy to be efficacious. Others argue that screening women first can be perceived as eugenic screening for breeding fitness (102).

Many studies have attempted to ascertain the optimal group for carrier screening: newborns, students, primary care recipients, pregnant women, couples planning to marry, or married couples (13,16,17,99,106). The studies often conflict: Each group offers advantages and disadvantages that vary by disease, available technology, and manifestations of the disease (ch. 6). Thus, the optimal population for CF carrier screening remains a topic of debate. Additionally, while CF mutations have a high prevalence in one racial group—Caucasians—the ethnic diversity of this population in the United States poses a challenge that did not exist for previous screening efforts. Beyond diversity per se, many Americans do not know specifics of their ethnic backgrounds, knowledge which might assist in more precisely assigning risks (96). Further, CF also occurs, albeit less frequently, in Asian Americans and African Americans. The extent to which routine screening within these two groups is appropriate is also an issue.

## *Public Education*

Consumer knowledge—both about medical aspects of the specific genetic condition and the meaning of carrier status—is key. Conditions that affect mental function, such as Tay-Sachs, appear to have a greater acceptance of carrier screening and prenatal diagnosis (20,22,26). The high acceptability of screening that resulted from public education for Tay-Sachs might be less informative for routine CF carrier screening, for example, although the mechanism of providing the education could be illustrative. Likewise, CF carrier screening might not serve as a good indicator of future screening for disorders such as fragile X syndrome (box B-3).

Stigmatization of carriers, widely cited as a negative outcome of genetic screening (27,30,67), is largely attributed to ignorance within both the community being screened and the general population. In contrast to acceptability, lessons about the importance of public education to avoid stigmatization can be gleaned from history. With Tay-Sachs screening, a concerted effort was made to educate community and religious leaders, who then helped educate the public through outreach, mailings, media, and word of mouth before any screening took place. Educational efforts were adjusted over time, and in general, data indicate the majority of carriers believe they were not stigmatized, although a small percentage of noncarriers expressed attitudes of superiority (13,40). Similarly, the Mediterranean β-thalassemia program demonstrates the effectiveness of massive public education through the local media, although some contend the massive public education bordered on propaganda and coercion (31). The sickle cell experience demonstrates how poor public outreach and lack of knowledge among professionals in the program lead to failure (54,66).

Regardless of the precise mechanism through which education is achieved, implementation of CF carrier screening can benefit from the knowledge that public education has had a positive effect on previous screening experiences. Lessons learned from Tay-Sachs, sickle cell, and β-thalassemia efforts can be applied to CF carrier screening to increase public knowledge and decrease stigmatization.

## *Counseling Carriers and At-Risk Couples*

CF carrier screening can be informed by the Tay-Sachs, sickle cell, and β-thalassemia experiences with genetic counseling. Lessons about informing patients, counseling carriers, and counseling carrier couples can be applied to CF carrier screening. The sickle cell experience often omitted counseling, and so individuals with sickle cell trait became anxious, not only because they were given incorrect information, but also because they had received inadequate counseling.

Tay-Sachs programs considered informed consent, counseling of carriers, and counseling carrier couples to be high priorities. Such informed consent and counseling aided in the understanding and adjustment of individuals who were identified as carriers. In contrast, with no apparent negative effect, β-thalassemia carrier screening in one pilot involving pregnant women was done on every blood sample drawn, with no separate informed consent (since providers felt they had patients' implicit consent for relevant diagnostic blood tests) (74). As is acknowledged to be essential, carriers and carrier couples received post-test counseling.

### *Box B-3—Fragile X Syndrome*

Recent advances elucidating the genetics of fragile X syndrome will have a great impact on future applications to screening. Fragile X syndrome is the most common form of inherited mental retardation. Its incidence is about 1 in 1,500 male and 1 in 2,500 female live births (58). It includes minor dysmorphic features such as an elongated face, large ears, prominent jaw, and macroorchidism (78). The genetics of this disorder are different from the autosomal recessive disorders of cystic fibrosis, Tay-Sachs disease, sickle cell anemia, and the thalassemias. Fragile X syndrome is an X-linked disorder—the gene for the disorder lies on the X chromosome, one of the two sex determining human chromosomes. (Females have two X chromosomes, while males have one X and one Y chromosome.)

Initially described in 1969, fragile X syndrome derives its name from the tendency for the tip of the X chromosome to break off or appear fragile (figure B-2) (49). Its mode of inheritance, however, is unlike other X-linked disorders such as hemophilia—it is neither recessive nor dominant. Unaffected men and women can transmit a fragile X chromosome without manifesting any symptoms of the syndrome or expressing the fragile site cytogenetically. Such transmitters, or carriers, can have children or grandchildren with fragile X syndrome (104). The mothers of all affected children are considered to be obligate carriers, and no affected offspring arise as a direct result of a new mutation (85). Among females, about half of the obligate carriers do not express the fragile X chromosome. About 50 percent of heterozygous females express the disorder to some extent, and 30 percent are mentally retarded. Approximately 20 percent of males who inherit the gene from their mothers are unaffected carriers. Furthermore, severity appears variable in different siblings even within the same family (58).

Physical signs of the disorder are neither specific nor constant, and generally appear after childhood, making diagnosis difficult. In the past, diagnosis generally occurred only after the birth of a second affected male (89), and involved expert, labor-intensive laboratory analysis (84). Once the X chromosome was shown to be fragile, linkage analysis of other family members could be performed, with prenatal tests also possible (33,86).

In 1991, researchers elucidated the unique inheritance of fragile X syndrome and developed a DNA probe to detect the fragile X site (58,104). Reliable and specific detection of all male or female carriers of a fragile X site is now possible (58,68). Similarly, prenatal analysis can be performed (58,68,90). Moreover, results from DNA assays appear to correlate with disease severity, and so can be used to predict clinical outcome (58). Fragile X syndrome will likely be one of the first disorders for which the primary diagnosis is based on the direct analysis of a mutation at the DNA level (68,84)

The increased knowledge of the genetics of fragile X syndrome has applications to genetic screening. Most families with fragile X syndrome are presently unknown, because routine screening of the entire developmentally disabled population has not been practical and because fragile X syndrome historically has been poorly diagnosed (89). Given the high sensitivity and specificity of the new fragile X assay—and the value of early diagnosis for genetic counseling purposes—the next step to consider is selected population screening (18). Although such screening would initially focus on developmentally disabled males and females, if the frequency of normal transmitting males is as high as the frequency of affected males, as one hypothesis predicts, an argument can be made for general population screening, as has been proposed in the United Kingdom (18).

With respect to genetic counseling needs for fragile X syndrome, one study examined interest in prenatal diagnosis and attitudes towards termination of affected pregnancies (51). Surveyed prior to the advent of the DNA assay, 81 percent of women said they would seek prenatal diagnosis, and 28 percent indicated they would terminate an affected pregnancy. There was no significant difference in responses between women who had affected children and those who did not. Issues the subjects considered most important for discussion with a genetic counselor included the availability of treatment, risk for having an affected grandchild, and expectations for the future functioning of children with fragile X syndrome (51).

SOURCE: Office of Technology Assessment, 1992.

**Figure B-2—Fragile X Chromosomes**

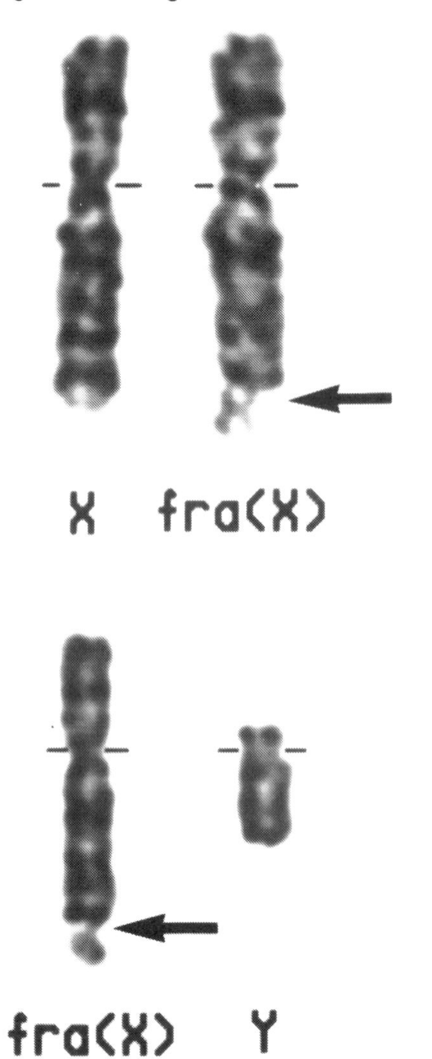


The fragile X site appears as a break or separation at the distal end of the long arm of X chromosomes (arrows). This computer-enhanced photomicrograph shows a carrier female (top) and an affected male (bottom).

SOURCE: The Denver Children's Hospital, 1992.

# APPENDIX B REFERENCES


1. Andermann, E., Scriver, C.R., Wolfe, L.S., et al., "Genetic Variants of Tay-Sachs Disease: Tay-Sachs Disease and Sandhoff's Disease in French Canadians, Juvenile Tay-Sachs Disease in Lebanese Canadians, and a Tay-Sachs Screening Program in the French Canadian Population," *Tay-Sachs Disease: Screening and Prevention*, M.M. Kaback (ed.) (New York, NY: Alan R. Liss, Inc., 1977).
2. Angastiniotis, M., "Cyprus: Thalassaemia Programme," *Lancet* 336:119-1120, 1990.
3. Benkendorf, J.L., Georgetown University Medical Center, Washington, DC, personal communication, December 1991.
4. Bernhardt, B.A., "Population Screening for the Cystic Fibrosis Gene," *New England Journal of Medicine* 324:61-62, 1991.
5. Boehm, C.D., Antonarakis, S.E., Phillips, J.A., et al., "Prenatal Diagnosis Using DNA Polymorphisms: Report on 95 Pregnancies at Risk for Sickle-Cell Disease or β-Thalassemia," *New England Journal of Medicine* 308:1054-1058, 1983.
6. Bowman, J.E., Department of Pathology, University of Chicago, personal communication, April 1992.
7. Bowman, J.E., and Murray, R.F., Jr., *Genetic Variation and Disorders in Peoples of African Origin* (Baltimore, MD: The Johns Hopkins University Press, 1990).
8. Cai, S.-P., Chang, C.A., Zhang, J.Z., et al., "Rapid Prenatal Diagnosis of β-Thalassemia Using DNA Amplification and Nonradioactive Probes," *Blood* 73:372-374, 1989.
9. Cao, A., Gossens, M., and Pirastu, M., "β-Thalassemia Mutations in Mediterranean Populations," *British Journal of Haematology* 71:309-312, 1989.
10. Cao, A., Rosatelli, C., Galanello, R., et al., "Prenatal Diagnosis of Inherited Hemoglobinopathies," *Indian Journal of Pediatrics* 56:707-717, 1989.
11. Cao, A., Rosatelli, C., Galanello, R., et al., "The Prevention of Thalassemia in Sardinia," *Clinical Genetics* 36:277-285, 1989.
12. Chan, V., Chan, T.K., and Todd, D., "Thalassemia Screening and Prenatal Diagnosis in Hong Kong," *American Journal of Human Genetics* 49(Supp.):79, 1991.
13. Childs, B., "The Personal Impact of Tay-Sachs Carrier Screening," *Tay-Sachs Disease: Screening and Prevention*, M.M. Kaback (ed.) (New York, NY: Alan R. Liss, Inc., 1977).
14. Clarke, J.T., "Screening for Carriers of Tay-Sachs Disease: Two Approaches," *Canadian Medical Association Journal* 119:549-550, 1978.
15. Clarke, J.T., Skomorowski, M.A., and Zuker, S., "Tay-Sachs Disease Carrier Screening: Follow-Up of a Case-Finding Approach," *American Journal of Medical Genetics* 34:601-605, 1989.
16. Clow, C.L., and Scriver, C.R., "Knowledge and Attitudes Toward Genetic Screening Among High-School Students: The Tay-Sachs Experience," *Pediatrics* 59:86-91, 1977.
17. Clow, C.L., and Scriver, C.R., "The Adolescent Copes With Genetic Screening: A Study of Tay-Sachs Screening Among High-School Students," *Tay-Sachs Disease: Screening and Prevention,*

M.M. Kaback, (ed.) (New York, NY: Alan R. Liss, Inc., 1977).

18. Connor, J.M., "Cloning of the Gene for the Fragile X Syndrome: Implications for the Clinical Geneticist," *Journal of Medical Genetics* 28:811-813, 1991.
19. Desnick, R.J., and Goldberg, J.D., "Tay-Sachs Disease: Prospects for Therapeutic Intervention," *Tay-Sachs Disease: Screening and Prevention*, M.M. Kaback (ed.) (New York, NY: Alan R. Liss, Inc., 1977).
20. Drugan, A., Greb, A., Johnson, M.P., et al., "Determinants of Parental Decisions to Abort for Chromosome Abnormalities," *Prenatal Diagnosis* 10:483-490, 1990.
21. Dumars, K., MediGene Facility, Inc., Fountain Valley, CA, personal communication, May 1992.
22. Faden, R.R., Chwalow, A.J., Quaid, K., et al., "Prenatal Screening and Pregnant Women's Attitudes Toward the Abortion of Defective Fetuses," *American Journal of Public Health* 77:288-290, 1987.
23. Fine, B.A., Northwestern University Medical School, Chicago, IL, personal communication, September 1991.
24. Fisher, L., Rowley, P.T., and Lipkin, M., Jr., "Genetic Counseling for β-Thalassemia Trait Following Health Screening in a Health Maintenance Organization: Comparison of Programmed and Conventional Counseling," *American Journal of Human Genetics* 33:987-994, 1981.
25. Githens, J.H., Lane, P.A., McCurdy, R.S., et al., "Newborn Screening for Hemoglobinopathies in Colorado: The First 10 Years," *American Journal of Diseases of Children* 144:466-470, 1990.
26. Golbus, M.S., Loughman, W.D., Epstein, C.J., et al., "Prenatal Diagnosis in 3,000 Amniocenteses," *New England Journal of Medicine* 300:157-163, 1979.
27. Goodman, M.J., and Goodman, L.E., "The Overselling of Genetic Anxiety," *The Hastings Center Report* 12:20-27, 1982.
28. Grebner, E.E., and Tomczak, J., "Distribution of Three α-Chain β-Hexosaminidase A Mutations Among Tay-Sachs Carriers," *American Journal of Human Genetics* 48:604-607, 1991.
29. Hechtman, P., Kaplan, F., Bayerlan, J., et al., "More Than One Mutant Allele Cause Infantile Tay-Sachs Disease in French Canadians," *American Journal of Human Genetics* 47:815-822, 1990.
30. Holtzman, N.A., *Proceed With Caution: Predicting Genetic Risks in the Recombinant DNA Era* (Baltimore, MD: The Johns Hopkins University Press, 1989).
31. Holtzman, N.A., Johns Hopkins University Hospital, Baltimore, MD, personal communication, December 1991.
32. Hubbard, R., and Henifin, M.S., "Genetic Screening of Prospective Parents and of Workers: Some Scientific and Social Issues," *International Journal of Health Services* 15:231-251, 1985.
33. Jenkins, E.C., Krawczun, M.S., Stark-Houck, S.L., et al., "Improved Prenatal Detection of fra(X) (q27.3): Methods for Prevention of False Negatives in Chorionic Villus and Amniotic Fluid Cell Cultures," *American Journal of Medical Genetics* 38:447-452, 1991.
34. Kaback, M.M., Nathan, T.J., and Greenwald, S., "Tay-Sachs Disease: Heterozygote Screening and Prenatal Diagnosis—U.S. Experience and World Perspective," *Tay-Sachs Disease: Screening and Prevention*, M.M. Kaback (ed.) (New York, NY: Alan R. Liss, Inc., 1977).
35. Kaback, M.M., Zeiger, R.S., Reynolds, L.W., et al., "Approaches to the Control and Prevention of Tay-Sachs Disease," *Progress in Medical Genetics* 10:103-134, 1974.
36. Kark, J.A., Posey, D.M., Schumacher, H.R., et al., "Sickle-Cell Trait as a Risk Factor for Sudden Death in Physical Training," *New England Journal of Medicine* 317:781-787, 1987.
37. Kazazian, H.H., "Molecular Pathology of the β-Globin Gene Cluster," *Trends in Neuroscience* 8:192-200, 1985.
38. Kazazian, H.H., "Thalassemia Syndrome: Molecular Basis and Prenatal Diagnosis in 1990," *Seminars in Hematology* 27:209-228, 1990.
39. Kazazian, H.H., Johns Hopkins University School of Medicine, Baltimore, MD, personal communication, December 1991.
40. Kenen, R.H., and Schmidt, R.M., "Stigmatization of Carrier Status: Social Implications of Heterozygote Genetic Screening Programs," *American Journal of Public Health* 68:116-1120, 1978.
41. Kovner, A., The National Tay-Sachs and Allied Diseases Association, Inc., Brookline, MA, personal communication, August 1991.
42. Kuliev, A.M., "The WHO Control Program for Hereditary Anemias," *Birth Defects: Original Article Series* 23(5B):383-394, March of Dimes Birth Defects Foundation, 1988.
43. Lebo, R.V., Cunningham, G., Simons, M.J., et al., "Defining DNA Diagnostic Tests Appropriate for Standard Clinical Care," *American Journal of Human Genetics* 47:583-590, 1990.
44. Lebo, R.V., Saiki, R.K., Swanson, K., et al., "Prenatal Diagnosis of α-Thalassemia by Polymerase Chain Reaction and Dual Restriction Enzyme Analysis," *Human Genetics* 85:293-299, 1990.

45. Lefkowitz, S., Chevra Dor Yeshorim, Center for the Prevention of Jewish Genetic Diseases, Brooklyn, NY, personal communication, March 1992.
46. Loukopoulos, D., "Screening for Thalassemia and HbS in the East Mediterranean," *American Journal of Human Genetics* 49(Supp.):79, 1991.
47. Lowden, J.A., "Role of the Physician in Screening for Carriers of Tay-Sachs Disease," *Canadian Medical Association Journal* 119:575-578, 1978.
48. Lowden, J.A., and Davidson, J., "Tay-Sachs Screening and Prevention: The Canadian Experience," *Tay-Sachs Disease: Screening and Prevention*, M.M. Kaback (ed.) (New York, NY: Alan R. Liss, Inc., 1977).
49. Lubs, H.A., "A Marker X Syndrome," *American Journal of Human Genetics* 21:231-244, 1969.
50. McDowell, G.A., Schultz, R.A., Schwartz, S., et al., "Presence of Both Ashkenazic Tay-Sachs Mutations in a Non-Jewish Inbred Population," *American Journal of Human Genetics* 45(Supp.):A9, 1989.
51. Meryash, D.L., and Abuelo, D., "Counseling Needs and Attitudes Toward Prenatal Diagnosis and Abortion in Fragile X Families," *Clinical Genetics* 33:349-355, 1988.
52. Merz, B., "Matchmaking Scheme Solves Tay-Sachs Problem," *Journal of the American Medical Association* 258:2636-2639, 1987.
53. Modell, B., and Petrou, M., "Review of Control Programs and Future Trends in the United Kingdom," *Birth Defects: Original Article Series* 23(5B): 433-442, March of Dimes Birth Defects Foundation, 1988.
54. National Research Council, Committee for the Study of Inborn Errors of Metabolism, *Genetic Screening: Programs, Principles, and Research* (Washington, DC: National Academy of Sciences, 1975).
55. National Tay-Sachs and Allied Disease Association, *Tay Sachs and the Allied Diseases: What Every Family Should Know*, brochure, 1987.
56. Navon, R., and Proia, R.L., "Tay-Sachs Disease in Moroccan Jews: Deletion of a Phenylalanine in the α-Subunit of α-Hexosaminidase," *American Journal of Human Genetics* 48:412-419, 1991.
57. Nora, J.J., and Fraser, F.C., *Medical Genetics: Principles and Practice* (Philadelphia, PA: Lea & Febiger, 1989).
58. Oberle, I., Rousseau, F., Heitz, D., et al., "Instability of a 550-Base Pair DNA Segment Abnormal Methylation in Fragile X Syndrome," *Science* 252:1097-1102, 1991.
59. O'Brien, J.S., Okada, S., Chen, A., et al., "Tay-Sachs Disease: Detection of Heterozygotes and Homozygotes by Serum Hexosaminidase Assay," *New England Journal of Medicine* 283:15-20, 1970.
60. Okada, S., and O'Brien, J.S., "Tay-Sachs Disease: Generalized Absence of a β-D-N-acetylhexosaminidase Component," *Science* 165:698-700, 1969.
61. Orkin, S.H., Little, P.F., Kazazian, H.H., et al., "Improved Detection of the Sickle Mutation by DNA Analysis," *New England Journal of Medicine* 307:32-36, 1982.
62. Paw, B.H., Tieu, P.T., Kaback, M.M., et al., "Frequency of Three *Hex A* Mutant Alleles among Jewish and Non-Jewish Carriers Identified in a Tay-Sachs Screening Program," *American Journal of Human Genetics* 47:698-705, 1990.
63. Potter, J.L., and Robinson, H.B., Jr., "Screening for the Tay-Sachs Carrier: A Compromise Program," *Clinical Chemistry* 27:523-525, 1981.
64. President's Commission for the Study of Ethical Problems in Medicine and Biomedical and Behavioral Research, *Screening and Counseling for Genetic Conditions: A Report on the Ethical, Social, and Legal Implications of Genetic Screening, Counseling, and Education Programs* (Washington, DC: U.S. Government Printing Office, 1983).
65. Reilly, P., *Genetics, Law and Social Policy* (Cambridge, MA: Harvard University Press, 1977).
66. Roberts, L., "One Worked: The Other Didn't (Genetic Screening Programs for Tay-Sachs and Sickle Cell Anemia)," *Science* 247:18, 1990.
67. Rosner, F., "Confidential Tay-Sachs Carrier Screening," *New York State Journal of Medicine* 89:585, 1989.
68. Rousseau, F., Heitz, D., Biancalana, V., et al., "Direct Diagnosis by DNA Analysis of the Fragile X Syndrome of Mental Retardation," *New England Journal of Medicine* 325:1673-1681, 1991.
69. Rowley, P.T., "Prenatal Screening for Hemoglobinopathies," *American Journal of Human Genetics* 49:466-467, 1991.
70. Rowley, P.T., Fisher, L., and Lipkin, M., Jr., "Screening and Genetic Counseling for β-Thalassemia Trait in a Population Unselected for Interest: Effects on Knowledge and Mood," *American Journal of Human Genetics* 31:718-730, 1979.
71. Rowley, P.T., Lipkin, M., Jr., and Fisher, L., "Screening and Genetic Counseling for Beta-Thalassemia Trait in a Population Unselected for Interest: Comparison of Three Counseling Methods," *American Journal of Human Genetics* 36:677-689, 1984.
72. Rowley, P.T., Loader, S., Sutera, C.J., et al., "Prenatal Hemoglobinopathy Screening: Receptivity of Southeast Asian Refugees," *American Journal of Preventive Medicine* 3:317-322, 1987.
73. Rowley, P.T., Loader, S., Sutera, C.J., et al., "Do Pregnant Women Benefit From Hemoglobinopathy

Screening?,'' *Annals of the New York Academy of Sciences* 565:152-160, 1989.

74. Rowley, P.T., Loader, S., Sutera, C.J., et al., ''Prenatal Screening for Hemoglobinopathies. I. A Prospective Regional Trial,'' *American Journal of Human Genetics* 48:439-446, 1991.
75. Rowley, P.T., Loader, S., and Walden, M., ''Toward Providing Parents the Option of Avoiding the Birth of the First Child With Cooley's Anemia: Response to Hemoglobinopathy Screening and Counseling During Pregnancy,'' *Annals of the New York Academy of Sciences* 445:408-416, 1985.
76. Saiki, R.K., Chang, C.-A., Levenson, C.H., et al., ''Diagnosis of Sickle Cell and β-Thalassemia With Enzymatically Amplified DNA and Nonradioactive Allele-Specific Oligonucleotide Probes,'' *New England Journal of Medicine* 319:537-541, 1988.
77. Sandhoff, K., Conzelmann, E., Neufeld, E.F., et al., ''The $G_{M2}$ Gangliosidoses'' *The Metabolic Basis of Inherited Disease*, C.R. Scriver, A.L. Beaudet, W.S. Sly, et al. (eds.) (New York, NY: McGraw Hill, 1989).
78. Schepis, C., Palazzo, R., Ragusa, R.M., et al., ''Association of Cutis Verticis Gyrata With Fragile X Syndrome and Fragility of Chromosome 12,'' *Lancet* 2:279, 1989.
79. Schneck, L., Friedland, J., Valenti, C., et al., ''Prenatal Diagnosis of Tay-Sachs Disease,'' *Lancet* 1:582-583, 1970.
80. Schulman, J.D., Genetics & IVF Institute, Fairfax, VA, personal communication, December, 1991.
81. Scriver, C.R., Bardanis, M., Cartier, L., et al., ''β-Thalassemia Disease Prevention: Genetic Medicine Applied,'' *American Journal of Human Genetics* 36:1024-1038, 1984.
82. Scriver, C.R., and Clow, C.L., ''Carrier Screening for Tay-Sachs Disease,'' *Lancet* 335:856, 1990.
83. Scriver, C.R., and Clow, C.L., ''Carrier Screening for Tay-Sachs Disease,'' *Lancet* 336:191, 1990.
84. Shapiro, L.R., ''The Fragile X Syndrome: A Peculiar Pattern of Inheritance,'' *New England Journal of Medicine* 325:1736-1738, 1991.
85. Shapiro, L.R., Wilmot, P.L., and Murphy, P.D., ''Prenatal Diagnosis of the Fragile X Syndrome: Possible End of the Experimental Phase for Amniotic Fluid,'' *American Journal of Medical Genetics* 38:453-455, 1991.
86. Shapiro, L.R., Wilmot, P.R., Shapiro, D.A., et al., ''Cytogenetic Diagnosis of the Fragile X Syndrome: Efficiency, Utilization, and Trends,'' *American Journal of Medical Genetics* 38:408-410, 1991.
87. Steinberg, M.H., and Embury, S.H., ''α-Thalassemia in Blacks: Interactions With the Sickle Hemoglobin Gene,'' *Birth Defects: Original Article Series* 23 (5A):43-48, March of Dimes Birth Defects Foundation, 1988.
88. Sullivan, L.W., ''Risks of Sickle-Cell Trait: Caution and Common Sense,'' *New England Journal of Medicine* 317:830-831, 1987.
89. Sutherland, G.R., ''Fragile Sites on Human Chromosomes: Demonstration of Their Dependence on the Type of Tissue Culture Medium,'' *Science* 197:25-26, 1977.
90. Sutherland, G.R., Gedeon, A., Kornman, L., et al., ''Prenatal Diagnosis of Fragile X Syndrome by Direct Detection of the Unstable DNA Sequence,'' *New England Journal of Medicine* 325:1720-1721, 1991.
91. ten Kate, L.P., and Tijmstra, T., ''Carrier Screening for Tay-Sachs Disease and Cystic Fibrosis,'' *Lancet* 336:1527-1528, 1990.
92. Thompson, J.S., and Thompson, M.W., *Genetics in Medicine* (Philadelphia, PA: W.B. Saunders Co., 1980).
93. Tocci, P.M., ''Seven Years Experience With Tay-Sachs Screening in Florida,'' *Journal of the Florida Medical Association* 68:24-29, 1981.
94. Triggs-Raine, B.L., Feigenbaum, A.S.J., Natowicz, M., et al., ''Screening for Carriers of Tay-Sachs Disease Among Ashkenazic Jews: A Comparison of DNA-Based and Enzyme-Based Tests,'' *New England Journal of Medicine* 323:6-12, 1990.
95. Tymstra, T., ''Prenatal Diagnosis, Prenatal Screening, and the Rise of the Tentative Pregnancy,'' *International Journal of Technology Assessment in Health Care* 7:509-516, 1991.
96. Valverde, K.D., Cystic Fibrosis Center, St. Vincent's Hospital and Medical Center, New York, NY, personal communication, December 1991.
97. Vecht, J., Zeigler, M., Segal, M., et al., ''Tay-Sachs Disease Among Moroccan Jews,'' *Israel Journal of Medical Science* 19:67-69, 1983.
98. Wang, S., Wang, L., and Zhang, B., ''A Survey of Hb Bart in Cord Blood and the α-Globin Gene in South China,'' *Birth Defects: Original Article Series* 23 (5A):23-30, March of Dimes Birth Defects Foundation, 1988.
99. Watson, E.K., Mayall, E., Chapple, J., et al., ''Screening for Carriers of Cystic Fibrosis Through Primary Health Care Services,'' *British Medical Journal* 303:504-507, 1991.
100. Weatherall, D.J., ''The Evolution of Thalassemia Screening,'' *American Journal of Human Genetics* 49(Supp.):79, 1991.
101. Weatherall, D.J., Clegg, J.B., Higgs, D.R., et al., ''The Hemoglobinopathies,'' *The Metabolic Basis*

*of Inherited Disease*, C.R. Scriver, A.L. Beaudet, W.S. Sly, et al. (eds.) (New York, NY: McGraw Hill, 1989).

102. Wilfond, B., University of Wisconsin Medical School, Madison, WI, personal communication, December 1991.
103. Willey, A.M., New York State Department of Health, Albany, NY, personal communications, January 1992, May 1992.
104. Yu, S., Pritchard, M., Kremer, E., et al., ''Fragile X Genotype Characterized by an Unstable Region of DNA,'' *Science* 252:1179-1181, 1991.
105. Yuen, J., and Hsia, Y.E., ''Consumer Attitudes Toward Thalassemia Screening: Heterozygotes of Multi-Ethnic Origin,'' *American Journal of Human Genetics* 49(Supp.):331, 1991.
106. Zeesman, S., Clow, C.L., Cartier, L., et al., ''A Private View of Heterozygosity: Eight-Year Follow-Up Study on Carriers of the Tay-Sachs Gene Detected by High School Screening in Montreal,'' *American Journal of Medical Genetics* 18:769-778, 1984.
107. Zhang, J.Z., Cai, S.P., and Kan, Y.W., ''Thalassemia Screening and Prenatal Diagnosis in South China,'' *American Journal of Human Genetics* 49(Supp.):80, 1991.

# Appendix C
# Acknowledgments

OTA would like to thank the members of the advisory panel who commented on drafts of this report, the contractors who provided material for this assessment, and the many individuals and organizations that supplied information for the study. In addition, OTA acknowledges the following individuals for their review of drafts of this report:

Naomi Aronson
Blue Cross and Blue Shield Association
Chicago, IL

Lori B. Andrews
American Bar Foundation
Chicago, IL

George J. Annas
Boston University School of Public Health
Boston, MA

Adrienne Asch
New York, NY

Sema K. Aydede
Istanbul, Turkey

Raj Barathur
Specialty Laboratories
Santa Monica, CA

Jonathan Beckwith
Harvard Medical School
Boston, MA

Michael Begleiter
Children's Mercy Hospital
Kansas City, MO

Clyde J. Behney
Health Program
Office of Technology Assessment

Judith Benkendorf
Georgetown University Medical Center
Washington, DC

David P. Bick
Genetics & IVF Institute
Fairfax, VA

Rob Bier
American Council of Life Insurance
Washington, DC

Barbara Bowles-Biesecker
University of Michigan Medical Center
Ann Arbor, MI

Paul R. Billings
California Pacific Medical Center
San Francisco, CA

Frank Blanchard
Howard Hughes Medical Institute
Bethesda, MD

Thomas F. Boat
University of North Carolina at Chapel Hill
Chapel Hill, NC

Mary Anne Bobinski
University of Houston Law Center
Houston, TX

Martin Bobrow
United Medical and Dental Schools of Guy's
and St. Thomas's Hospital
London, England

James E. Bowman
University of Chicago
Chicago, IL

David J.H. Brock
Western General Hospital
Edinburgh, Scotland

Keith Brown
GeneScreen, Inc.
Dallas, TX

Tony J. Beugelsdijk
Los Alamos National Laboratory
Los Alamos, NM

Peter Carpenter
Stanford University Center for Bioethics
Atherton, CA

Aravinda Chakravarti
University of Pittsburgh
Pittsburgh, PA

Gary Claxton
National Association of Insurance Commissioners
Washington, DC

Ellen Wright Clayton
The Vanderbilt Clinic
Nashville, TN

Francis S. Collins
Howard Hughes Medical Institute
Ann Arbor, MI

P. Michael Conneally
Indiana University School of Medicine
Indianapolis, IN

Elizabeth A. Conway
Blue Cross and Blue Shield Association
Washington, DC

James Corbett
MIB, Inc.
Westwood, MA

Virginia Corson
Johns Hopkins University Hospital
Baltimore, MD

Robert M. Cook-Deegan
Institute of Medicine
Washington, DC

George Cunningham
California Department of Health Services
Berkeley, CA

Peter T. D'Ascoli
St. Paul Ramsey Medical Center
St. Paul, MN

Neil Day
MIB, Inc.
Westwood, MA

Robert J. Desnick
Mt. Sinai School of Medicine
New York, NY

Daniel W. Drell
U.S. Department of Energy
Germantown, MD

Sherman Elias
University of Tennessee School of Medicine
Memphis, TN

C.M. Eng
Mt. Sinai School of Medicine
New York, NY

Mark Evans
American Medical Association
Chicago, IL

Lee Fallon
Genetics & IVF Institute
Fairfax, VA

Susan Dee Fernbach
Baylor College of Medicine
Houston, TX

Norman Fost
University of Wisconsin School of Medicine
Madison, WI

Michelle Fox
UCLA Medical Center
Los Angeles, CA

Robert Funk
GeneScreen, Inc.
Dallas, TX

Barry R. Furrow
Widener University School of Law
Wilmington, DE

Alan Garber
National Bureau of Economic Research, Inc.
Stanford, CA

Fred Gilbert
Cornell University Medical College
New York, NY

Robert C. Giles
GeneScreen, Inc.
Dallas, TX

Frank Greenberg
Institute for Molecular Genetics
Houston, TX

Gayle Greenberg
Cystic Fibrosis Foundation-Metropolitan
Washington, DC Chapter
Bethesda, MD

Robert M. Greenstein
University of Connecticut Health Center
Farmington, CT

Wayne W. Grody
UCLA School of Medicine
Los Angeles, CA

Joseph L. Hackett
U.S. Food and Drug Administration
Rockville, MD

Mark Hall
Arizona State University College of Law
Tempe, AZ

Barbara Handelin
IG Laboratories, Inc.
Framingham, MA

Peter S. Harper
University of Wales College of Medicine
Cardiff, Wales

Kiki Hellman
U.S. Food and Drug Administration
Rockville, MD

Patricia J. Hoben
Howard Hughes Medical Institute
Bethesda, MD

Neil A. Holtzman
Johns Hopkins University Hospital
Baltimore, MD

Frits A. Hommes
Medical College of Georgia
Augusta, GA

Juli A. Horwitz
Vivigen
Santa Fe, NM

Lauren S. Jenkins
Kaiser Permanente
San Jose, CA

James H. Jett
Los Alamos National Laboratory
Los Alamos, NM

Shirley Jones
Genetics & IVF Institute
Fairfax, VA

Eric T. Juengst
National Center for Human Genome Research
Bethesda, MD

Michael M. Kaback
University of California,
San Diego Medical Center
San Diego, CA

Deborah Kaplan
World Institute on Disability
Oakland, CA

Gail C. Kaplan
Vivigen
Santa Fe, NM

Nancy E. Kass
Johns Hopkins University
Baltimore, MD

Haig H. Kazazian, Jr.
Johns Hopkins University Hospital
Baltimore, MD

Bartha Maria Knoppers
University of Montreal Faculty of Law
Montreal, Canada

Michael R. Knowles
University of North Carolina at Chapel Hill
Chapel Hill, NC

Elinor Langfelder
National Center for Human Genome Research
Bethesda, MD

Jane S. Lin-Fu
Maternal and Child Health Bureau
Rockville, MD

Walter B. Maher
Chrysler Corp.
Washington, DC

Babetta L. Marrone
Los Alamos National Laboratory
Los Alamos, NM

Karla J. Matteson
University of Tennessee Medical Center-Knoxville
Knoxville, TN

F. John Meaney
Scottsdale, AZ

Abbey S. Meyers
National Organization for Rare Disorders
New Fairfield, CT

Arno S. Motulsky
University of Washington
Seattle, WA

Patricia D. Murphy
New York State Department of Health
Albany, NY

Thomas H. Murray
Case Western Reserve University
Cleveland, OH

Frank Nardi
MIB, Inc.
Westwood, MA

David M. Orenstein
Children's Hospital of Pittsburgh
Cystic Fibrosis Center
Pittsburgh, PA

David Orentlicher
American Medical Association
Chicago, IL

Jude C. Payne
Health Insurance Association of America
Washington, DC

Valois M. Pearce
Financial Concepts, Inc.
Santa Fe, NM

L.C. Pham
Roche Biomedical Laboratories
Research Triangle Park, NC

Robert J. Pokorski
North American Reassurance Co.
Westport, CT

Martin Pollard
Lawrence Berkeley Laboratory
Berkeley, CA

Stirling M. Puck
Vivigen
Santa Fe, NM

Kimberly A. Quaid
Indiana University School of Medicine
Indianapolis, IN

Harvie E. Raymond
Health Insurance Association of America
Washington, DC

C. Sue Richards
GeneScreen, Inc.
Dallas, TX

John Robertson
University of Texas School of Law
Austin, TX

Martin Rose
Genentech, Inc.
Washington, DC

Karen Rothenberg
University of Maryland School of Law
Baltimore, MD

Mark A. Rothstein
University of Houston Law Center
Houston, TX

Peter T. Rowley
University of Rochester Medical Center
Rochester, NY

Suzie Rubin
Biological Applications Program
Office of Technology Assessment

Marianne Schwartz
Rigshospitalet
Copenhagen, Denmark

Charles R. Scriver
McGill University
Montreal, Canada

Burton L. Shapiro
University of Minnesota Health Science Center
Minneapolis, MN

Ann C.M. Smith
Reston, VA

Bonnie Steinbock
University of Albany
Albany, NY

Edward Theil
Lawrence Berkeley Laboratory
Berkeley, CA

Helen Travers
IG Laboratories, Inc.
Miami, FL

Lap-Chee Tsui
Hospital for Sick Children
Toronto, Canada

Donald C. Uber
Lawrence Berkeley Laboratory
Berkeley, CA

Kathleen D. Valverde
St. Vincent's Hospital Cystic Fibrosis Center
New York, NY

Madge Vickers
Medical Research Council
London, England

Martha Volner
Alliance of Genetic Support Groups
Washington, DC

Judith L. Wagner
Health Program
Office of Technology Assessment

Ernestyne H. Watkins
National Institutes of Health
Bethesda, MD

Dorothy C. Wertz
Shriver Center for Mental Retardation, Inc.
Waltham, MA

Joan Weiss
Alliance of Genetic Support Groups
Washington, DC

Nancy S. Wexler
Hereditary Disease Foundation
Santa Monica, CA

Benjamin Wilfond
Arizona Health Sciences Center
Tucson, AZ

Robert Williamson
St. Mary's Hospital Medical School
London, England

Ann Willey
New York State Department of Health
Albany, NY

David R. Witt
Kaiser Permanente
San Jose, CA

# Appendix D
# List of Contractor Documents

For this assessment, OTA commissioned reports on various topics relevant to carrier screening for cystic fibrosis. The manuscripts of six of these contractors are available in a single volume (NTIS# PB 92-183185) from the National Technical Information Service, 5285 Port Royal Road, Springfield, VA 22161; (703) 487-4650.

Adrienne Asch, New York, NY, ''Carrier Screening, Cystic Fibrosis, and Stigma.''

Jeffrey L. Fox, Washington, DC, ''Cystic Fibrosis, Genetic Screening, and Insurance: Summary of OTA Workshop Proceedings, February 1, 1991.''

F. John Meaney, Scottsdale, AZ, ''CORN Report on Funding of State Genetic Services Programs in the United States, 1990.''

Thomas H. Murray, Center for Biomedical Ethics, Case Western Reserve University, Cleveland, OH, ''Genetics, Ethics, and Health Insurance.''

Mark V. Pauly, Leonard Davis Institute of Health Economics, University of Pennsylvania, Philadelphia, PA, ''The Cost-Effectiveness of Population Carrier Screening for Cystic Fibrosis: Draft Final Report;'' ''The Cost-Effectiveness of Population Carrier Screening for Cystic Fibrosis: Final Report;'' ''The Cost-Effectiveness of Population Carrier Screening for Cystic Fibrosis: Appendix to Final Report.''

Bonnie Steinbock, University of Albany, Albany, NY, ''Ethical Implications of Population Screening for Cystic Fibrosis: The Concept of Harm and Claims of Wrongful Life.''



327-046 - 92 - 10 : QL 3

# Appendix E

# List of Workshops and Participants

## Cystic Fibrosis, Genetic Screening, and Insurance
## February 1, 1991

Norman Fost, *Workshop Chair*
University of Wisconsin School of Medicine
Madison, WI

Rob Bier
American Council of Life Insurance
Washington, DC

Elizabeth A. Conway
Blue Cross and Blue Shield Association
Washington, DC

Barry R. Furrow
Widener University School of Law
Wilmington, DE

Robert M. Greenstein
University of Connecticut Health Center
Farmington, CT

Dennis Hodges
Group Health Association of America
Washington, DC

Nancy Kass
Johns Hopkins University
Baltimore, MD

Angèle Khachadour
Hastings College of the Law
San Francisco, CA

Jude C. Payne
Health Insurance Association of America
Washington, DC

Robert J. Pokorski
North American Reassurance Co.
Westport, CT

*Contractors:*
John M. Boyle
Schulman, Ronca, & Bucuvalas, Inc.
Silver Spring, MD

R. Alta Charo
University of Wisconsin Law School
Madison, WI

## Cystic Fibrosis, Genetic Tests, and Self-Insurance Under ERISA
## August 6, 1991

Angèle Khachadour, *Workshop Chair*
Hastings College of the Law
San Francisco, CA

Sonia Muchnik
Washington Business Group on Health
Washington, DC

Nancy Kass
Johns Hopkins University
Baltimore, MD

Ann Burdell
Genetics & IVF Institute
Fairfax, VA

Walter B. Maher
Chrysler Corp.
Washington, DC

Deborah Chollet
Georgia State University
Atlanta, GA

Mark A. Rothstein
University of Houston Law Center
Houston, TX

Gary Claxton
National Association of Insurance Commissioners
Washington, DC

Valois M. Pearce
Financial Concepts, Inc.
Santa Fe, NM

Mary Anne Bobinski
University of Houston Law Center
Houston, TX

# Appendix F
# Acronyms and Glossary

## Acronyms

-/- —negative CF carrier/negative CF carrier (couple)
+/- —positive CF carrier/negative CF carrier (couple)
+/+ —positive CF carrier/positive CF carrier (couple)
621+1G→T—a CF mutation
A —adenine
AAP —American Academy of Pediatrics
ABMG —American Board of Medical Genetics
ACOG —American College of Obstetricians and Gynecologists
ADA —Americans With Disabilities Act of 1990
AFP —alpha-fetoprotein
AHCPR —Agency for Health Care Policy and Research
AIDS —acquired immunodeficiency syndrome
AMA —American Medical Association
AMP —adenosine monophosphate
APKD —adult polycystic kidney disease
ARMS —amplification refractory mutation system
ASHG —American Society of Human Genetics
ASO —allele-specific oligonucleotide
ATP —adenosine triphosphate
BABI —blastomere analysis before implantation
Bart's —St. Bartholomew's Hospital (London)
BC/BS —Blue Cross and Blue Shield
C —cytosine
$Ca^{2+}$ —calcium ion
cAMP —cyclic adenosine monophosphate
CAP —College of American Pathologists
CDHS —California Department of Health Services
CF —cystic fibrosis
CF Trust —Cystic Fibrosis Research Trust (United Kingdom)
CFF —Cystic Fibrosis Foundation
CFR —Code of Federal Regulations
CFTR —cystic fibrosis transmembrane conductance regulator
CHAMPUS—Civilian Health and Medical Program of the Uniformed Services
$Cl^-$ —chloride ion
CLIA —Clinical Laboratory Improvement Amendments of 1988
CORN —Council of Regional Networks for Genetic Services
CVS —chorionic villus sampling
ΔF508 —delta F508 (most prevalent CF mutation)
ΔF508+6-12—delta F508 plus six to 12 additional CF mutations
ΔI507 —a CF mutation
DHHS —U.S. Department of Health and Human Services
DNA —deoxyribonucleic acid
DNase —deoxyribonuclease
DOD —U.S. Department of Defense
DOE —U.S. Department of Energy
EEOC —U.S. Equal Employment Opportunity Commission
ELSI —Ethical, Legal, and Social Issues Programs (NIH or DOE)
ERISA —Employee Retirement Income Security Act of 1974
FCC —Federal Communications Commission
FDA —U.S. Food and Drug Administration
FFDCA —Federal Food, Drug, and Cosmetic Act of 1938
FR —Federal Register
G —guanine
G542X —a CF mutation
G551D —a CF mutation
GAO —General Accounting Office
GI —gastrointestinal
GP —general practitioner (United Kingdom)
Hb —hemoglobin
HCFA —Health Care Financing Administration
HexA —hexosaminadase A
HHMI —Howard Hughes Medical Institute
HIAA —Health Insurance Association of America
HMO —health maintenance organization
HSRC —Health Services Research Committee (MRC; United Kingdom)
HSRP —Health Services Research Panel (MRC; United Kingdom)
HUGA-1 —Human Genome Analysis System (Japan)
IDE —Investigational Device Exemption
IRT —immunoreactive trypsin
IV —intravenous
kb —kilobase(s); 1,000 base pairs
L —liter
LCR —ligase chain reaction
MCH —Maternal and Child Health (Federal block grant)
MCV —mean corpuscular volume
MDA —Medical Device Amendments of 1976
MDR —medical devices reporting (1984 regulation)

mg —milligram
MHSS —Military Health Services System
MIB —Medical Information Bureau, Inc.
mmol —millimole
MRC —Medical Research Council (United Kingdom)
mRNA —messenger ribonucleic acid
MSAFP —maternal serum alpha-fetoprotein
N1303K —a CF mutation
$Na^+$ —sodium ion
NCHGR —National Center for Human Genome Research (NIH)
NHLBI —National Heart, Lung, and Blood Institute (NIH)
NHS —National Health Service (United Kingdom)
NICHD —National Institute of Child Health and Human Development (NIH)
NIDDK —National Institute of Diabetes and Digestive and Kidney Diseases (NIH)
NIH —National Institutes of Health
NRC —National Research Council
NSF —National Science Foundation
NSGC —National Society of Genetic Counselors
NTD —neural tube defect
NTSAD —National Tay-Sachs and Allied Diseases Association
OTA —Office of Technology Assessment
PCR —polymerase chain reaction
PMA —premarketing approval application
PPO —preferred provider organization
PT —physical therapy
R553X —a CF mutation
RAC —Recombinant DNA Advisory Committee (NIH)
RFA —request for applications
RFLP —restriction fragment length polymorphism
RNA —ribonucleic acid
SMDA —Safe Medical Devices Act of 1990
SPRANS —Special Projects of Regional and National Significance
T —thymine
UCLA —University of California, Los Angeles
W1282X —a CF mutation

# Glossary of Terms

**Adverse selection:** The tendency of persons with poorer than average health expectations to apply for or continue insurance to a greater extent than persons with average or better health expectations. Also known as "antiselection."

**Amino acid:** Any of a group of 20 molecules that combine to form proteins in living things. The sequence of amino acids in a protein is determined by the genetic code.

**Amniocentesis:** The most widely used technique of prenatal diagnosis. Cells shed by the developing fetus are extracted from a sample of amniotic fluid withdrawn from the expectant mother's uterus at about 16 weeks of gestation by means of a hypodermic needle. The cells are cultured and then tested for chromosomal defects. In addition, scientists can now analyze the DNA of these cells directly, identifying specific genetic errors.

**Allele:** Alternative form of a genetic locus (e.g., at a locus for eye color there might be alleles resulting in blue or brown eyes); alleles are inherited separately from each parent.

**Allele-specific oligonucleotide (ASO) probe:** Probes that are able to exactly match the nucleotide sequence of a portion of a gene, detecting even single-base differences.

**Automation:** Technology, such as robotics, developed to increase the speed, volume, and accuracy of routine DNA diagnostic procedures.

**Autoradiogram:** An x-ray film image showing the position of radioactive substances. Sometimes called "autorad."

**Autoradiograph:** See *autoradiogram.*

**Autoradiography:** A process for identifying radioactively labeled molecules or fragments of molecules.

**Autosome:** Chromosome not involved in sex determination. In a complete set of human chromosomes, there are 44 autosomes (22 pairs).

**Base pair:** Two complementary nucleotides held together by weak bonds. Two strands of DNA are held together in the shape of a double helix by the bonds between base pairs. The base adenine pairs with thymine, and guanine pairs with cytosine.

**β-thalassemia:** An autosomal recessive disorder affecting the red blood cells, resulting in anemia, infections, growth retardation, and other complications. β-thalassemia predominantly occurs among individuals of Mediterranean, Middle Eastern, Asian Indian, Chinese, Southeast Asian, and African descent.

**Blot:** See *Southern blot.*

**Buccal:** Relating to the inside of the cheek. A buccal swab collects cells from the inside of the cheek for CF mutation analysis.

**Carrier:** An individual apparently normal, but possessing a single copy of a recessive gene obscured by a dominant allele; a heterozygote.

**Cell:** The smallest component of life capable of independent reproduction and from which DNA can be isolated.

**Chest physical therapy (chest PT):** A cornerstone of CF therapy that moves the mucus blocking major air passages out of the lungs. One form of chest PT is bronchial drainage, during which an individual claps on the chest or back of the patient who is usually lying on a table or over a couch, to loosen mucus that the patient coughs up.

**Chorionic villus sampling (CVS):** A method of prenatal diagnosis undertaken as early as the 9th week of pregnancy. Fetal cells from chorionic villi (protrusions of a membrane, called the chorion, that surround a fetus during early development) are suctioned out through the uterus and their DNA is analyzed.

**Chromosomal aberration:** An abnormal chromosomal complement resulting from the loss, duplication, or rearrangement of genetic material.

**Chromosome:** A threadlike structure that carries genetic information arranged in a linear sequence. In humans, it consists of a complex of nucleic acids and proteins.

**Cloning:** The process of asexually producing a group of cells (clones), all genetically identical to the original ancestor. In recombinant DNA technology, the process of using a variety of DNA techniques to produce multiple copies of a single gene or segment of DNA.

**Community rating:** A method of determining premium rates based on the allocation of total costs without regard to past group experience. Community rating is required of federally qualified health maintenance organizations.

**Complementary DNA (cDNA):** DNA synthesized from a messenger RNA template; the single-strand form is often used as a probe in physical mapping.

**Confidentiality:** A fundamental component of the health care provider-patient relationship in which the professional has the duty to keep private all that is disclosed by the patient.

**Consanguineous:** Related by blood or origin, rather than by marriage.

**Cystic fibrosis (CF):** A life-shortening, autosomal recessive disorder affecting the respiratory, gastrointestinal, reproductive, and skeletal systems, as well as the sweat glands. CF is caused by mutations in the CF gene that affect the gene product, cystic fibrosis transmembrane conductance regulator. Individuals with CF possess two mutant CF genes.

**Cystic fibrosis carrier:** An individual who possesses one CF mutation and one normal CF gene. CF carriers manifest no symptoms of the disorder. See *carrier*.

**Cystic fibrosis carrier screening:** The performance of tests on persons for whom no family history of CF exists to determine whether they have one aberrant CF gene and one normal CF gene. Compare *cystic fibrosis screening*.

**Cystic fibrosis screening:** The performance of tests to diagnose the presence or absence of the actual disorder, in the absence of medical indications of the disease or a family history of CF. This type of diagnostic screening usually involves newborns, but is rare except in Colorado and Wisconsin. Compare *cystic fibrosis carrier screening*.

**Cystic fibrosis transmembrane conductance regulator (CFTR):** The CF gene product, which regulates chloride ($Cl^-$) conductance and might be a $Cl^-$ ion channel, the structure that governs $Cl^-$ entry and exit in the cell. CFTR produced by a mutant CF gene is frequently impaired, resulting in the medical manifestations of CF in affected individuals.

**ΔF508:** A three base pair deletion in the CF gene that results in a faulty CF gene product (i.e., a flawed CFTR). This mutation results in the deletion of one amino acid, phenylalanine, at position number 508 in CFTR. ΔF508 is the most common mutant allele among the more than 170 mutations identified in the CF gene.

**Deoxyribonucleic acid (DNA):** The molecule that encodes genetic information. DNA is a double-stranded helix held together by weak bonds between base pairs of nucleotides.

**Discrimination:** Differential treatment or favor with a prejudiced outlook or action.

**Dominant:** An allele that exerts its phenotypic effect when present either in homozygous or heterozygous form.

**Dot blot:** A variation of Southern blotting that involves placing DNA into discrete spots on a nylon membrane. A probe or probes can be hybridized to the membrane and diagnosis made rapidly.

**DNA:** See *deoxyribonucleic acid*.

**DNA analysis:** A direct examination of the genetic material, DNA, to reveal whether a individual has a CF mutation. Also known as *DNA test* and *DNA assay*.

**DNA band:** The visual image, e.g., on a autoradiogram or an ethidium bromide stained gel, that represents a particular DNA fragment.

**DNA probe:** Short segment of DNA labeled with a radioactive or other chemical tag and then used to detect the presence of a particular DNA sequence through hybridization to its complementary sequence.

**DNA sequence:** Order of nucleotide bases in DNA.

**Double helix:** The ladder-like shape formed by two linear strands of DNA bonded together.

**Electrophoresis:** Technique used to separate molecules such as DNA fragments or proteins. Electric current is passed through a gel and the fragments of DNA are separated by size. Smaller fragments migrate farther than larger pieces.

**Enzyme:** A protein that acts as a catalyst, speeding the rate at which a biochemical reaction proceeds, without being permanently altered or consumed by the reaction so that it can act repeatedly.

**Epidemiology:** The scientific study of the distribution and occurrence of human diseases, health conditions, and their determinants.

**Eugenics:** Attempts to improve hereditary qualities through selective breeding.

**Exocrine glands:** Glands that secrete into ducts or onto specific organ surfaces. Exocrine glands are classified as serous (producing a watery substance) or mucous (producing a viscous substance). CF affects both types, increasing the salt content of serous secretions and diminishing the salt content of mucus secretions. Mucous exocrine glands from individuals with CF then produce thicker than normal secretions leading to obstruction of the glands' ducts. Examples of exocrine glands include lacrimal (tear) glands, sweat glands, and part of the pancreas.

**Exons:** The protein-coding DNA sequences of a gene. Compare *introns*.

**Frequency:** The number of occurrences of a given allele within a given population.

**Gel:** The semi-solid matrix (e.g., agarose gel or acrylamide) used in electrophoresis to separate molecules.

**Gene:** The fundamental physical and functional unit of heredity. A gene is an ordered sequence of nucleotide base pairs to which a specific product or function can be assigned.

**Gene mapping:** Determining the relative locations of different genes on chromosomes.

**Gene therapy:** The deliberate administration of genetic material into the cells of a patient with the intent of correcting a specific genetic defect.

**Genetic code:** The sequence of nucleotides, coded in triplets along the mRNA, that determines the sequence of amino acids in protein synthesis. The DNA sequence of a gene can be used to predict the mRNA sequence, and this genetic code can in turn be used to predict the amino acid sequence.

**Genetic counseling:** A clinical service involving educational, informational, and psychosocial elements to provide an individual (and sometimes his or her family) with information about heritable conditions. Genetic counseling is performed by genetics specialists, including physicians, Ph.D. clinical geneticists, genetic counselors, nurses, and social workers.

**Genetic screening:** The analysis of samples from asymptomatic individuals with no family history of a disorder, groups of such individuals, or populations.

**Genetic testing:** The use of specific assays to determine the genetic status of individuals already suspected to be at high risk (e.g., family history or symptoms) for a particular inherited condition.

**Genetics:** The study of the patterns of inheritance of specific traits.

**Genome:** All the genetic material in the chromosomes of a particular organism; its size is generally given as its total number of base pairs. The human genome is 3.3 billion base pairs.

**Genotype:** The genetic constitution of an organism, as distinguished from its physical appearance, or phenotype.

**Health maintenance organization (HMO):** A health care organization that serves as both payer and provider of comprehensive medical services, provided by a defined group of physicians to an enrolled, fee-paying population.

**Hemoglobin (Hb):** A protein that carries oxygen in red blood cells. Sickle cell and thalassemia mutations affect hemoglobin.

**Heterozygote:** A heterozygous individual, such as a CF carrier.

**Heterozygous:** Having two different alleles at a particular locus.

**Homozygote:** A homozygous individual.

**Homozygous:** Having the same alleles at a particular locus.

**Hybridization:** The process of joining two complementary strands of DNA, or of DNA and RNA, together to form a double-stranded molecule.

**Immunoreactive trypsin (IRT) test:** An assay that measures levels of pancreatic trypsin, a digestive enzyme. As a protocol for newborn CF screening, a drop of blood is isolated on a card, dried, and chemically analyzed to detect elevated levels of the enzyme. It is not intended to be a diagnostic test.

**In vitro:** Literally, "in glass," pertaining to a biological process or reaction taking place in an artificial environment, usually a laboratory.

**In vivo:** Literally, "in the living," pertaining to a biological process or reaction taking place in a living cell or organism.

**Introns:** DNA sequences interrupting the protein-coding DNA sequences of a gene that are transcribed into mRNA, but are spliced out of the mRNA before the mRNA is translated into protein. Compare *exons*.

**Karyotype:** A photomicrograph of an individual's chromosomes arranged in a standard format showing the number, size, and shape of each chromosome.

**Ligase chain reaction (LCR):** An in vitro process that amplifies only the region of DNA directly underneath the known sequence to make millions of copies of this sequence. Mutations differing by a single base can be easily detected with this technique.

**Linkage:** The proximity of two or more markers (e.g., genes, RFLP markers) on a chromosome. The closer together the markers are the lower the probability that they will be separated during meiosis, and hence the greater the probability that they will be inherited together.

**Linkage analysis:** The process of studying DNA markers to trace the transmission of a particular gene and a specific mutation in a particular family. This can be performed only in families with a living CF affected member or from DNA samples stored from a deceased affected family member.

**Locus:** A specific, physical position on a chromosome.

**Marker:** A stretch of DNA with a known location on a chromosome that is used as a point of reference when mapping another locus. Markers can be important to linkage analysis and diagnosing genetic disease.

**Meiosis:** The process of reduction of genetic material and cell division in the diploid progenitors of sex cells. Meiosis results in four, rather than two, daughter cells, each with a single set of chromosomes.

**Messenger RNA (mRNA):** A class of RNA produced by transcribing the DNA sequence of a gene. An mRNA molecule is involved in translating instructions from the DNA sequence into proteins.

**Mutation:** Changes in the composition of DNA.

**Nucleotide:** The unit of DNA consisting of one of four bases—adenine, guanine, cytosine, or thymine—attached to a phosphate-sugar group. The sugar group is deoxyribose in DNA. In RNA, the sugar group is ribose and the base uracil substitutes for thymine.

**Open enrollment:** A health insurance enrollment period during which coverage is offered regardless of health status and without medical screening. Open enrollment periods are characteristic of some BC/BS plans and HMOs.

**Phenotype:** The appearance of an individual or the observable properties of an organism that result from the interaction of genes and the environment. Compare *genotype*.

**Polymerase chain reaction (PCR):** An in vitro process through which repeated cycling of the reaction reproduces a specific region of DNA between two sites, yielding millions of copies from the original.

**Polymorphism:** The existence of more than one form of a genetic trait.

**Preexisting condition:** A condition existing before an insurance policy goes into effect and commonly defined as one which would cause an ordinarily prudent person to seek diagnosis, care, or treatment.

**Proband:** The individual in a family first identified as manifesting a given heritable trait.

**Probe:** A short segment of DNA tagged with a reporter molecule, such as radioactive phosphorus ($^{32}P$), used to detect the presence of that particular complementary DNA sequence.

**Protein:** A biological molecule whose structure is determined by the sequence of nucleotides in DNA. Proteins are required for the structure, function, and regulation of cells, tissues, and organs in the body.

**Rated premium:** A premium with an added surcharge that is required by insurers to cover the additional risk associated with certain medical conditions. Rated premiums usually range from 25 to 100 percent more than the standard premium.

**Recombinant DNA technology:** Processes used to form a DNA molecule through the union of different DNA molecules, but often commonly used to refer to any techniques that directly examine DNA.

**Recessive:** An allele that exerts its phenotypic effect only when present in homozygous form, otherwise being masked by the dominant allele.

**Reliability:** The ability of a test to accurately detect that which it was designed to detect and to do so in a consistent fashion.

**Replication:** The synthesis of new DNA from existing DNA. PCR is an in vitro technology based on principles of replication.

**Restriction endonuclease:** An enzyme that has the ability to recognize a specific DNA sequence and cut it at that sequence.

**Restriction enzyme:** See *restriction endonuclease.*

**Restriction fragment length polymorphism (RFLP):** Variations in the size of DNA fragments produced by a restriction endonuclease at a polymorphic locus.

**Restriction fragment length polymorphism (RFLP) analysis:** DNA technique using single-locus or multilocus probes to detect variation in the DNA sequence by revealing size differences in DNA fragments produced by the action of a restriction enzyme. See *restriction fragment length polymorphism.*

**Reverse dot blot:** Blotting in which allele-specific oligonucleotides (ASOs) are immobilized on the membrane. Key segments of the individual's unknown DNA are then amplified, labeled, and hybridized to the probes on the membrane.

**Ribonucleic acid (RNA):** A chemical found in the nucleus and cytoplasm of cells that plays an important role in protein synthesis and other chemical activities of the cell. The structure of RNA is similar to that of DNA. There are several classes of RNA molecules, including messenger RNA, transfer RNA, ribosomal RNA, and other small RNAs, each serving a different purpose.

**RNA:** See *ribonucleic acid.*

**Sensitivity:** The ability of a test to identify correctly those who have a disease.

**Sex chromosomes:** The X and Y chromosomes in human beings that determine the sex of an individual. Females have two X chromosomes in somatic cells; males have an X and a Y chromosome.

**Sickle cell anemia:** An autosomal recessive disorder affecting red blood cell flow through the circulatory system, causing complications in numerous other organ systems. Sickle cell anemia predominantly occurs in individuals of African descent.

**Sickle cell trait:** The heterozygous state of sickle cell anemia; sickle cell carrier status.

**Single-gene disorder:** Hereditary disorder caused by a single gene (e.g., cystic fibrosis, Tay-Sachs disease, sickle cell anemia).

**Somatic cells:** Any cells in the body except reproductive cells and their precursors.

**Southern blot:** The nylon membrane to which DNA has adhered after the process of Southern blotting.

**Southern blotting:** The technique for transferring DNA fragments separated by electrophoresis from the gel to a nylon membrane, to which DNA probes that detect specific fragments can then be applied.

**Specificity:** The ability of a test to identify correctly those who do not have the characteristic which is being tested.

**Stigmatization:** Branding, marking, or discrediting because of a particular characteristic.

**Sweat test:** An assay used to confirm CF that measures levels of sodium ($Na^+$) and chloride ($Cl^-$) ions. These ions appear in high concentrations in patients with CF. Sweating is induced by running a low electric current through a pilocarpine-soaked gauze pad on the individual's arm or back. The amounts of $Na^+$ and $Cl^-$ in the sweat can then be determined to confirm or question a diagnosis of CF.

***Taq* polymerase:** DNA polymerase—the enzyme used to form double-stranded DNA from nucleotides and a single-stranded DNA template—isolated from the bacterium *Thermus aquaticus*, which normally lives in hot springs. *Taq* polymerase can withstand the high temperatures required in the repeating cycles of polymerase chain reaction (PCR).

**Tay-Sachs disease:** A lethal autosomal recessive disorder affecting the central nervous system which results in mental retardation and early death. Tay-Sachs disease predominantly occurs among Jews of Eastern and Central European descent and populations in the United States and Canada descended from French Canadian ancestors.

**Transcription:** The synthesis of mRNA from a sequence of DNA (a gene); the first step in gene expression. Compare *translation.*

**Translation:** The process in which the genetic code carried by mRNA directs the synthesis of proteins from amino acids. Compare *transcription.*

**Underwrite:** The process by which an insurer determines whether and on what basis it will accept an application for insurance.

**Wrongful birth:** A malpractice claim in which the parents assert that failure to receive timely, accurate information robbed them of the opportunity to avoid conception or birth of an affected child.

**Wrongful life:** A malpractice claim in which an affected child asserts he or she was harmed by a failure to give the parents an opportunity to avoid conception or birth. The claimant argues that to never have existed would be better than to exist with severe disabilities.

# Index

## Other Related OTA Reports

- *Artificial Insemination Practice in the United States: Summary of a 1987 Survey*
  OTA-BP-BA-48, August 1988; 112 p.
  NTIS order #PB89-139903
- *Biotechnology in a Global Economy*
  OTA-BA-494, October 1991; 282 p.
  GPO stock # 052-003-01258-8
- *Genetic Monitoring and Screening in the Workplace*
  OTA-BA-455, October 1990; 262 p.
  NTIS order #PB91-105940
- *Genetic Witness: Forensic Uses of DNA Tests*
  OTA-BA-438, July 1990; 204 p.
  NTIS order #PB90-259110
- *Healthy Children: Investing in Our Future*
  OTA-H-344, February 1988; 310 p.
  NTIS order #PB88-178454
- *Human Gene Therapy*
  OTA-BP-BA-32, December 1984; 116 p.
  NTIS order #PB85-206076
- *Infertility: Medical and Social Choices*
  OTA-BA-358, May 1988; 402 p.
  NTIS order #PB88-196464
- *Medical Monitoring and Screening in the Workplace: Results of a Survey*
  OTA-BP-BA-67, September 1990; 96 p.
  GPO stock #052-003-01255-3
- *Mapping Our Genes—The Genome Projects: How Big, How Fast?*
  OTA-BA-373, April 1988; 232 p.
  NTIS order #PB88-212402
- *Medical Testing and Health Insurance*
  OTA-H-384, August 1988; 224 p.
  NTIS order #PB89-116958
- *New Developments in Biotechnology: Public Perceptions of Biotechnology*
  OTA-BP-BA-45, May 1987; 136 p.
  NTIS order #PB87-207544
- *Technologies For Detecting Heritable Mutations in Human Beings*
  OTA-H-298, September 1986; 156 p.
  NTIS order #PB87-140158

NOTE: Reports are available from the U.S. Government Printing Office, Superintendent of Documents, Dept. 33, Washington, DC 20402-9325, (202) 783-3238, and/or the National Technical Information Service, 5285 Port Royal Road, Springfield, VA 22161-0001, (703) 487-4650.

## Superintendent of Documents **Publications** Order Form

P3

Order Processing Code:
* **6322**

Charge your order. It's Easy!

**To fax your orders (202) 512–2250**

☐ **YES**, please send me the following:

_____ copies of ***Cystic Fibrosis and DNA Tests: Implications of Carrier Screening (308 pages)***
S/N 052-003-01291-0 at $16 each.

The total cost of my order is $_____. International customers please add 25%. Prices include regular domestic postage and handling and are subject to change.

(Company or Personal Name) (Please type or print)

(Additional address/attention line)

(Street address)

(City, State, ZIP Code)

(Daytime phone including area code)

(Purchase Order No.)

**May we make your name/address available to other mailers?** YES ☐ NO ☐

**Please Choose Method of Payment:**

☐ Check Payable to the Superintendent of Documents

☐ GPO Deposit Account ☐☐☐☐☐☐☐ – ☐

☐ VISA or MasterCard Account

☐☐☐☐☐☐☐☐☐☐☐☐☐☐☐☐☐☐☐☐

☐☐☐☐ (Credit card expiration date)

***Thank you for your order!***

(Authorizing Signature)

8/92

Mail To: New Orders, Superintendent of Documents
P.O. Box 371954, Pittsburgh, PA 15250–7954

ISBN 0-16-037986-5
90000
9 780160 379864